Book Series on
Complex Metallic Alloys – Vol. 2

PROPERTIES AND APPLICATIONS
OF COMPLEX INTERMETALLICS

Series on Complex Metallic Alloys

Series Editor: Jean Marie Dubois

Book Series on
Complex Metallic Alloys – Vol. 2

PROPERTIES AND APPLICATIONS OF COMPLEX INTERMETALLICS

edited by

Esther Belin-Ferré

Laboratoire de Chimie Physique-Matière et Rayonnement
Centre National de la Recherche Scientifique, Université Pierre et Marie Curie, France

NEW JERSEY · LONDON · SINGAPORE · BEIJING · SHANGHAI · HONG KONG · TAIPEI · CHENNAI

Published by

World Scientific Publishing Co. Pte. Ltd.

5 Toh Tuck Link, Singapore 596224

USA office: 27 Warren Street, Suite 401-402, Hackensack, NJ 07601

UK office: 57 Shelton Street, Covent Garden, London WC2H 9HE

British Library Cataloguing-in-Publication Data
A catalogue record for this book is available from the British Library.

PROPERTIES AND APPLICATIONS OF COMPLEX INTERMETALLICS
Book Series on Complex Metallic Alloys — Vol. 2

ISBN-13 978-981-4261-63-0
ISBN-10 981-4261-63-7

Printed in Singapore by World Scientific Printers

FOREWORD

This volume is the second of a series of books issued from the lectures delivered at the Euro-School on Materials Science, organised yearly in Ljubljana, by the European Network of Excellence Complex Metallic Alloys (CMA) under contract NMP3-CT-2005-500145.

The second session of the CMA Euro-School, was mostly dedicated to physical properties and applications of complex metallic alloys. These are introduced in the first chapter which complements the presentation already given by the same author in the first volume of the series. The second chapter is dedicated to the growth of single crystals and the problems encountered using various growing techniques. Then, four chapters report about thermoelectric properties, electronic structure and magnetism, they are followed by two more theoretical sections about the calculation of physical properties. The two next chapters are dedicated to hydrogen storage by complex metallic alloys and their applications. During this second session of the CMA Euro-School, a lecture was given about the discovery of bulk metallic glasses. It is reported in the last chapter of this volume.

Grateful thanks go to the European Commission for providing financial support. All authors whose contributions have made possible editing this volume are also warmly acknowledged.

Esther Belin-Ferré
Paris, March 2009

CONTENTS

CHAPTER 1

COMPLEX, METALLIC, AND SO DIFFERENT

Jean-Marie Dubois

Institut Jean Lamour (FR 2797b CNRS-INPL-UHP)
Ecole des Mines, Parc de Saurupt, 54042 Nancy, France
E-mail: dubois@mines.inpl-nancy.fr

Complex Metallic Alloys (or CMAs for short) include quasicrystals as the ultimate state of complexity in metallic crystals. Phase selection rules and crystallographic complexity are addressed with the help of examples taken from selected Al-Mg and Al-TM (TM=transition metals) systems. Specific properties (transport, surface) are illustrated with the aim to present CMAs as typical examples of smart materials, which allow us to take advantage of compromise properties usually not encountered in conventional metallic alloys. Inverse nano-structuration is introduced as a new unifying concept, helping us to better understand what makes CMAs so different.

This chapter is dedicated to Esther Belin-Ferré on the occasion of her 70th birthday.

1. Introduction

The acronym CMA, for Complex Metallic Alloys, embodies simultaneously a concept, a family of intermetallic compounds, and the name of a European Network of Excellence. The concept deals with the phenomena that take place is an essentially metallic medium when the range of the interactions become small, or very small, in comparison to the period of the system, although the system is well ordered. The family of compounds called CMAs comprises a large number of crystals known already for decades, but also a very large number still to be discovered in ternary, quaternary, etc. intermetallics containing metals alloyed with metalloids (N, C, Si, etc.), rare earths, chalcogenides, etc. It is usually admitted that oxygen, and therefore oxides, do not enter this category. The most famous CMAs are quasicrystals,[1] which were first pointed out 25 years ago,[2] but other well-known CMAs are the so-called approximants, for their resemblance with quasicrystals,[3] or clathrates (Fig. 1) and skutterudites for their thermoelectric properties.[4]

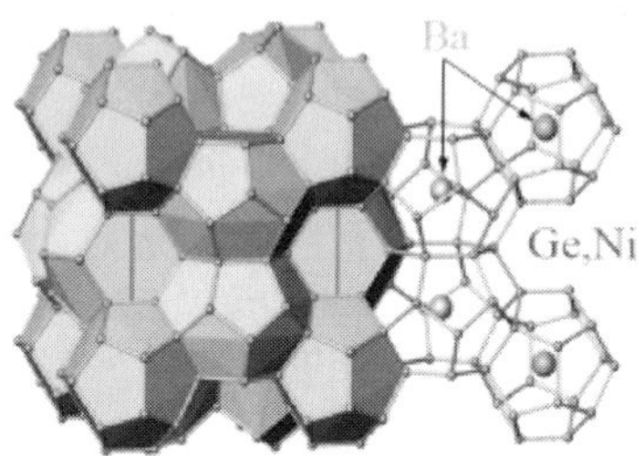

Fig. 1. A clathrate compound, also called a cage compound, as a typical example of a CMA. Electrons propagate nearly freely within the "cages" (shown open is the right hand side part of the figure) whereas phonons are hindered by the heavy atoms located at their centre. This results in an efficient decoupling between electronic and heat conduction, thus making the material appropriate for thermoelectricity generation. A large fraction of the present volume is dedicated to this topic. (Courtesy of Y. Grin, MPG Dresden, *CMA*).

The *CMA* network of excellence[*] is dedicated on the one hand to discovering new CMAs, preferably with attractive properties in view of technological applications, and on the other hand to disseminating the knowledge gained on those compounds toward academia, industry, and the Grand Public. It is in this frame that the European School on Materials Science of the *CMA* network operates every year, with a major deliverable that is the volume series to which the present book belongs.

A brief introduction to CMAs was given in Vol. I of this series.[5] In the present chapter, we review in more details the fundamentals that underline the phase selection, growth and preparation of single-grain as well as multi-grained CMAs. We pay attention to the major property of the broader category of CMAs, namely compounds of aluminium with transition metals (TM) that are characterized by the presence at the Fermi energy of a pseudo-gap, thus reducing conductivity more or less in proportion to the complexity of the crystal lattice. We will show in the

---

[*] In this chapter, the acronym CMA (or often, CMAs when several compounds are mentioned) is being used to label crystalline compounds. Italics are used to refer to the European network of excellence *CMA*.

course of this chapter that this reduction of the density of electronic states is also responsible for major surface properties of CMAs, i.e. solid-solid adhesion in vacuum and wetting by polar liquids.

The presentation of this chapter is oriented towards considering CMAs as smart materials, materials able to compensate a deficit on one property by the enhancement of another property. This is typically given by a few CMAs, which are both hard and present a low coefficient of friction against steel, in contrast with hard steels that offer only high friction when sliding against each other. Few potential technological niches that may take advantage of the so far surprising properties of CMAs will be mentioned in due place in the course of the chapter. More could have been quoted as well, but the Reader will find the relevant information in this series of books,[5] or elsewhere.[1]

2. Historical Background

The foundation of the field goes back to the early works of Linus Pauling, the most famous chemist ever and the only individual so far who was awarded an unshared Nobel Prize in Stockholm (for chemistry) and few years later, another unshared Nobel Prize in Oslo (for peace). When he was still a PhD student, Pauling studied cluster-forming crystals like $Al_{12}Mo$ (Fig. 2). Crystal chemistry of the time being was powerful enough to resolve such a structure, although it contained 2×13 atoms per unit cell. He solve many more complex alloys and derived most of his chemistry rules from such studies, which he compiled in his 'Bible' for Chemists, first edited in 1939.[6]

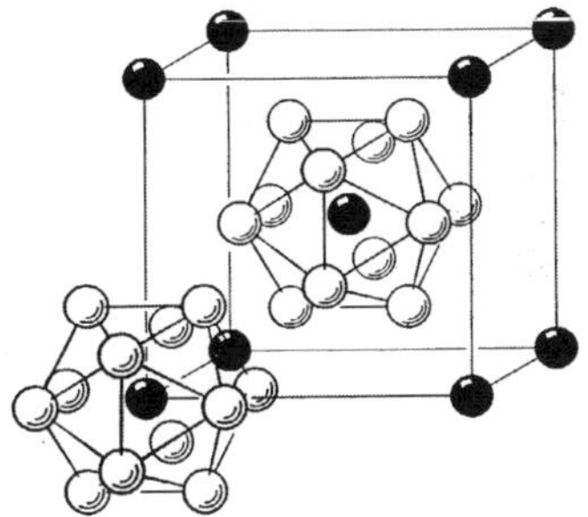

Fig. 2. The two nodes of the bcc lattice of $Al_{12}Mo$ CMA are both occupied by a 12-Al icosahedron, with centre occupied by Mo (black dots).

                                 Jean-Marie Dubois

His students continued the work of the Master until the mid of the sixties when Samson attacked the β-Al_3Mg_2 compound, some times referred to as the Samson phase.[7] This compound was far more difficult to solve than all previous crystals studied by Pauling's team. Nevertheless, Samson went through all difficulties and using the most up-to-date X-ray diffraction techniques available, as well as cardboard modelling (Fig. 3), he came to a very good agreement between experimental and calculated structure factors. The complexity of the compound revealed itself: the cubic unit cell is huge, with a lattice parameter a = 2.824 nm; it contains 1832 possible atomic positions, but many are simply unoccupied, reducing to only 1168 fully or partially occupied positions; well-defined clusters of atoms are formed, of icosahedral point group symmetry, but simultaneously, partial occupancy of sites leads to some disorder in the regions separating clusters. This structure was revisited in recent years by *CMA* scientists, using modern diffraction techniques applied to single grain specimens.[8] Amazingly, the work of Samson appeared very accurate, except for details of second rank. We will come back to the Samson phase later in this chapter.

Fig. 3. Cardboard model used by Samson to represent the crystal structure of β-Al_3Mg_2 in the 60's (Taken from Ref. 7; courtesy of W. Steurer, ETH-Zürich, *CMA*).

Unfortunately, this was the maximum that crystallography could achieve with the tools available at that time. Possibly, the Master was in the mean time also focusing his interests to other topics, like DNA.[9] His team stopped research on intermetallics and abandoned the field it had pioneering in the early 30's. For twenty years, nothing happened, and no one apparently showed any interest in complex intermetallics. The truth is that these materials are not only complex, therefore nearly impossible to solve with sufficient accuracy without having single crystals of high quality, which was not the case in Pauling's time, but they are also very brittle at room temperature. This is detrimental to any mechanical application in e.g. aeronautics or the automotive industry, except if the intermetallics is a very diluted, nanosized precipitate dispersed in a ductile matrix. The benefit of such a nano-precipitation in aluminium alloys was well identified, following the works of Guinier in France[10] and Preston in the US.[11] Another way to circumvent the intrinsic brittleness of Al-based compounds is to obtain them as metallic glasses, preferably by rapid solidification from the melt. This is the way I followed with my PhD supervisor by the end of the 70's, giving rise to the first patent on aluminium-based glasses.[12] As counter-examples used to define the concentration domain of glass-forming ability, I mentioned several compounds that do not amorphize under the conditions used in my study. Few compositions, like $Al_{69}Cu_{17}Fe_{10}Mo_1Si_3$ (at. %) turned out four years later to yield substantial amounts of a stable quasicrystal, which indeed requires very high cooling rates to get amorphous, a process that we had no access to in 1980.

The revolution came with Dany Shechtman, who was following the same track when he was working at the National Bureau of Standards (now, the NIST) about at the same time. He studied the Al-Mn system with the intention to increase as much as possible the solubility of manganese in the aluminium matrix. On a day of April 1982, with a melt-spun Al75Mn25 (per weight) alloy, he observed for the first time a specific diffraction pattern and noticed immediately its 10-fold symmetry. A long way started then for Shechtman, until he had assembled enough electron microscopy pieces of evidence to convince himself that he was facing a genuine effect and not an artefact like multiple twinning and could place with three co-authors a report in Phys.

Rev. Lett.[13] This article introduced the concept of a crystal (diffraction peaks are narrow, and therefore the material must be well-ordered like a crystal) which nevertheless is not compatible with translation periodicity in 3-dimensional space (the actual point group symmetry is 5-fold symmetric). Independent theoretical confirmations followed at the end of 1984 and beginning of 1985, as well as the discovery of other aperiodic crystals with 10-fold, 8-fold and 12-fold point symmetries, also considered as impossible by classical crystallography until then (see Ref. 1 for more details and quotation of the original papers).

The labour of Shechtman, and of many others including the present author,[14] was by far not completed at this stage because the two-century old paradigm that an ordered piece of solid matter must fit translational order was that much anchored in the spirit of many crystallographers, and especially in the mind of the most reputed one, that it took years till our community could talk about quasicrystals without facing criticism, when not irony, or accusation of cheating experimental data. The most drastic attacks[15] came from the Master, Linus Pauling, who had established his conviction decades ago when he and his students solved increasingly complex crystals, yet always trusting conventional crystallography. He claimed that multiple twinning was the artefact hidden behind the "apparent five-fold symmetry" of quasicrystals. Despite the more and more convincing pieces of evidence that Shechtman and others could produce and show him, it is plausible that Pauling passed away without modifying his mind regarding what he believed was the only acceptable paradigm for crystals, and for any solid that shows sharp diffraction lines.

Mathematicians did not follow this track and enormously benefited from the discovery of quasicrystals, as is beautifully explained by Senechal in her gorgeous book on aperiodic crystals.[16] A few metallurgists also did not trust Pauling and, based on their acquaintance with icosahedral order in liquid and amorphous metals, tried to understand better how quasicrystals form, and whether they are actually metastable (as suggested by Shechtman's experiments), or could correspond to a ground state of the system and be stable crystals. Out of several attempts, Tsai in Japan[17] produced the most successful ones within a short period of years. He discovered first the stable Al-Cu-Fe quasicrystals, and then a full series of so far unidentified icosahedral and decagonal quasicrystals (again, see Ref. 1 for details). More recently, his group at NIMS, Tsukuba and Tohoku University, Sendai, in Japan found new aperiodic compounds, especially the binary Cd-Yb icosahedral crystal the structure of which could be solved with extraordinary accuracy.[18]

Amazingly enough, I could recognize the compositions secured in the patent I had filed in 1982 when I read Tsai's initial paper on Al-Cu-Fe upon a visit I paid him in 1987 in Sendai. That quasicrystals could be stable changed the (industrial) game because it was clear to me that they could most presumably undergo thermal excursions during processing and service without loosing the essential part of their attractive properties. At that time, I knew already two such properties of potential interest for technology, namely reduced sticking against meat and reduced friction against steel. We will go back to both properties in the second part of this chapter. All this was secured in a patent dated 1988,[19] but could not be published before a few years until a whole family of patents was secured.[1] My early papers attracted some interest by metallurgists first, physicists and chemists later, especially from Iowa State University.[20] Many more compounds were identified, the so-called approximants, because they resemble quasicrystals, but are periodic compounds with a large, sometimes a very large, unit cell. We pointed out a few compounds of technological relevance in the Al-Cr-Fe(-Cu) systems, solved the Al-TM skeleton of the orthorhombic lattice, and set-up a process to produce it at industrial level (1000 kg/day) by gas atomization, followed by plasma spraying to prepare large areas of coatings.[21] Independently in Germany, Urban and his co-workers studied many other compounds and especially focused at the fundamentals of mechanical properties (plasticity, dynamics of dislocations) in aperiodic crystals.[22] A clear understanding of plasticity arose, around the key concept of metadislocations that is addressed later in this chapter. The joint effort of Urban, Schlapbach and myself lead ultimately to the birth of the *CMA* network, which is now in charge of European research in this area of condensed matter science and engineering. There, quasicrystals are viewed as the ultimate degree of complexity in a crystal made essentially of metals.

3.   Complexity in Real and Reciprocal Space

Complexity is often understood as a difficulty to fully account for the ensemble of atomic positions using simple terms, and few sentences, in contrast with many metals that may be described just a close packed array of spheres, or at most as small unit cells with only few atoms. It is true that many CMAs exhibit so large unit cells, with tens, if not hundreds of atom positions, that they are difficult to account for in few sentences. In this respect, the list of independent positions is large, and must be associated with another list of identical length that contains occupancy factors often smaller than 1. In Sections 3.1 to 3.3, we try to clarify the notion of complexity in alloys, based on the example of the Samson phase and of complex crystals of the same family. We end with a definition of a CMA in reciprocal space that will become more relevant in the coming sections dedicated to electronic properties and phase stability.

3.1.   *The example of compounds of Al, Mg and Zn*

Aluminium, magnesium, and zinc, are fairly simple metals that close pack in the most trivial ways available: Mg is hcp, whereas Al and Zn are fcc. As seen before, the mixture of 3 Al atoms with 2 Mg atoms yields an extremely complex crystal structure that was revised recently.[8] Why is that so is a fundamental question of great relevance, which we shall try to answer partially in Section 5.3 below.

The Bravais lattice of β-Al_3Mg_2 is bcc with a = 2.8242 nm, the lattice parameter. The actual composition is $Al_{61.43(3)}Mg_{38.6(2)}$ at. %. It is decorated by a number of large clusters of atoms, with near icosahedral point symmetry, leaving few atoms in between to fill-in space. The atoms positions are listed in Ref. 8. The most abundant decoration clusters are presented in Fig. 4. As a matter of fact, the structure undergoes a phase transition at 487 K.[8] Below that temperature, the crystal structure exhibit a rhombohedral distortion of the initial cubic lattica with parameters a = 1.9936 nm and c = 4.9110 nm.

Addition of Zn to Al and Mg leads to the formation of the so-called Bergman phase[23,24] in the vicinity of the $Al_{30-40}Mg_{40}Zn_{30-20}$ composition area (at. %). The Bergman phase is characterized by a cubic cell of lattice parameter a = 1.43 nm, which contains 163 atoms. Similarly to the Samson phase, atoms lie on concentric shells, with icosahedral symmetry (Fig. 5). Linkage between clusters is insured by so-called "glue atoms", here Al and Zn, but not Mg, which must be there since icosahedra cannot fill-in 3D Euclidian space. The influence of such atoms on physical properties is essential, as we shall see later in this chapter.

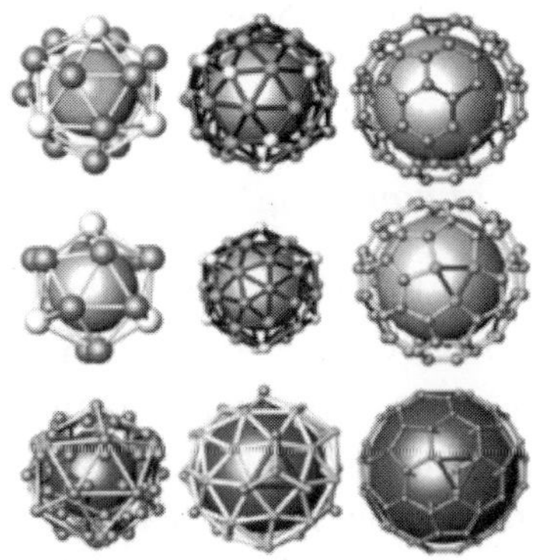

Fig. 4. Main clusters of icosahedral symmetry that build-up the crystal structure of b-Al$_3$Mg$_2$. The clusters in each row are embedded into each other, like Russian dolls of respective diameter 0.4, 0.6 and 0.8 nm, from left to right. The top row is for Mg-centred clusters, middle for Al or Mg centres and the bottom row is the so-called Samson cluster (Taken from Ref. 8; courtesy of M. Feuerbacher, Jülich Res. Centre, *CMA*).

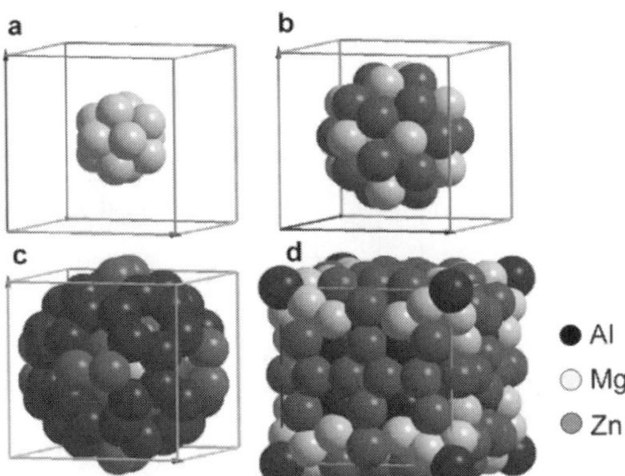

Fig. 5. Atomic decoration of the 3 concentric shells (a, b, c) that decorate the bcc lattice (d) of the Bergman phase[24] (Courtesy U. Mizutani, Nagoya Univ.).

Finally, the $MgZn_2$ compound corresponds to the formation of a C-14 hexagonal Laves phase[25] in which the basic clusters are smaller. Around the smallest atom (Zn), distorted icosahedra are found whereas around the larger (Mg), it is the so-called Friauf polydron that forms.[26] The model used by Samson (Fig. 3) for β-Al_3Mg_2 is made of individual Friauf polyhedra consisting of truncated tetrahedral units, with triangular and hexagonal facets. Accordingly, the size of the unit cell is smaller than in the two previous examples, as shown by its lattice parameters a = 0.522 nm and c = 0.8567 nm.

3.2. *Hierarchy, groups of atoms and clusters*

The most salient feature of the atomic structure of the three compounds introduced above is the presence of icosahedral clusters, or more precisely of clusters based on icosahedral symmetry. A typical example taken from the Bergman phase was already featured in Fig. 5. All Laves phases, of generic composition A_1B_2, where A and B stand for two different chemical species (typically, two metals like Mg and Cu, or Zn and Mg, with hard sphere diameters in the ration 1.2-1.3), are based on clusters, often called the Friauf polyhedron,[26] which form hierarchical icosahedra when packed together.

The importance of the hierarchical order defined by the successive shells of atoms in a cluster was recognized long ago by Pauling and his collaborators.[6] An early illustration of the atomic packing in the Bergman phase is shown in Fig. 6.

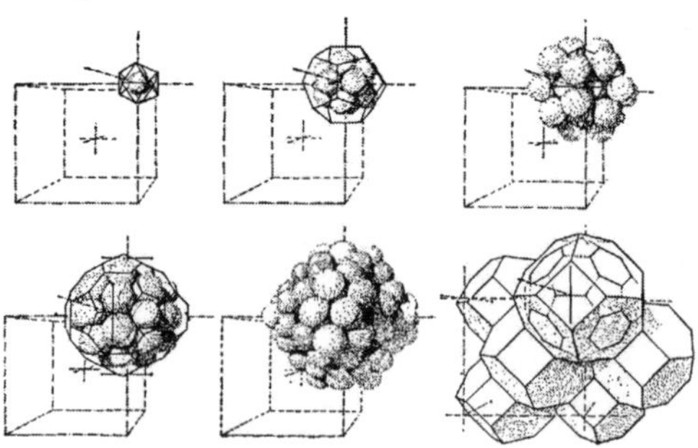

Fig. 6. Artistic presentation by Pauling[6] of the embedded shells of atoms forming the basic cluster in the Bergman phase.

Pushing the reasoning further, Janot[27] could reproduce the main characteristics of icosahedral order up to the long range, just assuming that the basic cluster in i-AlCuFe or in i-AlPdMn forms because it traps almost the right number (or magic number) of electrons that satisfies Schrödinger equation and therefore corresponds to a particularly favourable energy state. In those two icosahedral CMAs, the basic cluster is a Mackay icosahedron (MI) formed by three successive shells of 1) 12 Al atoms placed on a small icosahedron, 2) followed by a larger dodecahedron of 12 Mn atoms embedded in 3) a slightly larger icosidodecahedron of 30 Al+Pd atoms. As a matter of facts, the central shell must be Al-defective, but this is too technical for our present purpose. A cut through Janot's model within a plane perpendicular to a 5-fold axis is shown in Fig. 7.

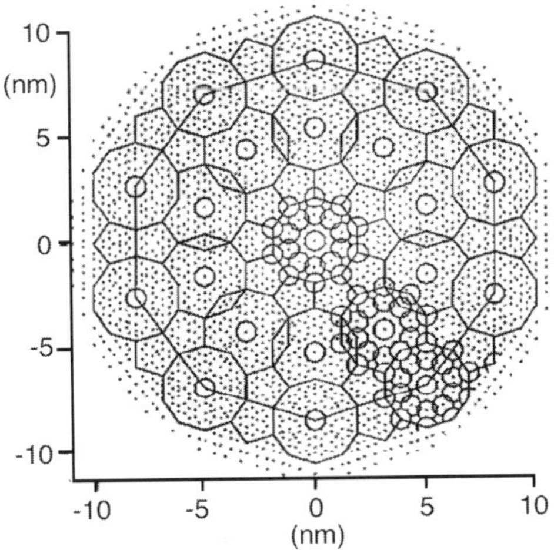

Fig. 7. Simplified presentation of the propagation of icosahedral order in Janot's model.[27] The first level of the hierarchy corresponds to the small circles drawn only in the bottom right hand side of the figure for simplicity (the small circle represents the trace of the equatorial atoms contained in the initial MI cluster). Around each of the MIs located on the largest decagon form two layers of MIs, placed respective to each other as atoms are placed in the MI. This mechanism may go on to infinity, thus mimicking icosahedral long-range order (Courtesy of C. Janot).

The building principle assumes simply that the magic number of electron states is slightly smaller than the charge observed at the first level of the hierarchy. The excess charge compared to the ideal magic number may thus propagate the icosahedral order of the initial MI cluster to the next level. Propagation of icosahedral order stops at this stage if the magic number for the two first levels of the hierarchy is satisfied. If not, the same mechanisms may continue to the next step, etc. It turns out that at a precisely defined composition, the excess charge corresponds to the MI arrangement propagating to infinity. Since the MI belongs to the icosahedral point group of symmetry (except for details in the central shell arrangement), true aperiodic order results in 3-dimension (3D) space (again, glue atoms are necessary to fill-in the space left unoccupied between the hierarchical clusters).

The extremely good agreement found between predicted compositions of the icosahedral CMAs and experimental ones, as well as a few remarkable predictions of transport properties as functions of temperature are in strong support of Janot's model. It is straightforward to apply the same model, but at a different composition that restricts the range of the hierarchy to a limited number of steps: two steps along all three directions of 3D space corresponds to the so-called 1/1 1/1 1/1 approximant (typically, the Bergman phase). Figure 8 features such an approximant, constructed with a basic cluster that is not a MI but a more complex unit.[28]

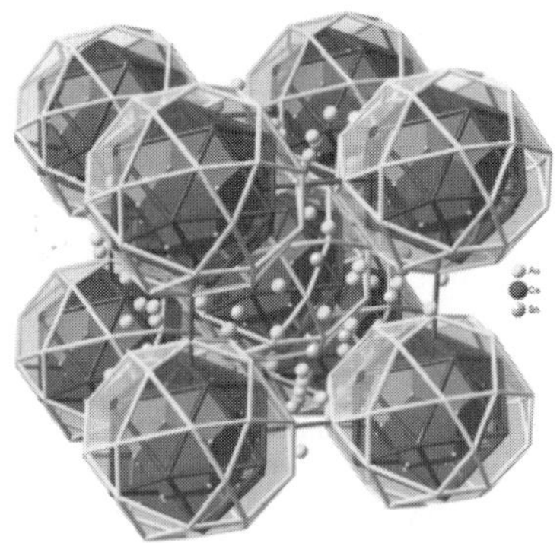

Fig. 8. Sketch of the cubic approximant structure of $Ce_{15}Au_{65}Sn_{20}$.[28] A central shell of Sn atoms on a pentagonal dodecahedron is surrounded by a larger icosahedron of Ce atoms, itself surrounded by a large icosidodecahedron of Au atoms. The cubic unit cell (a = 1.5190 nm) contains two such clusters located on the bcc nodes and glue atoms (shown as small spheres), ending with 161 atoms per unit cell. Notice the resemblance with the Bergman phase, except for the building cluster.

Here, the unit is a central pentagonal dodecahedron, embedded in an icosahedron, itself embedded in a large icosidodecahedron. Glue atoms that are not associated to vertices of the building clusters are visible as small spheres in the interval between basic clusters. Once more, Nature has chosen a reasonable compromise between filling Euclidian 3-dimensional space with large icosahedral clusters, and leaving no open space in between. The clusters however should not be taken as rigid entities that remain unchanged over a broad temperature range. On the contrary, as we see in the next sub-section, they may move upon atom popping over short interatomic distances. Altogether, structural complexity appears as intimately related to two opposite trends, the formation of well-defined and large clusters, most often of icosahedral symmetry, and the occurrence of disorder and defects. The resulting entropic contribution must undoubtedly play an important role in minimizing the free energy of the system. Whether entropy is the contribution that stabilizes complex architectures and favours them against simpler ones resulting uniquely from enthalpic contributions is a very important question. To the best of the knowledge of the author, this has not yet been solved to an entirely satisfactory level of comprehension and is still frequently a matter of discussion among experts.

3.3. *The key role played by disorder and defects*[b]

All types of defects that are observed in normal metallic crystals are also observed in CMAs: vacancies, dislocations, staking faults, etc. Systematic studies of atomic mobility have essentially shown that diffusion coefficients of the constitutive species are not different at high temperature from the ones observed in crystals of nearby composition, but smaller unit cell size (for a review of those results, see Ref. 1 and references therein).

An essential difference is however manifest in quasicrystals and large

---

[b] A defect is something that is missing compared to the ideal architecture of the lattice; disorder is something that is at the wrong place. Defects contribute to disorder the lattice because they relax the lattice and allow ideally placed objects (e.g. atoms) to move to the wrong place. A real crystal at finite temperature contains defects in statistical equilibrium. It may also be disordered, but still in thermodynamic equilibrium at $T > 0$ K.

unit cell CMAs. It can be pointed out using various techniques, especially inelastic neutron scattering, which proves that a small fraction of the constituent atoms pop over short distances along specific directions in the crystal.[29] Radioactive tracer diffusion techniques also show that mobility of at least one of the constituents is still active at rather low temperatures, at a level that is above the one extrapolated from high temperature range.[30] The existence of such a mobility had been predicted on the basis of high-dimensional crystallography[31] and happens to be in qualitative agreement with the best (and most difficult to perform) diffusion experiments.[30,32] Making few words out of a long story, we may summarize the essential facts in two figures.

Figure 9 illustrates the enormous influence of those defects on physical properties. The diagram shows the specific heat (divided by temperature) that is observed versus temperature (squared) for a highly perfect quasicrystal (bottom line) and the same material, but prepared highly defective by fast cooling from high temperature (top line).[33] Similar differences can be observed as well on transport properties and mechanical properties.

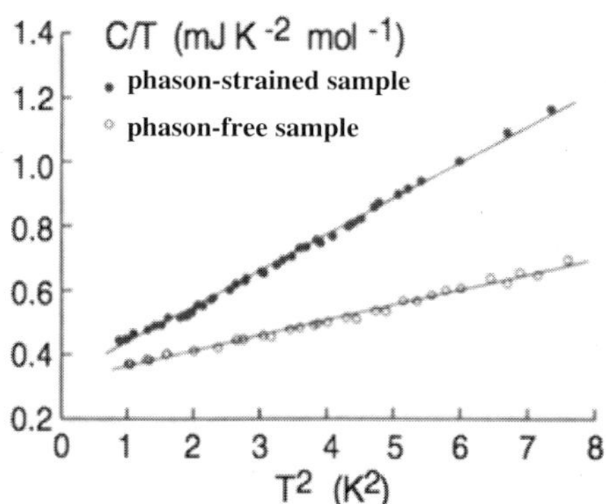

Fig. 9. Specific heat divided by temperature versus squared temperature as observed for a sample of i-Al-Cu-Fe when prepared upon slow cooling from high temperature (open dots, bottom line) and by fast cooling from the same temperature (solid dots, top line). (Courtesy of P. Garoche, Uni. Paris-Sud, Orsay).

Figure 10 presents a direct visualisation of the effect of these specific defects proper to quasicrystals and high-order approximants, namely the mobility of full clusters as it can be observed using an appropriate set-up of scanning transmission electron microscopy.[34] The left part of the figure was taken at room temperature. Three decagonal clusters are made more visible: the two right ones overlap partially, whereas the left ones only share an edge. At high temperature, here 1100K, the cluster on the right has disappeared, but not the two edge-sharing ones. Nevertheless, the underlying pentagonal network, which links the small, bright clusters is even more visible than at 300K.

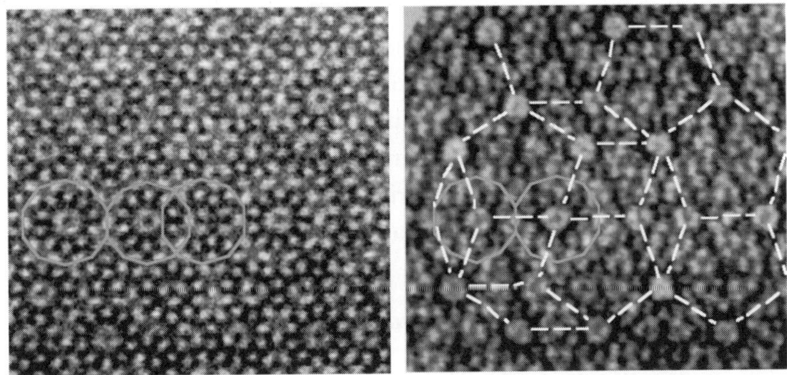

Fig. 10. High-resolution images showing how the quasi-periodic lattice of decagonal $Al_{72}Ni_{20}Co_8$ transforms from 300K (left) to 1100K (right). For clarity of the illustration, three decagonal clusters (diameter: 2 nm) are made more visible on the left. The one located on the right is not visible any more at 1100K in the right hand side part of the figure (Courtesy: E. Abe, NIMS, Tsukuba).

So far, we did not coin a specific term to label those defects. In the case of quasicrystals, they are called phason flips because they appear as a discontinuous change of an atom position upon a continuous change of the atom coordinates in complementary space (or perpendicular space).[27] In periodic CMAs, like the Samson phase, they correspond to sites with a partial occupancy. In both cases, they are associated with short bond distances between two sites that cannot be occupied simultaneously, something like a half vacancy. This explains why we prefer to rank them among defects. They are able to cause a lot of disorder, as shown by the data of Fig. 10, and are responsible for mass transport in the lattice (atomic diffusion on top of normal vacancy-mediated mobility; plastic deformation as demonstrated by Feuerbacher *et al.*[22,35]) because they introduce free volume in the lattice like disorder does in metallic glasses.

The question arises then to know to which extent these specific defects may be eliminated from CMA lattice, or equivalently, to know whether a perfect quasicrystal may be void of any such defects in its fundamental state. So far, following the most accurate diffraction experiments that were performed on single grains of icosahedral quasicrystals,[36] it seems that total absence of diffuse scattering in the interval between Bragg peaks can never be observed, whatever the quality of the heat treatment applied to the sample. Diffuse scattering signs the presence of disorder, and its variation with the reciprocal vector along a simple direction in q-space may help identifying the type of disorder. This is too technical for our present purpose. Nonetheless, and as a matter of fact, it appears that entropy, or disorder, is closely related to complexity in CMAs and is intrinsically related to the formation of well-identified clusters of icosahedral symmetry.

3.4. *Definition of a CMA in reciprocal space*

We may now settle an essential question, which is that of the definition of what we call a CMA. First, it is a compound, or a phase, or an alloy, essentially made of metals. This does not mean that the alloy is a metal itself, nor an alloy characterized by metallic properties, because most often the metallic character of the alloy species has become poor or much weaker than in the pure metal constituents. In scarce cases like Al_2Ru, it has turned to semi-conducting. It simply means that the major part of the constituents belongs to sp or d metals (Al, Ga, Sn, Fe, Ni, Pd, W, Rh, Re, etc.), possibly alloyed with semi-conductors (Ge, Si), chalcogenides (Se) and/or rare earths. In few cases, the situation is reversed: the major constituent is the semi-conductor like in clathrates. Oxides, although some may be structurally very complex, are excluded from the CMA family because they present no metallic behaviour whatsoever (except in few cases at very high temperature). The broad variety of chemical combinations that may be synthesized out of about 80 metals in the periodic table participates to the complexity of the compounds considered in this book. This also means that the potential for discovering new ternary, quaternary, etc. CMAs is enormous.

Second, many CMAs do not require lots of independent positions to be accounted for. Very frequently instead, a distribution of occupancy factors must be considered in order to match the chemical disorder inherent to the compound. This is the case for instance of the superstructures of the β-CsCl–type cubic phase that forms in Al-Cu alloys.[37] The basic unit cell is the 2-atom, body-centred cubic unit cell of parameter 0.29 nm. Depending on the Al/Cu composition ratio, substitution vacancies order in the lattice and increase dramatically the size of the unit cell. The largest superstructure known so far forms at composition $Al_{36}Cu_{48}V_{12}$ (V = vacancy), with a unit cell volume 47.7 times larger than that of the conventional β-phase (Fig. 11).

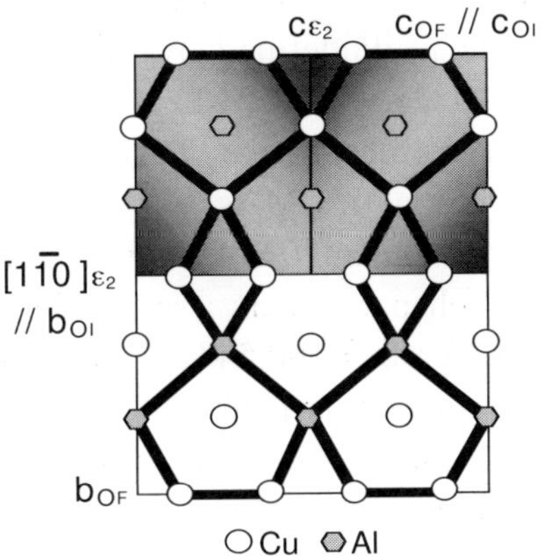

Fig. 11. Chemical ordering on the ε_2-Al_3Cu_2 lattice may produce various lattices with increasing sizes of the unit cell upon introducing vacancies (V, not shown in the drawing). Here is shown one 110 layer. The unit cell of ε_2-Al_3Cu_2 is represented by the top left, grey rectangle. Another cell, called O_I, of larger size and composition $Al_8Cu_{12}V_4$ is represented by the two top grey rectangles. By continuing the process until composition $Al_{36}Cu_{48}V_{12}$, one forms the O_F compound with unit cell 50 times larger than that of the original B2 lattice of the Al-Cu β-phase (After Dong *et al.*[37]).

The example of the superstructures of the β-cubic phase points out the difficulty to describe accurately complexity in real space due to the need to introduce a function that adequately fits the chemical disorder in the lattice, although atomic positions may be easily accounted for by a simple Bravais lattice like the body-centred cubic one. It is more relevant to call 'complex' an alloy, or a compound, essentially made of metals as presented above, whose reciprocal space exhibits complexity within the

Jones zone.[c] A measure of complexity of the crystal is then supplied by the number of peaks that fall inside the Jones zone or equivalently, by the inverse of the reciprocal distance between the first diffraction peak and the origin of reciprocal space: the most complex CMA in a series presents a diffraction peak that falls at the shortest distance to the origin of reciprocal space. In many Al-based compounds, this is a quasicrystal (with in principle 0 distance of the first diffraction peak from the origin of reciprocal space). As an immediate consequence, metallic glasses are not considered to belong to the group of CMAs, although it is often observed that metallic glasses of specific composition crystallise in a CMA, because they show no sharp Fourier component is their diffraction pattern. The Jones zone is ill defined by the broad main halo in the diffraction pattern and the pre-peak, when it exists, does not fall that close to the origin of reciprocal space. Their physics however is often very much reminiscent of that of CMAs.[38]

As an example for a typical CMA, Fig. 12 illustrates the case of triclinic $Al_{11}Mn_4$ (unit cell parameters a = 0.5087 nm; b = 0.8848 nm; c = 0.5052 nm; a = 89.72°; b = 100.54°; c = 105.37°; unit cell volume: 0.215 nm^3). The major diffraction peaks fall in the vicinity of the wave vector q = 30 nm^{-1} (q = $4\pi \sin \theta/\lambda$ with θ the Bragg angle and λ the wavelength) whereas many other peaks of variable intensity are observed inside the range [0, q]. The average distance between opposite centres of the facets of the Jones zone is often labelled K_P, so that here we have $K_P \approx 30$ nm^{-1}.

The number of peaks and their Miller indices in the vicinity of q = K_P reflects the degree of symmetry of the Jones zone. The larger the symmetry of the Jones zone, the closer the shape will be from a sphere. Such a resemblance to a sphere is actually achieved in many CMAs, for instance, quasicrystals but also γ-brass phases, etc. Furthermore, electronic concentration is very often naturally selected so that a close matching between Jones zone and Fermi surface is observed, which fulfils the Bragg condition $K_P = 2$ k_F (where k_F is the Fermi vector). Such

---

[c] The Jones zone is the Brillouin zone constructed with the most intense Fourier components in reciprocal space. For a simple crystal, it is identical to the first Brillouin zone. The notions of Brillouin zone, Fermi surface, etc. are introduced in Vol. I.[5]

a selection is responsible for the opening of Hume-Rothery gaps and therefore for the enhanced stability of the compound. In $Al_{11}Mn_4$, the Fermi vector amounts to $k_F = 14.25$ nm^{-1} if the contribution to the valence band by Al is taken equal to +3 electrons and that of Mn is assumed negative like in many other transition metals such as Fe, Ru, Re (but not Cu or Pd, see Ref. 1 for more information on this point) and equal to –3 electrons according to its position in the periodic table along the 3d-metal series. Hence, it is observed that $2k_F \approx K_P$, a result which is indeed traditionally associated with the formation of a CMA in a given system (see Section 5.1 hereafter).

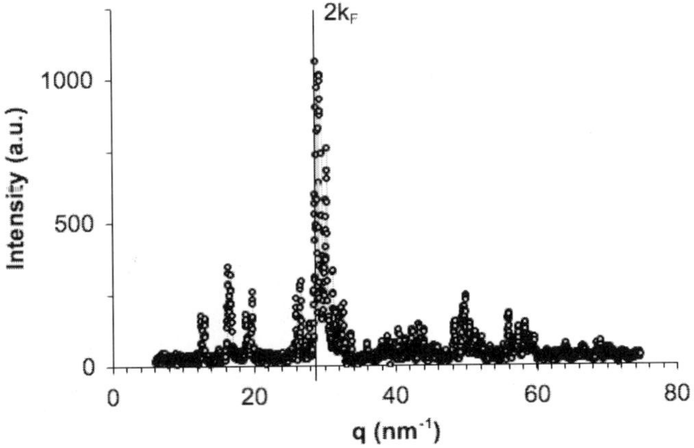

Fig. 12. Example of a typical CMA in reciprocal space (X-ray powder average). The most intense peaks, which define the Jones zone, are found in the vicinity of $q \equiv K_P \approx 30$ nm^{-1}. It turns out that the diameter of the Fermi surface is very close to this value with $2k_F = 28.5$ nm^{-1}.

4. Metallurgy and Surface Chemistry of CMAs

In this section, we address a number of the contingencies that make life difficult to all those who wish to study, or use, CMAs. These are preparation, stability of the samples in their environment, and mechanical properties. Most of the material necessary to this end was already presented in Vol. I[5] or elsewhere.[1] As a consequence, we shall reduce the presentation to a brief sketch of the basic knowledge required to start with.

4.1. *Preparation methods*

Single crystal growth is beyond doubt the most important technique to start with in view of growing single grains of excellence lattice perfection and well-defined composition. Reliable conclusions regarding physical properties can only be reached this way. The goal is nonetheless very difficult to reach because on the one hand, a precise knowledge of the phase diagram that contains the CMA of interest is a mandatory prerequisite to grow the crystal, and on the other hand because favourable thermodynamic conditions must be found within the diagram landscape to achieve successfully the growth, which is not always possible because of the occurrence of a cascade of peritectic reactions. The *CMA* network has triggered an intense research on the growth of complex single crystals, see as an example the Al-Cr-Fe grains obtained by the Czochralsky technique in Munich by Peter Gille[39] (Fig. 13).

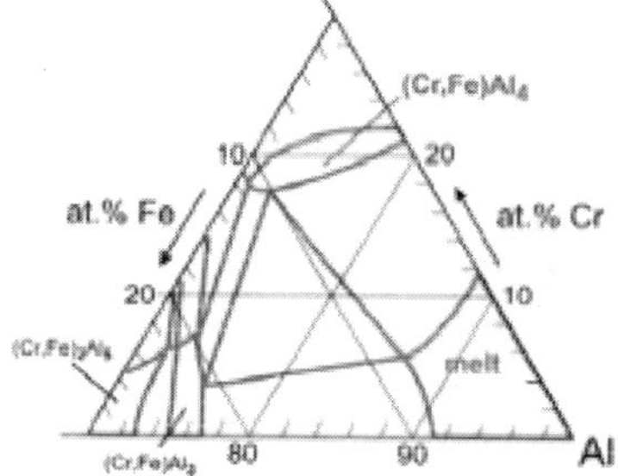

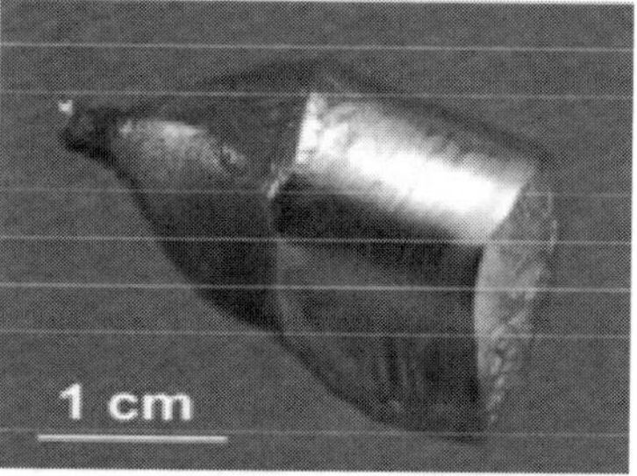

Fig. 13. Centimetre-size single grain (right) obtained by Czochralsky growth in the Al₄(Cr,Fe) region (left).(Courtesy of P. Gille, Uni. München, Munich).

A second technique,[40] by far easier to set-up, but which supplies only multi-grain materials, is sintering powders of appropriate average composition (Fig. 14). Bulk specimens with very low, when not zero, porosity may be prepared this way in the shape of large pellets a few centimetres in diameter. Most of the results reported in the following of this chapter on densities of states, friction, wetting, adhesion, etc. were obtained using such samples. The technique consists basically in three steps: i- melting of the pure constituents according to standard metallurgical routes, ii- grinding the ingot (if necessary, after removing corrupted parts due to oxidation for instance) to a mesh size typically in the range 25-50 μm and iii- sintering under uniaxial pressure in a carbon or molybdenum crucible. Alternative processing, using e.g. isostatic pressure were also tested with essentially identical results.[41] The final bulk material must then be polished to expose the CMA material to experiment. It may also be shaped, machined, etc. to fit the constraints of the experiment.

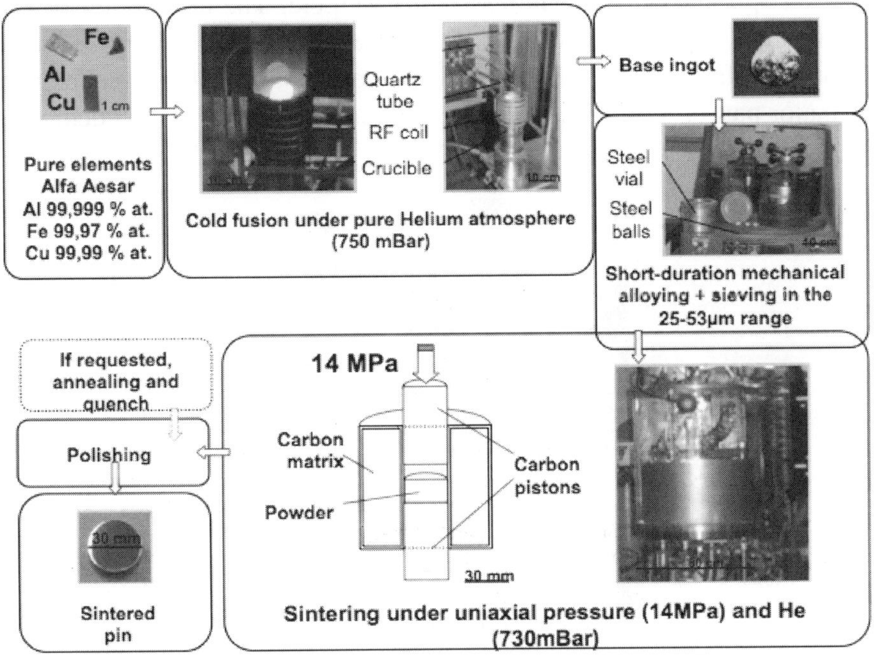

Fig. 14. Schematics of the sintering route used to prepared bulk, multi-grained specimens of CMAs. The successive steps are presented from top left to bottom left: alloying of the pure constituents, grinding and sieving, sintering under protective atmosphere and uniaxial pressure and final preparation of the bulk sample (Courtesy: P. Brunet, IJL, Nancy).

Interestingly, the two first steps of the process may be by-passed when using atomized powders, which are also the basic ingredient used by thermal spray processes like plasma torch spraying.[1] A lot of progress was done in recent years on surface coating technologies applied to CMAs. An illustration is given in Fig. 15, showing how stacks of multi-layers of the individual species Al, Cu and Fe ultimately lead to the formation of an icosahedral AlCuFe film by thermal annealing. The layers are first deposited one after the other on a substrate (here a WC-Co hard material), in the appropriate Al-Fe-Cu sequence. Mixing of the constituents, which are in the stoichiometric ratio by controlling the thickness of the individual layers, is obtained here by thermal anneal.

Many other techniques may be used: PVD, MO-CVD, laser assisted plasma spray, electron beam assisted PVD, etc. The description of this technology is too technical for our present purpose. It is however an issue in so far that the technological development of CMAs is concerned. A major part of Vol. III in the same series of textbooks will be dedicated to these topics.

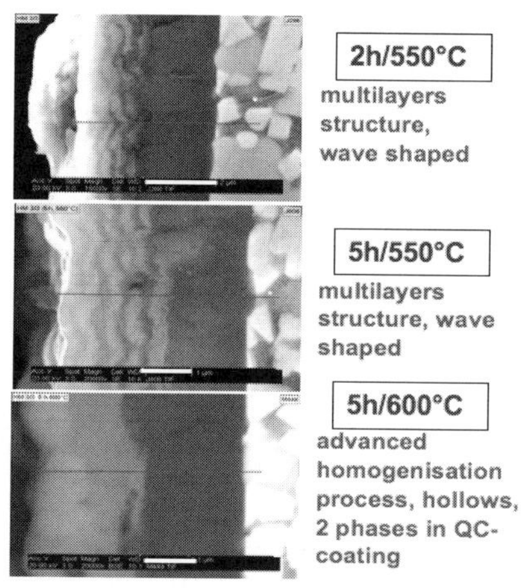

Fig. 15. Thin film of an AlCuFe icosahedral CMA prepared by physical vapour deposition of the individual constituents followed by thermal annealing. The substrate is a WC-Co material (right side part of the image), covered by a bound coat that shows up in the picture as a dark layer. Three main steps of the process are captured, see the annealing parameters given in the right hand side of the figure. The length scale of the scanning electron microscopy images is given in each figure by a white bar. The one at the top stands for 2 μm, whereas the two other ones are for 1 μm. The initial layers are clearly visible in the two top images (the heavier the element, the darker it appears in the figure). At the bottom, thermal mixing is completed, but unfortunately the film is not icosahedral since two phases appear within the thickness of the layer. Bonding to the underlying coating is however of very good quality (Courtesy: M. Cekada, JSI, Ljubljana; unpublished results).

4.2. *Corrosion, oxidation and interaction with chemical atmosphere*

These topics were studied as from the beginning of quasicrystals research. The essential two conclusions are i- that oxidation at relatively low temperature affects only aluminium, just like it does for pure aluminium[42] and ii- corrosion resistance is not determined by the crystal structure, but by the composition only.[1]

This does not mean however that nothing interesting may be found. On the contrary, specific compositions like that of the large orthorhombic unit cell of the O1-AlFeCr compound[43] shows outstanding resistance to pitting corrosion in the presence of halide species, which is even better than that of stainless steels. We will come back to this material in Section 8.

Interaction of CMAs surfaces with gaseous species will be addressed specifically in Vol. III, to be associated with the third session of the *CMA EuroSchool*.[5] The potential of CMAs was analysed in view of applications to catalysis[44] long ago and will be part of Section 8 hereafter.

Aside oxidation, nitridation is another important reaction, frequently used for surface hardening, improvement of wear resistance and corrosion properties. To the best of author's knowledge, nitridation of quasicrystals has never been studied except recently by Kenzari *et al.*[45]

Figure 16 shows the X-ray diffraction pattern of a sample of *i*-$Al_{59}Cu_{25.5}Fe_{12.5}B_3$ powder (at. %, noted in this sub-section hereafter *i*-AlCuFeB) after annealing for 9h at 625°C under 9.10^{-2} mbar of vacuum. It can be indexed as an icosahedral phase containing a small amount of β-cubic, CsCl type, phase with a ≈ 2.9 Å. This extra phase coexists with the quasicrystalline phase in gas-atomized powders because it takes part in the peritectic reaction that forms the icosahedral phase and is trapped out of equilibrium by the quenching process associated with atomization. It is also well known that the β-phase appears upon oxidation at high temperature (typically 500 to 800°C in air) of quasicrystalline particles.[42] The growth of the β-phase is due to the formation of an amorphous oxide layer of increasing thickness at the surface of the particles, which causes Al depletion in the *i*-phase. Inevitably, if the particles loose too much of aluminium, and in agreement with the phase diagram, the *i*-phase must transform to β-$Al_{50-x}(Cu,Fe)_{50+x}$. Figure 16 shows the main diffraction

24 *Jean-Marie Dubois*

peaks of the *i*-phase (18/29; 20/32) and β-phase (110) after annealing the powder at 625°C for 9 hours either in primary vacuum, in air or in pure nitrogen (N_2 in the following). First, we see that similarly to what happens in the presence of oxygen, a phase transformation occurs under N_2. The growth of the β-phase with a lower Al content than the *i*-phase suggests the formation of an aluminium nitride surface layer. Secondly, as can be seen from the intensity ratio of the β-(110) peak to the *i*-(18/29) peak, the transformation rate under nitrogen appears much larger compared to the growth rate observed in other atmospheres (here, low vacuum and air).

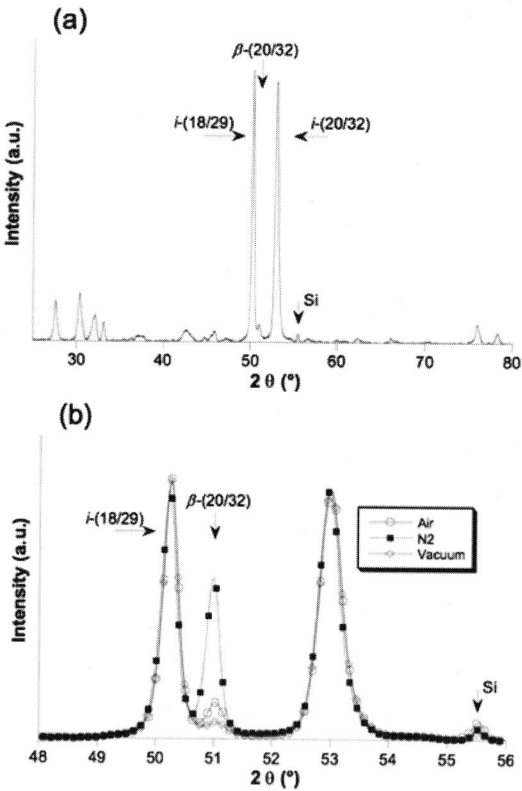

Fig. 16. X-ray diffraction patterns (λ=KαCo) of a) i-AlCuFeB powder annealed in vacuum (10^{-2} mbar) at 625°C during 9 h and b) measured on the same powder, but after annealing in different atmospheres (vacuum, air or nitrogen) at 625°C for 9 h. Silicon peaks for internal calibration are marked by arrows. Only peaks with the largest intensity are indexed (Courtesy: S. Kenzari, IJL, Nancy).

Figure 17 shows the weight gain of i-AlCuFeB powders measured by thermo-gravimetry analysis during thermal treatments performed at different temperatures and in the same atmospheres. When the thermal treatment is performed in air, the weight gain kinetics shows a parabolic behaviour, with almost no evolution anymore when the temperature is varied from 625°C to 700°C. This behaviour is the consequence of the rapid passivation of the surface by an amorphous oxide layer. The growth of the oxide layer implies Al diffusion from the bulk toward the oxide-air interface explaining the parabolic kinetics, in agreement with a previously published article.[42] As a result, the weight increment in air is in any case lower than 1% under the experimental conditions used in the present study. No crystalline oxide can be detected by XRD.

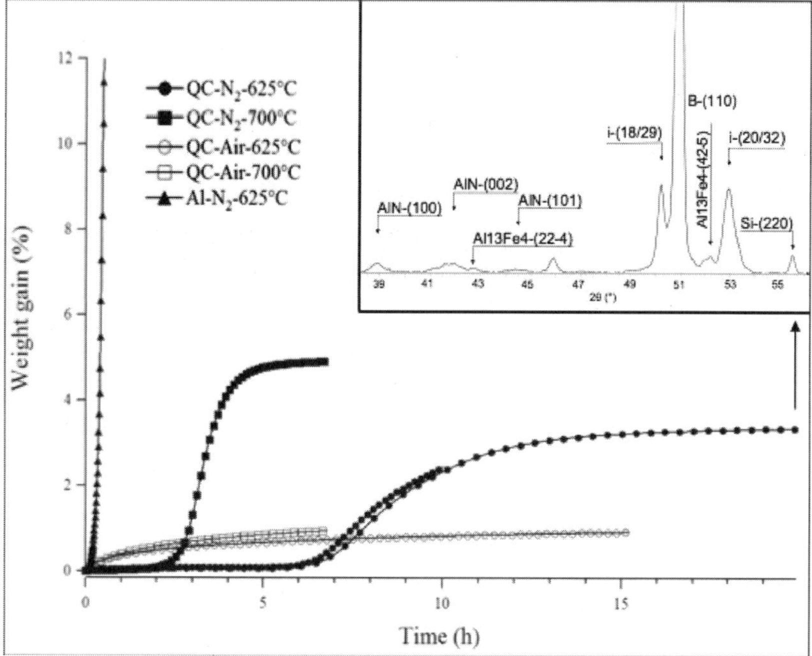

Fig. 17. Weight gain measured by TGA on the i-AlCuFeB powder during heat-treatment in various conditions as indicated in the top left part of the figure (QC stands for i-AlCuFeB; Al for fcc aluminium). Observe the marked difference in transformation kinetics, from parabolic behaviour in the presence of oxygen to S-shaped, Avrami type under N_2. Inset: XRD pattern recorded after annealing the powder at 625°C for 20 h in N_2, showing the presence of i-, β- and λ-phases together with hexagonal AlN (Courtesy: S. Kenzari, IJL, Nancy).

When the thermal treatment is performed in nitrogen, the weight gain is much larger and exceeds 3.3% or 4.9% at 625°C or 700°C respectively, due to the formation of a nitride at the surface. Several observations can be made from Fig. 17. First, the weight gain kinetics increases with temperature, which can be inferred to the non-passivating character of the nitride layer. Secondly, we observe a delay before the reaction starts. This incubation time decreases with increasing temperature, from 7 h at 625°C to 2 h at 700°C. For the purpose of comparison, the weight gain of an pure Al powder (99.8% in weight) measured under N_2 atmosphere at 625°C is also reported in Fig. 17. No incubation time is observed in this case and the nitridation kinetics appears much faster than for the *i*-phase placed in air, with a mass gain exceeding 16% after just 1 h.

This difference may be due to the different nature of the native oxide layer present on Al and on the *i*-phase, respectively, before the experiment begins. The alumina layer formed on Al is already crystallized at pretty low temperature (ca. 400°C) whereas it is still amorphous up to 625°C on the quasicrystalline phase[42] (Fig. 18).

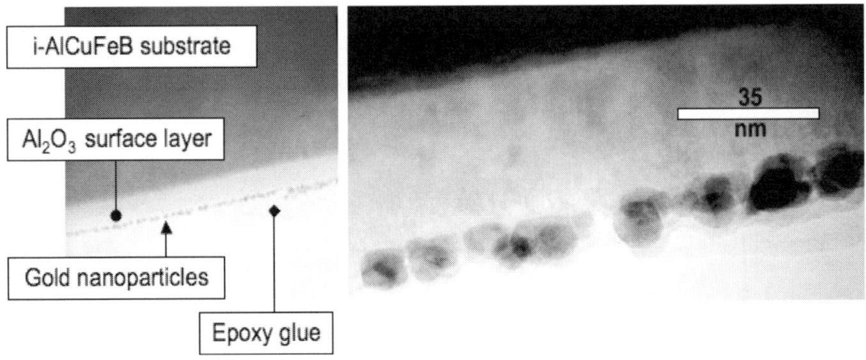

Fig. 18. Transmission electron micrograph of a thin sample cut perpendicular to the oxidized surface of an AlCuFeB icosahedral substrate held at 500°C for 144h. Despite the long duration of the heat treatment, the oxide later formed at the surface is still amorphous (right). Gold nanoparticles were dispersed on the initial sample in order to mark the top surface of the oxide (left) (Courtesy: G. Bonhomme, LSG2M, Nancy and St Gobain CREE, Cavaillon).

Aluminium diffusion through the crystallized layer (especially, at grain boundaries) should be faster than for an amorphous layer, explaining the rapid growth of the AlN layer on Al particles. In the case of *i*-AlCuFeB powders, we can speculate that the incubation time corresponds to the time needed to disrupt the amorphous oxide layer. A possible scenario includes the diffusion of atomic nitrogen through the oxide, followed by the formation of aluminium nitride or oxinitride at the interface that would crack the oxide layer. Once this amorphous layer is mechanically disrupted, Al diffusion is faster allowing the growth of the AlN layer.

The phases contained in the i-AlCuFeB powder after heat treatment under N_2 have been investigated by XRD. When the weight gain is about 2%, only icosahedral and β phases are detected and the XRD pattern is similar to the one presented in Fig. 16, indicating that the volume fraction of the AlN layer is low. For higher weight gain, two other phases are detected: hexagonal AlN and monoclinic λ-$Al_{13}Fe_4$ (insert of Fig. 17). The formation of the λ-phase is in accordance with the phase diagram where $i + \beta + \lambda$ phases can be stabilized due to a depletion in aluminium of the *i*-phase. Compared to the oxidation reaction, two main differences are observed: the nitrided layer is crystallized whereas the oxidized layer is amorphous under similar conditions and phase transformations during oxidation are always limited to the formation of the β phase. Figure 19 shows a few SEM images of two AlCuFeB particles glued by a distinctive layer. Associated EDS maps (Fig. 19a) and profiles (Fig. 19b) of this zone clearly demonstrate that this layer contains mostly aluminium and nitrogen elements and that the nitride grows at the surface of the particles. Therefore, aluminium diffusion controls the $i \rightarrow \beta + \lambda$ transformation.

Conclusively at this stage, comparison of nitradation[45] to oxidation experiments[42] proves useful since it sheds light on the role of a passivating (under oxygen) and non-passivating (under N_2) layer on the surface transformation kinetics. In both cases, the progressive depletion of Al in the bulk of the i-AlCuFeB particles results in the appearance of the crystalline cubic β-Al_{50-x} $(Cu,Fe)_{50+x}$ phase first, followed by the monoclinic λ-$Al_{13}Fe_4$ compound for longer thermal treatments, in accordance with the phase diagram.

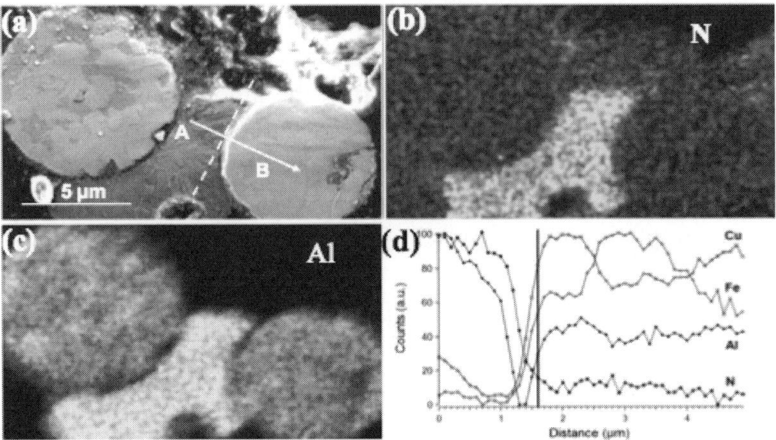

Fig. 19. Scanning electron microscopy (SEM) images and concentration profile observed after annealing the i-AlCuFeB powder in N_2 for 9h at 625°C: (a) Secondary electron SEM image of two AlCuFeB particles glued by a thick nitride layer; (b, c) Corresponding EDS maps for Al and N of the same zone. (d) EDS elemental profiles taken along the segment [A-B] as indicated by the arrow in (a). The straight vertical line corresponds to the interface between the particle and the nitride layer, as indicated by the dashed line in (a) (Courtesy: S. Kensari, IJL, Nancy).

4.3. *Atom transport*

Careful studies of atomic transport were performed essentially for quasicrystals and compared to previous knowledge of diffusion in common metals and solid compounds of aluminium and transition elements, most often with a short lattice period.[46] Figure 20 summarizes this data, showing immediately that atom transport is by essence not different from the one encountered in metals and simple binary compounds, i.e. is mediated by vacancies. To this end, see for instance the binary Al-Fe compounds led in the left hand side of the figure. The only departure from conventional analysis is observed in the central set of data (dotted lines), which show the existence of two diffusion regimes, one above 750K and one below.[30] So far, only very little work was dedicated to this finding, which was however observed on two diffusing species, Pd and Au, and not on the transition metal itself (here Mn). Aluminium is extremely difficult to study as such because of the very low activity of its radioisotope and in fact the experiment was never attempted. The author and his colleagues in Nancy made another indirect trial based on an oxidation experiment managed *in situ* in an Auger scanning analyser.[47] The final result is drawn in the right upper part of Fig. 20 (dashed line). The absolute value of the diffusion coefficient is set at an arbitrary level and has no meaning in itself. The two-regime mechanism is however clearly visible and yields activation energies of 2.2 eV and 0.6 eV, which fit with vacancy-mediated transport and phason-assisted mobility, respectively, as already explained earlier in this section.

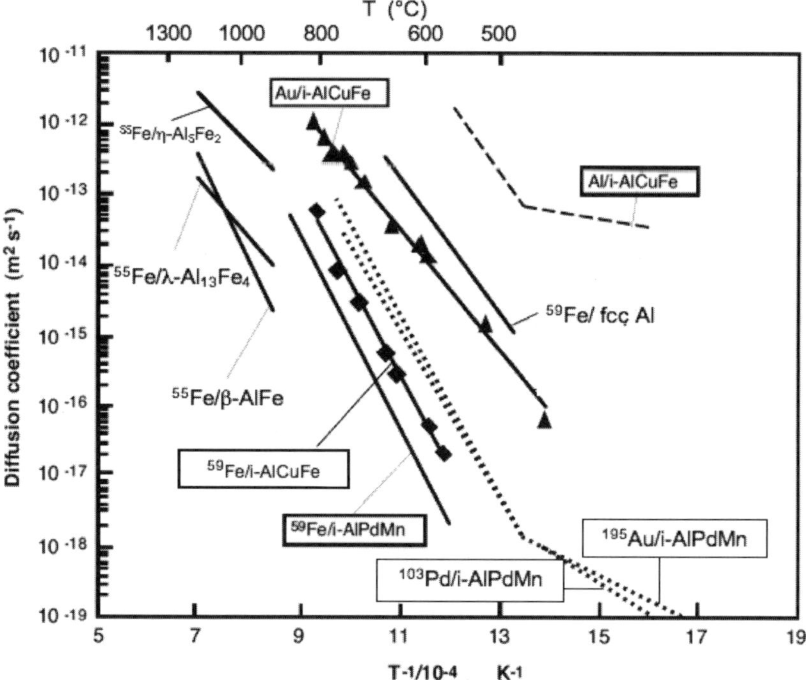

Fig. 20. Diffusion coefficient versus inverse temperature as deduced from radio-tracer experiments. Each label gives the tracer used versus (/) substrate. Observe that the slope of each line, which is proportional to the activation enthalpy of diffusion is essentially the same for all experiments, including Fe-diffusion in fcc Al or Al-Fe crystalline compounds. The dashed line at the top right side of the figure was deduced from an oxidation experiment managed *in situ* in an Auger analyser. Only slopes and position in temperature of the regime change are meaningful.

4.4. *Essential mechanical properties*

Aluminium-based CMAs, and most of the icosahedral and decagonal quasicrystals discovered so far, are very brittle at room temperature. A comparison is made in Table 1 between their most salient mechanical characteristics at room temperature, and the ones of conventional metals and non-metals.[48] Clearly, the low to very low toughness of these materials embodies their main drawback against technological applications. As a consequence, applications are restricted to technology niches, which we will discuss in Section 8 of this chapter.

Table 1. Comparison between essential mechanical properties at room temperature of two icosahedral and one periodic CMAs with conventional metals and non-metals.

	Hardness	Fracture toughness K_{IC}	Young's modulus	Density
	(H_v)	$(MPa\,m^{1/2})$	(GPa)	(g/cm^3)
CMAs				
Icosahedral Al-Pd-Mn (single grain)	700-900	0.3	200	5.1
Icosahedral Al-Cu-Fe (polygrain)	800-1000	1.6	68-85	4.6
β-Al$_3$Mg$_2$	350	<0.5	68	2.3
Common Metals				
Copper	100	30	130	8.9
Aluminium	25	40	70	2.7
High-carbon steel	880	<50	200	7.8
Low-carbon steel	120	50	200	7.8
Non-Metals				
Yttria-doped zirconia	1000	3	100-200	5.6
Alumina	1950	4.7	300	4.0
Diamond	>5000	3-4	1200	3.5

The situation looks entirely different at elevated temperature, say above 900K for aluminium-TM CMAs. Under compressive stress, the materials become plastic and flow with no work hardening after an initial narrow elastic stage during which strain is proportional to stress, following Hooke's law.[49] Representative curves of the stress-strain relationship[50] are presented in Fig. 21. They are typical for all large unit cell and quasicrystalline CMAs studied so far, and are entirely different from the one that can be recorder with a conventional alloy like a steel or a light alloy (Al-, Ti-, Mg-basis, …). In those alloys, the straight elastic regime observed at low deformation is followed by a work-hardening regime during which both stress and strain first increase on average, often by small successive jumps, before a final drop preceding fracture. CMAs do not show this behaviour at all.[50] The elastic regime is followed by an abrupt yield drop until a nearly flat plateau is reached during which a dynamical state is established with a balance between hardening and softening processes (Fig. 21). Relaxation experiments may be performed, as shown in the figure, to probe the thermodynamic parameters associated with the raise of plasticity.[50] Once again, this type of plasticity is not met in metals and alloys, especially those made of aluminium or magnesium,[51] and resembles more that of silicon.

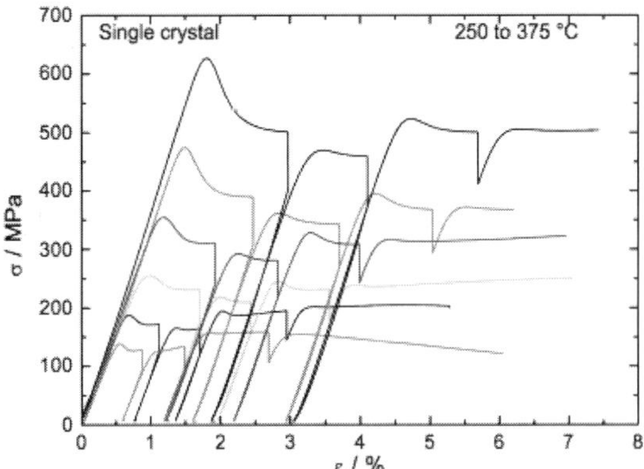

Fig. 21. True stress-true strain curves obtained in compression at a strain rate of 10^{-4} s^{-1} for β-Al$_3$Mg$_2$ samples. The two black curves are for single crystals and the grey ones for multi-grain materials. The temperature of the test was varied between 250°C (top curve) up to 375°C (bottom curve). Observe the linear elastic regime on the left side of the curves followed by the plastic regime, with no work hardening, on the right (Courtesy: M. Feuerbacher, Res. Centre Jülich. *CMA*).

Understanding the origin of this specific type of stress-strain curves was the matter of a discussion among experts. Transmission electron microscopy shows direct evidence that dislocations mediate plasticity, like they do in normal crystals.[52] Their dynamics must however be specific in order to interpret correctly the work-softening stage (as opposed to common work-hardening) described above. A first model, initially worked out in Jülich[53], suggested that the rigid clusters examined in previous sections, oppose a supplementary energy dissipation term to the glide of dislocations. Fair agreement with experimental data was found and it as used by the present author as a rationale to understand better the mechanical properties of quasicrystals.[1] It turns out however that a more careful study of the mobility of dislocations, especially during *in situ* straining, proves that dislocations move by climbing, not by gliding. The movement is due to the occurrence of phason-flips and leads to plasticity mediated by meta-dislocations.

4.5. *Metadislocations*

The notion of metadislocation is probably one of the most exciting, and unifying, concepts that have emerged from two decades of research on quasicrystals and CMAs.[54] It is illustrated in Fig. 22 for the case of a deformed single crystal of Ψ-Al-Pd-Mn. The Ψ-Al-Pd-Mn phase is a superstructure of an approximant of the decagonal phase called ξ'–Al-Pd-Mn (*Pnma*; a = 2.354, b = 1.656 and c = 1.234 nm[55]).

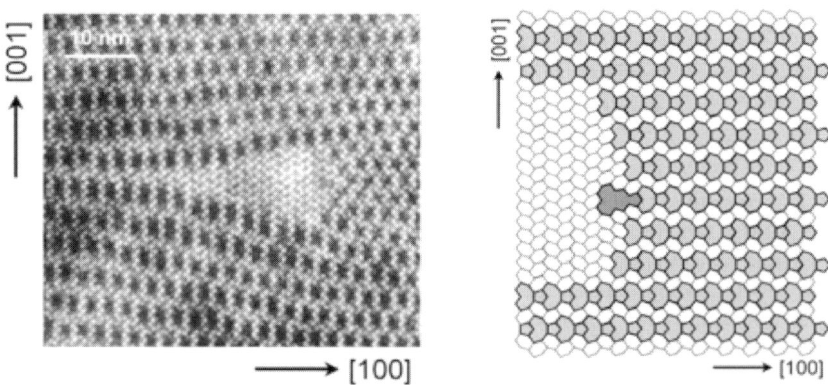

Fig. 22. Core of a metadislocation observed by high-resolution transmission electron microscopy (TEM; left) in a thin foil of Ψ-AlPdMn after mechanical straining. The image is interpreted (right) using a tiling model of the Ψ-AlPdMn. The dark grey area in the centre of the figure corresponds to the same area visible in the TEM and to a Burger's vector of 0.183 nm. Deformation is accounted for by an intergrowth of the tiling units of Ψ-AlPdMn with flattened hexagons in alternating orientations, which are characteristic of the parent ξ'-AlPdMn phase (Courtesy of M. Feuerbacher, Res. Centre Jülich, *CMA*).

The left hand side part of Fig. 22 presents an experimental observation of a dislocation core: six extra layers of tiling units are inserted in the right, yet leading to a Burger's vector that is an order of magnitude larger than an usual one. On the right of the figure is proposed an interpretation of the high-resolution image that is based on a tiling model of the Ψ-Al-Pd-Mn structure, mostly visible in the right side of Fig. 22. The raise of deformation is accounted for by assuming a local intergrowth of the Ψ-AlPdMn host phase with its parent structure, the ξ'-AlPdMn phase. The later is characterized by a tiling of flattened hexagons in alternating orientations (white units), whereas the former exhibits a more complex scheme based on small pentagons and banana-shaped tiles (light grey units). It was shown by Beraha *et al.*[56] that a transition from one type of units to the other is easily achieved by phason-flips, which therefore happen to act as the major vehicle for plastic deformation in CMAs.

5.　Phase Selection

The variety of solid structures encountered in compounds made of metals that go from simple, close packed and periodic to non-periodic arrangements of atoms, raises questions about why Nature builds so sophisticated and complex systems. More generally, the question of the stability of a given phase or compound was addressed long ago by Hume-Rothery[57] and others.[58] What essentially matters is the formation of a gap in the electron density of states due to the scattering of electrons at the Fermi energy by the Bragg planes of the crystal. When this arises, a gap opens at the Fermi level and a substantial fraction of the electronic states see their energy decreased towards the top of the valence band, which in turn stabilises that specific architecture and selects the corresponding phase. Friedel and Denoyer[59] in France, few others in different countries[1] noticed early after the discovery of quasicrystals that the mechanism predicted by Hume-Rothery applies fully to these new compounds and may explain why they can be readily observed. Section 5.1 hereafter will discuss this point. Reality however turns out to be more subtle as was elegantly described by Mizutani and his co-workers in their throughout and persuasive study[60] of γ-brass phases (Section 5.2).

Finally, experimental observation of the formation of d-like states just below the Fermi energy in Al-Mg, Al-Mg-Zn and Zn-Mg compounds, knowing that none of Al and Mg constituents yield significant d-states in their density of states, leads us in Section 5.3 to propose a supplementary mechanism to explain the stability of these materials.

Clearly enough, knowledge of the electronic structure is one of the most important means that may help to approach an answer. The electronic structure can be investigated by various methods, among which Soft X-ray Spectroscopy techniques.[61] In strong contrast to photoemission spectroscopy, which investigates the total DOS (but requires knowledge of cross sections for a better understanding of the total DOS), for the valence band (VB), Soft X-ray Emission Spectroscopy (SXES) probes separately the distribution of partial occupied states around each chemical species in the solid, averaging data over the various atomic sites. At a given energy, the intensity of the emitted x-rays is proportional to the probed partial density of states (DOS). Therefore, comparison of the shapes of the distribution curves for a given spectral character around a specific component in various materials informs about changes of the electronic interactions from one compound to another. Thus, as we show below, we have access – separately – to the $3s+3d$ sub-bands and $3p$ sub-bands of – separately - Al and Mg in an Al-Mg alloy (unfortunately, quantum mechanics does not allow us separate $3s$ from $3d$ states). We will extensively use such results in the following of this section. Calibration details, including positioning the Fermi edge on the x-ray transition energy scale, may be found elsewhere.[62] Because the transition probabilities involved in the x-ray transition process that gives rise to the signal are not known, all data can only be normalised for each sub-band, individually. Total densities of states are therefore not accessible.

SXES delivers partial DOS corresponding to a single species in a compound and therefore, each spectral curve refers to the atom that produces the signal. We insist on this: the spectral character pointed out in a series of SXES experiments belongs to the source atom, in this sub-section either Al or Mg, and not to the other alloy constituents. SXES intensities need to be expressed in arbitrary units. To this end, the most frequently applied scheme in the SXES community is a normalization to maximum intensity, instead of a normalisation to the total integrated intensity, which as a matter of facts makes little difference. After normalisation, intensities obtained from various samples, but in the same experimental conditions, may be compared.

5.1. *Hume-Rothery rules*

Three main factors control the structure of an alloy: the electrochemical factor, the size factor and the valence electron factor. Hume-Rothery intermetallic compounds are mainly determined by the third factor, i.e., by the valence electron concentration. As explained earlier, Jones[58,63] was the first to point out that Hume-Rothery compounds are stable because their Fermi surface interacts with important Bragg diffraction planes.

Stability of the crystal structure implies the existence of a minimum of the free energy F = U-TS, where U is for the enthalpy, or internal energy of the system, T is temperature and S entropy. To a very large extent, the internal energy U depends on atomic size ratios, electronegativity differences and a band structure term. Like in simple Hume-Rothery alloys, many complex metallic phases including Al-TM compounds exhibit small differences in electronegativity and atomic size. In this case, it is the band structure term E_b and its minimization that dominate the stability of the system. This band structure term writes:

$$E_b = \int^{E_F} \varepsilon n(\varepsilon) d\varepsilon \qquad (1)$$

where ε stands for the electronic energy and $n(\varepsilon)$ for the number of states with energy ε.

The mechanism by which minimization of the band structure term can be understood is the following. In a nearly-free electron approximation, the scattering of the electronic waves by a Bragg plane defined by reciprocal lattice vector K (Fig. 23a) opens a gap in the dispersion relation $\varepsilon(k)$ at k = K/2 (Fig. 23b). The magnitude of the gap scales like V(K), the Fourier component of the potential. Integrating over all directions in reciprocal space, Bragg scattering produces a depletion in the density of states $n(\varepsilon)$. Such a depletion will fall at the Fermi energy

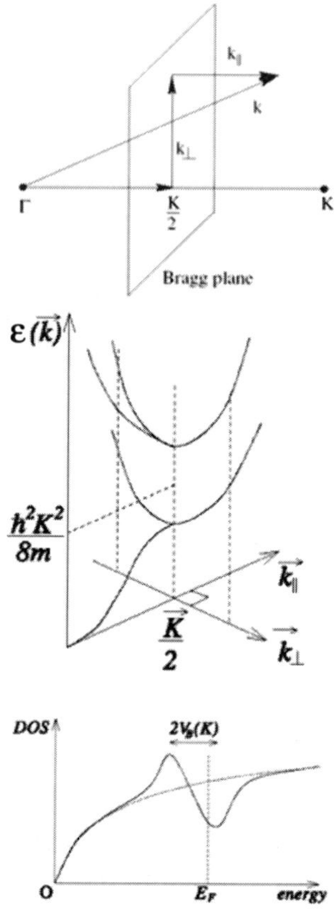

Fig. 23. a) Bragg plane corresponding to a reciprocal lattice vector K. b) Dispersion relation for a nearly free electron in the vicinity of the Bragg plane with potential V_K. c) Density of states in the nearly free approximation as compared to the free electron case. Reprinted with permission from G. Trambly de Laissardière.[64]

if and only if the Fermi sphere is tangent to the Bragg plane (Fig. 23c). In this case, namely when $2k_F \sim K$, the band term is lowered as compared to the free electron case.

The criterion $2k_F \sim K$ implies a condition on the number of valence electron per atom e/a defined by:

$$(e/a) = \int_{-\infty}^{E_F} N(E).dE \qquad (2)$$

The Hume-Rothery rule implies that the e/a ratio is such that the Fermi sphere of radius k_F is tangent to the Jones zone, which is the

Brilouin zone constructed from Bragg planes with strongest scattering potential (see Fig. 12). Bragg planes corresponding to strong scattering potentials can be identified with the most intense peaks in X-ray diffraction patterns, as diffracted intensities depend on V_K as well. Accordingly, the depletion in $n(\varepsilon)$ is strong (large V_K), the band term is minimized and the compound is stabilized by an electronic effect. The dip produced in the density of states at E_F is called a pseudo-gap.

Initially, Hume-Rothery rules were discovered for simple *sp* alloys but they also apply in *sp-d* alloys, like the Al_9Co_2 or Al_3Ni aluminides. The most emblematic system in this respect is Cu-Zn, which forms well defined compounds at composition $Cu_{50}Zn_{50}$, with a B2 CsCl-type of structure, and e/a=3/2, and the Cu_5Zn_8 γ-brass type, with 52 atoms per unit cell and e/a=21/13. Another example is the γ-Al_9Cu_4 compound, also with e/a=21/13. Raynor[65] showed that the structure of these alloys is related to e/a, with the Fermi surface being close to Bragg planes corresponding to intense diffraction peaks. A negative valence was given to the transition element, to account for the coupling between *sp* and *d* states, which was ascribed as a charge transfer from the conduction band to the unfilled *d* band. However, it was soon realized that the notion of charge transfer would lead to unphysical electrostatic energy. More recently, Trambly de Laissardière *et al.*[64] showed that *sp-d* coupling contributes to deepen the pseudo-gap and enhances the number of *sp* states below the Fermi level. This introduces a more acceptable interpretation of the negative valence associated with transition metals, without requiring the notion of charge transfer.

An interpretation of the former Hume-Rothery rule was based by Blandin in 1960s on pseudo-potential theory.[58] In this model, the band structure term contributing to the total energy is described as a sum of pair interactions $\Phi(r_{ij})$ between atoms presenting Friedel oscillations for large inter-atomic distances r_{ij}. Then, the electron band term is minimized if the atoms sit in minima of the pair interactions, which implies a correlation between r_{ij} and the Fermi wavelength $\lambda_F = 2\pi/k_F$, equivalent to the Hume-Rothery condition $2k_F \sim K$.

In 1987, Friedel and Dénoyer[59] identified the newly discovered icosahedral quasicrystals like Al-Li-Cu or Al-Cu-Fe as a new class of Hume-Rothery alloys. Despite the lack of periodicity, their diffraction pattern shows sharp peaks and a Jones zone can be constructed from the most intense Bragg planes. Due to the high symmetry of the icosahedral quasicrystal, the multiplicity of Bragg planes is large and therefore the Jones zone is nearly spherical, thus producing a large number of contact points with the Fermi sphere, all with strong interaction potential. This is shown in Fig. 24 for both a quasicrystalline structure and a crystalline approximant. The higher the symmetry of the Jones zone, the closer it fits with the Fermi sphere. Therefore one expects a stronger depletion of the density of states at the Fermi level in quasicrystals than for simple crystals of lower symmetry. Thus quasicrystals and related complex metallic alloys are considered as a particularly stable case of Hume-Rothery alloys.

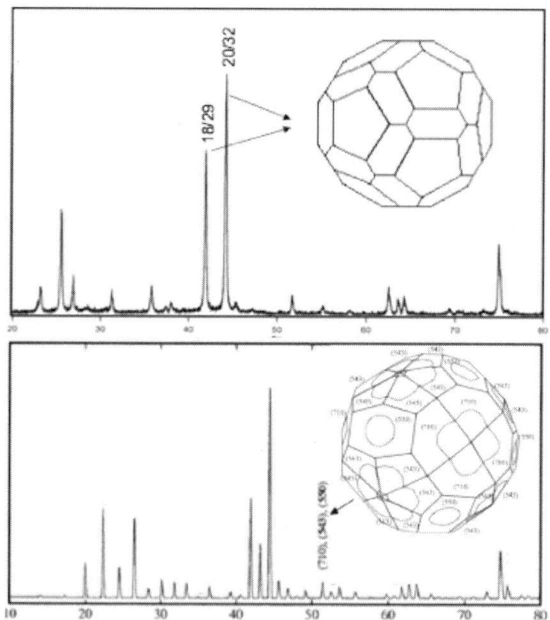

Fig. 24. a) X-ray diffraction pattern of the icosahedral Al-Cu-Fe quasicrystal and corresponding Jones zone with 42 facets constructed from the (18/29) and (20/32) main reflections. b) X-ray diffraction pattern of the Al-Cu-Ru-Si 1/1 approximant and the most prominent Brillouin zone consisting of the 48-fold {543}, 12-fold {550} and 24-fold {710} Bragg planes.

The Hume-Rothery stabilization mechanism implies that alloys of a given structure will only form at a specific chemical composition. This has been noticed for example in the Al-Cu-Fe system. Assuming a valence of -2 for Fe, +1 for Cu and +3 for Al, it is found that the quasicrystalline phase occurs along a line of constant $e/a = 1.86$. In the close vicinity of this line, related complex ternary phases were also discovered. Also, Tsai[66] used the e/a criterion to discover many stable quasicrystalline phases in Al-TM1-TM2 systems. All these phases are isostructural to the icosahedral Al-Cu-Fe phase and they all have the same e/a ratio (see next sub-section). More recently, Tsai and his group discovered a new family of quasicrystal and approximant in the CdYb system, with an $e/a = 2.0$, and a different structure type.[67] Again, isoelectronic substitutions led to the discovery of many quasicrystalline and giant unit cell crystals in Cd-RE (rare earth), Cd-Ca, Ag-In-RE, Au-In-RE, etc.[68]

5.2. *More on specific Al-TM CMAs*

We examine in this subsection the case of a few Al-TM complex compounds from the point of view of Hume-Rothery interactions and their composition dependence. One consequence of the appliaction of the Hume-Rothery rule is that the compound must form at a specified value of e/a. This is directly related to the chemical composition of the compound:

$$e/a = \Sigma\, n_i\, c_i = \text{constant} \tag{3}$$

where n_i is the contribution of element i to the valence band and c_i is the atomic fraction of element i. In the case of the AlCuFe quasicrystal, the ideal composition is found around $Al_{62.3}Cu_{24.9}Fe_{12.8}$ and the valence contributions n_i for elements Al, Cu and Fe are, respectively: +3, +1 and -2 according to Gratias *et al.*[69] Hence, $e/a = 1.862$, and since $c_{Al}+c_{Cu}+c_{Fe} = 1$, equation (1) is equivalent to:

$$c_{Fe} = 0.228 - 0.4c_{Cu} \tag{4}$$

This relation means that there is only one degree of freedom for the composition changes if e/a must keep constant. Therefore it can be expected that the composition zone of the AlCuFe icosahedral phase tends to extend according to Eq. (4), as pointed out by Gratias *et al.* in the experimental diagram.[69]

What is interesting is that this straight line also passes through or near crystalline approximants, as is illustrated in the schematic Al-Cu-Fe concentration diagram of Fig. 25: λ-$Al_{13}Fe_4$, Ψ-$Al_{62.3}Cu_{24.9}Fe_{12.8}$, ϕ-$Al_{10}Cu_{10}Fe$ and ζ-Al_3Cu_4. We expect that the radius of the Fermi sphere k_F will also keep constant if the atomic density ρ does not change much along this line. As a matter of fact, this is the case for the λ, Ψ and ϕ phases whose ρ values are respectively 67.7, 66.7 and 67.2 at.nm^{-3}. The atomic density ρ of the ζ-Al_3Cu_4 phase is higher (72.8 at.nm^{-3}), so that its $2k_F$ value is also relatively larger (31.8 nm^{-1}).

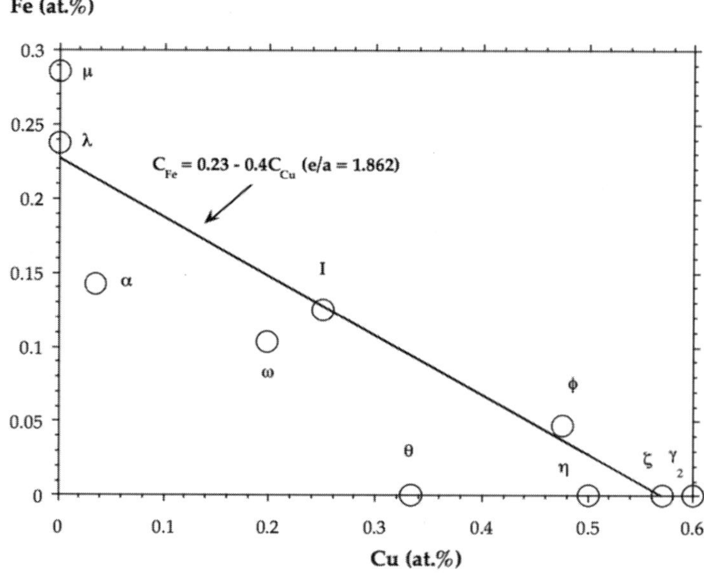

Fig. 25. Schematic concentration diagram of the Al-Cu-Fe system. The e/a-constant line, with e/a = 1.86, passes through or near four phases: λ-$Al_{13}Fe_4$, Ψ-$Al_{62.3}Cu_{24.9}Fe_{12.8}$, ϕ-$Al_{10}Cu_{10}Fe$ and ζ-Al_3Cu_4. This diagram is drawn according to the data reported in Bradley and Goldschmidt[70] and Faudot *et al.*[71]

The values of $2k_F$ for these Al-Cu-Fe phases are compared with the positions K of the most intense Bragg peaks in Fig. 26. The ζ-Al3Cu4 phase is not presented here because of lack of structure data. We see immediately that the intense reflections of the three phases lying on the e/a-constant line are quite near their respective Fermi diameters, thus satisfying the $2k_F \approx K$ condition. In comparison, for the other phases, either only few intense reflections are near their respective $2k_F$ values (i.e. θ–Al_2Cu and ω-Al_7Cu_2Fe phases), or these reflections are relatively remote from the Fermi vector (e.g. the γ-Al_2Cu_3 phase with $2k_F = 31.8$ nm^{-1}). In particular, the θ-Al_2Cu phase is a well-known tetragonal phase whose structure is mainly determined by close packing. An exception is found for the μ-Al_5Fe_2 phase, whose e/a value is quite inferior to that of Ψ-AlCuFe but whose intense peaks are all near the Fermi diameter due to its high atomic density ($\rho = 71.8$ at.nm^{-3}). It is interesting to note that this phase is the first phase formed during solidification from the liquid state in the $Al_{65}Cu_{20}Fe_{15}$ alloy[72], before the appearance of the λ-$Al_{13}Fe_4$ and Ψ phases.

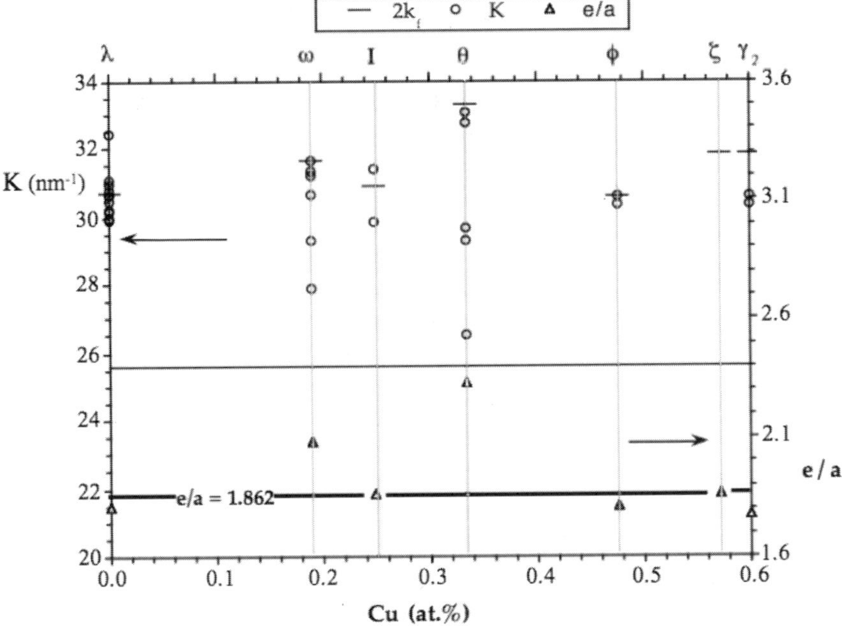

Fig. 26. Position of the intense peaks of phases λ, I, θ, ω, ϕ, ζ and γ (open circles) as well as of the corresponding diameter of the Fermi sphere $2k_F$ (short horizontal lines) plotted as a function of Cu content. As a comparison, the outer electron concentration per atom e/a (open triangles) is shown in the lower part of the figure for each phase. The bold line corresponds to the constant e/a = 1.86 ratio.

5.3. *The case of γ-brass type CMAs*

The so-called γ-brass phase belongs to the family of CMAs, but is one of its simplest member due to a cubic unit cell containing only 52 atoms (Fig. 27). It is therefore an excellent choice for performing complete sets of calculations of the band structure, using *ab initio* methods. This is the reason why it was selected by Prof. Mizutani and his group at the University of Nagoya for a throughout and deep study of the mechanism that stabilizes this compound.[24,73] Furthermore, a broad variety of binary compounds adopt this structure type, which in turn allows varying the nature of the electronic states found in the system. This phase is dealt with in details by Mizutani in this series of books.[73] We will therefore only summarize his most important results.

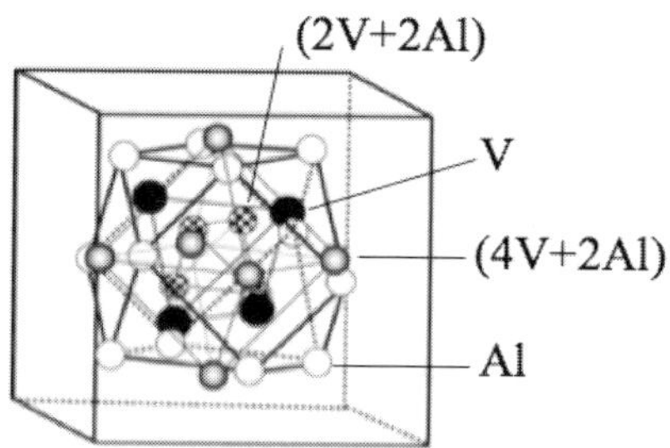

Fig. 27. Experimentally determined positions of the atoms surrounding the centre of the cubic cell in γ-Al_8V_5. In total, there are two such atom shells per unit cell, i.e. 52 atoms in the unit cell.

A first-principle computational method called FLAPW (for Full Potential Linearized Augmented Plane Wave method) was used in a systematic study of a large number of binary compounds that crystallize in the γ-brass structure type. For many such compounds, but not all, Mizutani could prove that stability occurs because the electron density matches the Hume-Rothery criterion, e/a = 21/13. This is the case of a few tens of binary compounds, among which γ-Al_4Cu_9, Cu_5Zn_8, Ni_2Zn_{11}, etc. They could confirm that the stability of those CMAs of g-brass crystal structure originates from interaction of a spherical Fermi sphere touching the {411} and {330} facets of the Jones zone. This corresponds to reciprocal vectors such that $|G|^2 = h^2+k^2+l^2 = 18$ in units of $(2\pi/a)^2$ where a is the γ-cubic unit cell edge length and h, k and l, the conventional Miller indices. The occurrence of a pseudo-gap below E_F is characteristic of these compounds.

Noticeable exceptions exist however, like the γ-brass Co_2Zn_{11}, Al_5V_8 or Ag_5Li_8 compounds, which cannot be stabilized by a similar mechanism, although their structure type is the same. Using computational techniques, Mizutani's group could show for instance that the γ-Al_5V_8 compound is stabilized by a more complex mechanism, combining on the one hand orbital hybridization between Al$3p$ and V$3d$ states, and on the other hand a resonance of V$3d$ states with a series of lattice planes, including not only the $|G|^2=18$ set, but also a full range of planes such that $14 \leq |G|^2 \leq 30$. The V$3d$ resonance was found "to play a key role in condensing the bonding states below the Fermi level to gain electronic energy".[74] We will examine a similar case in the next sub-section.

The case of γ-Ag_5Li_8 is different.[75] First, the effective e/a = 1 ratio that corresponds to stability departs very significantly from the ratio e/a = 21/13 assigned by the Hume-Rothery rule. Second, based on a careful screening of the effect of electronic hybridization between specific atomic pairs, it could concluded that some Ag-Ag and Ag-Li pairs are critically important to produce orbital hybridization in the Ag$4d$ band. These states in turn favour the stability of the γ-type structure against its bcc counterpart.

5.4. *The case of Al-Mg(-Zn) alloys*

We now examine a case that cannot be entirely studied on a computer due to the very large size of the unit cell, namely the Samson β-Al_3Mg_2 phase already introduced in Section 3.1 above. For comparison purposes, we report also on data gained with the Bergman phase (also introduced above in Section 3.1) and a simpler phase of Laves type, Zn_2Mg. The XES technique was extensively applied to the investigation of the valence band (VB) structure in this series of Al and Mg-based compounds, at both edges of Al and Mg. Special attention was paid to Al s,d and Mg s,d distributions in the pure fcc Al and hcp Mg metals, respectively. As explained earlier, s and d states are obtained together. Because transition probabilities favour d states with respect to s ones, we focus at those states and come to the conclusion that complexity of the crystal structure is linked to the occurrence of *d*-like states below, but close to, the Fermi energy. This conclusion is of special relevance since the individual constituents (except Zn), namely fcc Al and hcp Mg, show very little *d*-states in their respective band structures.

To achieve this conclusion, we first locate *d*-like states on the binding energy scale in Al-TM and Mg-TM samples. These states form due to *sp-d* hybridization and are readily observed on the partial DOS of the *sp*-element. Then, we focus at Al-Mg(-Zn) alloys, and show that they exhibit a pronounced *d*-like peak just below the Fermi energy. Simple computations in a fcc unit cell with 4 average atoms ((3Al+1Mg)/4 combined in 1 single atom) confirm the *d* character of those states, including when a Au_3Cu order is enforced to mimic the β-Al_3Mg_2 compound. We assume that these *d*-like states most probably originate from the opening of Van Hove singularities at the top of the valence band, via Bragg scattering of *sp* electron waves by dense atomic planes (unfortunately, computer simulations cannot be achieved with the most complex crystal to confirm this hypothesis). We then point out that the abundance of *d*-like states compared to *s*-states depends on the crystal complexity: the more complex, the more abundant.

Accordingly, a supplementary term is added to the stabilization mechanism. It results from self-hybridization between *s-p* states and *d*-like states. This new term favours the selection of the most complex phase instead of a simpler packing, in strong contrast with what is usually assumed in conventional metallic crystals.

5.4.1. *Locating d-like states in Al-TM based alloys*

We start first with a brief overview of the partial DOS in a typical Al-TM CMA. This will help us in the following section to locate *d*-states on the binding energy scale and with respect to the position of the Fermi edge and the other partial DOS. To exemplify, Fig. 28 shows a set of partial distribution curves in the binding energy scale representing the valence band of stable Al_3Ni. This is a compound of the cementite family, with Ni atoms inserted in tri-capped triangular prisms of Al. In the Al *3p* DOS, a significant proportion of Al states lie in a narrow

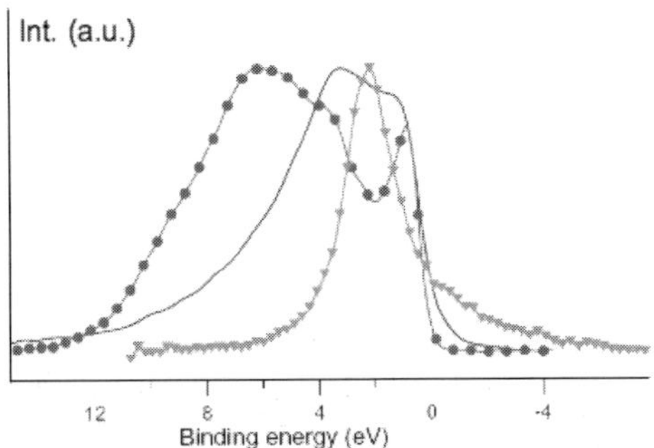

Fig. 28. Partial densities of states measured by XES for Al$_3$Ni. The solid line is for Al$3p$ states, the line with triangles for Ni$3d$ states and the dotted line for Al $3s$ and $3d$ states, which are measured together.

energy range of about 1eV below E$_F$. DOS calculations have shown that Al $3s$ states extend between 4 and 11 eV below E$_F$ whereas Al $3d$ states are hybridized to Al $3p$ states in the upper part of the valence band below E$_F$.[76] This is also true for many other systems including approximants of quasicrystals, for which valence band calculations[77,78] have pointed out that Al $3s,d$ states close to E$_F$ are essentially of d character. According to DOS calculations available for Al-MT systems, these states have a localised, or for short a d-like, character.

Conclusively at this stage, we have shown experimentally that in stable complex structures d-like states arising from hybridization effects of Al states with TM states are present close to E$_F$, at the top of the valence band.

5.4.2. *Alloys based on Al, Mg, and possibly containing Zn*

Fcc Al and hcp Mg are free-electron metals with essentially sp electronic states, and also a faint proportion of d-like states located close to their valence band edge. Actually, according to Papaconstantopoulos,[78] the valence band of Al contains 1.55 s, 1.38 p and 0.07 d electronic states (total: 3 e$^-$/at.) whereas in hcp Mg, there are 0.77 s, 1.12 p and 0.11 d (total: 2 e$^-$/at.) electronic states. In both Al and Mg elemental constituents, and in the expected energy range for the d-like states below the Fermi energy, X-ray emission spectroscopy (XES) yields a pretty sharp peak that arises in the experimental partial s,d DOS on top of a smooth and broad massif (Fig. 29). With a closer view at the data (inset of Fig. 29), and starting from Fermi energy, which is set to E = 0 eV in the figure, going down to increasing binding energies as is the habit in XES, we recognize on the Al *3s,d* curve, first a very narrow peak superimposed on top of a slightly broader peak, itself placed on top of the broad density that extends over more than 10eV.

The very narrow peak in aluminium originates mainly from a many-body effect following the x-ray emission process.[79] It is shown enlarged in the inset of Fig. 29 for the Al *3s,d* DOS. For hcp Mg, it can only hardly be distinguished from the *3d* component. Its integrated intensity is negligible and it will be disregarded in the following comparisons between Al and Mg curves. It does also not appear in the DOS of the binary or ternary compounds. The other two components of the DOS of the pure metals are the partial densities of *3d* states, i.e. the narrow peak about 0.5-1.5 eV broad, and the partial *3s* DOS, the much broader bump extending to large binding energies in Fig. 29. Not shown in Fig. 29 is the *3p* DOS that is measured separately.

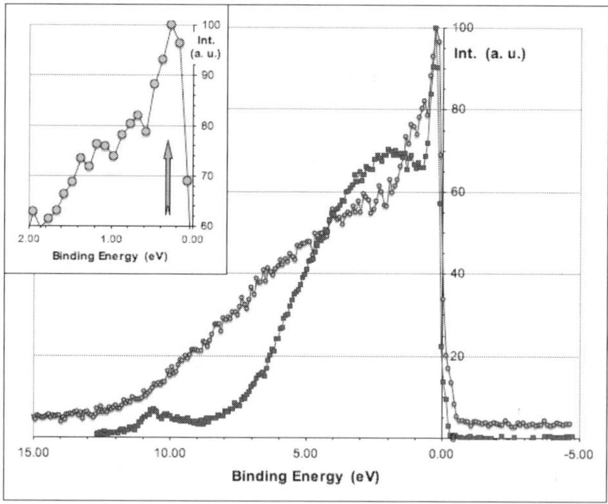

Fig. 29. Partial 3s,d DOS measured by XES for pure Al (dots, light grey line) and Mg (squares, dark grey line). The inset shows an enlargement of the narrow peak observed for fcc Al below the Fermi energy. The very narrow part involving the many-body effect is indicated by an arrow in the inset. In Mg, the peaks due to *3d* states and to the many-body effect overlap and cannot be separated. Courtesy: E. Belin-Ferré, *CMA*.

The compounds for which we have investigated the Al and/or Mg $3s,d$ distributions are the stable Samson β-Al_3Mg_2 phase, actually of composition $Al_{61.5}Mg_{38.5}$ and one metastable quasiperiodic counterpart of composition $Al_{61}Mg_{39}$, noted β'-Al-Mg in the following, which was claimed an aperiodic cubic crystal structure.[80] In the present context, we will consider it as a defected, superstructure version of the low temperature Samson phase.[8] We also studied the Laves phase $MgZn_2$,[25] and finally stable crystalline Bergman[23] phases of composition $Mg_{39.5}Al_{30}Zn_{30.5}$ (Al edge) and $Mg_{39.5}Al_{40}Zn_{20.5}$ (Mg edge). For convenience in the following, we will refer to these compounds by β- and β'-Samson, Bergman, Laves, and fcc-Al or hcp-Mg. It should be noticed already that the number of atoms in each unit cell decreases along the series, from 1168, 586, 162, 12 to 4 and 2.

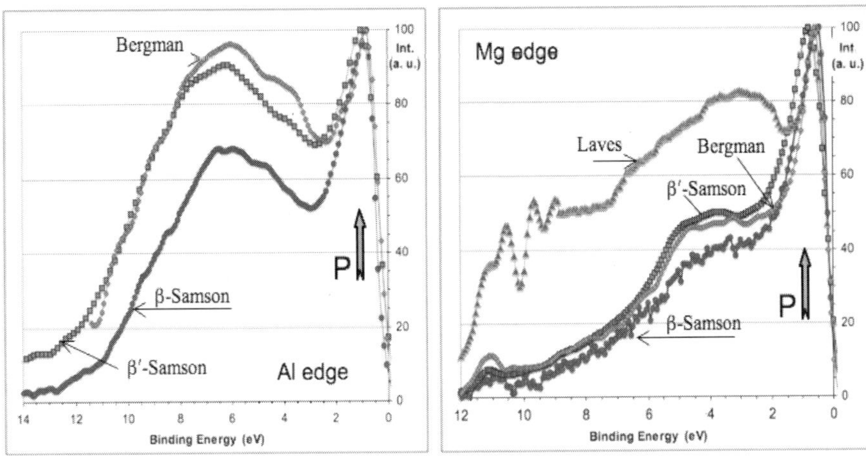

Fig. 30. Experimental $3s,d$ DOS measured at the Al edge (left) and Mg edge (right) for the β-Samson phase, its low temperature β'-allotropic form, the Bergman phase and Laves phase (as indicated in the figure). Oscillations visible at high binding energy on the Laves phase curve is an experimental artefact that does not affect the discussion in text. Courtesy: E. Belin-Ferré, *CMA*.

For all the compounds, the Al $3s,d$ as well as the Mg $3s,d$ curves consist in a broad peak near E_F (denoted P hereafter) followed by a large marked bump whose maximum intensity is set around 4 eV below P (Fig. 30). This emphasises a complete redistribution of the $3s,d$ states in the compounds. In an attempt to put this observation on a more quantitative basis, we have measured the energy extent W of the base of peak P in each of the compounds. This was done first by extrapolating the part of the curve corresponding to the bump on the low energy side of peak P, and second by measuring the energy distance between this point and the point on the other side of peak P at the same intensity. Measuring furthermore the height H (intensity difference between maximum intensity of the peak and its base line) supplies us with the approximate area A=WxH/2 of the peak P (which is sufficiently accurate for the present purpose). The extent of the spectral distribution of the peak, of the order of $0.9 \leq W/2 \leq 1.2$ eV, signals a well-defined peak and therefore the occurrence of localized, or equivalently *d*-like, states. We will come back to this point in the following. Similarly, it is easy to estimate the area of the DOS underneath the broad bump. This data will be used in the following subsection. Figure 31 shows now the Al and Mg *3p* DOS measured in the β-Samson phase (for the present purpose, we do not need to present to whole set of data). Peak P is also shown for the sake of the discussion (small dots for Al and solid line for Mg).

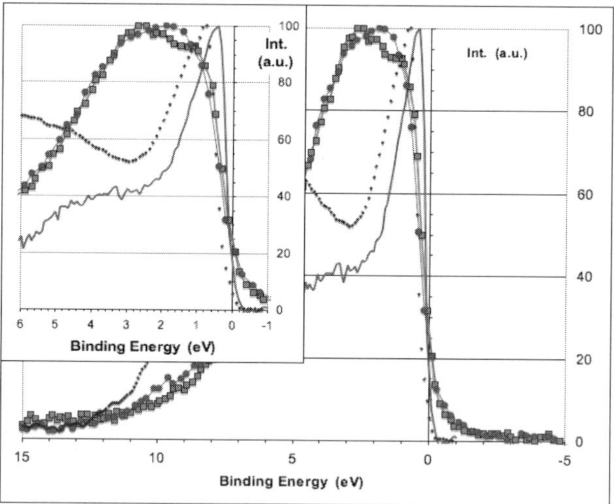

Fig. 31. Al *3p* (squares) and Mg *3p* (circles) distributions measured in the β-Samson phase together with the respective *3s,d* DOS (small dots for Al and solid line for Mg). The inset presents a blow up of the energy region between Fermi energy and 6 eV. Courtesy of E. Belin-Ferré, *CMA*.

Quite clearly, the positions of the components of the band structure are strongly correlated: observe for instance the coinciding position of the maximum of the Al$3p$ band with that of the local minimum of the Al $3s,d$ band. Similarly, the maximum of the Mg $3p$ band superimposes on a local minimum of the Al $3p$ band, and the maximum of the Al $3d$ peak goes with the shallow, secondary maximum of the Al $3p$ component at about 1eV from E_F. From this analysis, it can be concluded that i-bonding Al $3p$ states split the Al $3s,d$ band into bonding Al $3s$ states and anti-bonding Al $3d$ states, ii- Al $3d$ states are more bonding than Mg $3d$ states, iii- Al $3d$ states hybridize with Al $3p$ states and, iv- Mg $3d$ states also hybridize with Mg $3s$ states.

The experimental intensity measured at a specific energy by XES is proportional to the product of the DOS at that energy, multiplied by the transition probability between initial and final states involved in the photon emission process, and convoluted by a broadening function that is resulting from finite time life of the core hole and spectrometer resolution. The transition probabilities depend on the spectral character of the DOS, and are usually unknown. They depend on the local potential experienced by the probe atom and do not vary with energy. Usually, it is therefore not possible to derive quantitative data from the integrated intensities of the partial DOSs. In the present case however, we can assume that the potentials are not excessively different in the close packed metal and, for the same species, in the equally dense Samson phase. This in turn allows us to compare the areas under the peak P in the pure metal and in the compound for a given emission edge and try to derive a qualitative estimate of the number of states involved, knowing the abundance of such states in the pure metal. This approximation however involves drastic assumptions that can hardly be verified by theoretical means: a) the transition probabilities do not vary with energy, b) they are identical in the metal and in the compound, which certainly introduces quite some uncertainty on the numbers reported below, c) peak P is due to d-like states: its position falls within a narrow range of no more than 2 eV below the Fermi energy, and its width at half-maximum intensity of about 1.0 to 1.5 eV (or less) clearly points out localised states. For this reason, we call it a peak of d-like states, although the elemental constituents, Al and Mg, carry very little

                    Jean-Marie Dubois

true *3d* states (2.3% and 5.5% in a total of 3 e⁻/at. and 2 e⁻/at., respectively, for Al and Mg). This label is consistent with the following results, but a numerical study, which cannot be applied yet to β-Al$_3$Mg$_2$ would be necessary to confirm this point.

Within this approach, if the peak P for Al represents 0.07 *d* states per atom, then the peak P in the β-Samson phase is for 0.18 *d*-like st./at. Similarly, for Mg, one obtains 0.7 *d*-like st./at. in the compound. Going one step further, it is possible to refer the area under the remaining part of the *3s,d* DOS (the broad bump in Fig. 30) and estimate the number of *3s* states involved. Within the crude assumptions emphasized above, one finds that the ratio between transition probabilities for *s* and for *d* states is nearly equal to 1 for the Al edge, but is as high as 3.2 for the Mg edge. Both values make sense and were used to build Table 2. The table also gives the number of *3p* states as the difference between the total number of states (2.62 st./at. in β-Al$_{61.5}$Mg$_{38.5}$) and the sum of *3s* and *3d* states.

Table 2. Number of *3s*, *3p* and *3d* states in the pure metals and in the ordered β-Samson and disordered β'-Samson phases, determined as indicated in text. Owing to the assumptions made, caution must be taken when considering these numbers.

	Al			Mg		
Compound	*3s* states	*3p* states	*3d* states	*3s* states	*3p* states	*3d* states
Pure metal	1.55	1.38	0.07	1.12	0.77	0.11
β-Samson	0.68	1.76	0.18	0.13	1.77	0.72
β'-disordered	0.97	1.49	0.14	0.19	1.74	0.67

Compared to the pure metals, the results in Table 2 show dramatic changes of the band populations. In the ordered Samson phase, the *3d* band yields now about 2.5 times more *d*-like states that experience the creation of the core hole on Al sites during the XES process and 6.5 on Mg sites. Simultaneously, the number of *3s* states has been reduced, whereas that of *3p* states is increased. This result is in agreement with a previous observation that *3p* states at E$_F$ account for the transport properties of many Al-TM quasicrystals and CMAs.[81] A small difference is observed between β and β' forms of the Samson phase, which is discussed below. Altogether, the data in Table 2 are in line with the qualitative conclusions drawn above for orbital hybridization between *d*-like and *sp* states.

5.4.3. *A supplementary mechanism for phase selection and stability?*

Let us now consider more quantitatively the peak P and see how it evolves in the various samples of this study. It is visible in Fig. 30 that the area underneath peak P is a function of the nature of the sample: the larger the unit cell size, the higher and broader the peak. This observation can be put on a more quantitative basis as shown in Fig. 32, again within the assumption that the transition probability between initial and final states involved in XES is independent of the specimen for a given emission edge.

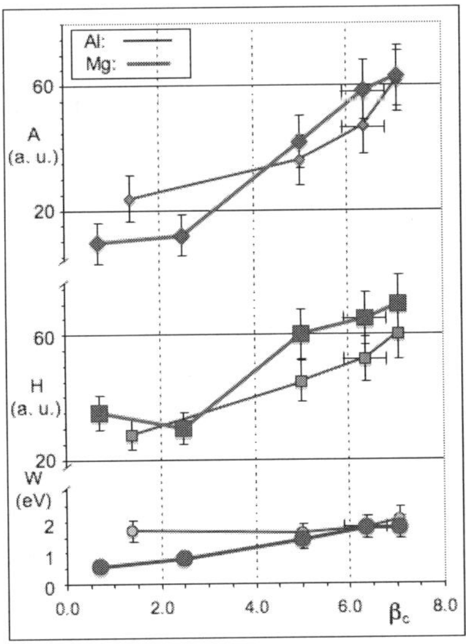

Fig. 32. Evolution of the full width W, height H and area A of the *d*-like peak (noted P) as a function of the complexity index defined in text. Large symbols are for the Mg partial *3d*-like DOS and small symbols for the Al *3d*-like DOS.

The x-axis in Fig. 32 is organised along increasing complexity of the samples. To put numbers in, the data is ordered according a complexity index defined as $\beta_c = Ln(N_{uc})$, where N_{uc} is the number of atoms in the unit cell. Up to a constant, this is identical to Ln (number of degrees of freedom) of the compound, which is itself a measure of the entropy of the system (many other choices are of course possible). N_{uc} is well known for all compounds, except for the disordered β' phase ($293 \leq N_{uc} \leq 2 \times 293$ at. due to the existence of planar stacking faults), which explains the large error bar on β_c shown in Fig. 32.

Happily enough, the smooth evolution of the curves in Fig. 32 confirms the validity of the assumptions made. More importantly, it points out a correlation between complexity of the crystals and formation of d-like states, presumably by orbital hybridization at the expense of s states as concluded already in the previous section. This has an immediate, and fundamental consequence. The d-like states form just below the Fermi energy, and they hybridize with the other states of different spectral character. Therefore, they contribute to stabilize the structure. Furthermore, their contribution to stability increases in proportion, or at least in proportion, to complexity. In other words, this mechanism selects the most complex compound preferably to a more simple crystal structure. This surprising result goes just opposite to common sense arguments based on the knowledge of conventional metals. It is very much in agreement with the conclusion of Mizutani et al.[74], but for a compound (γ-Al_5V_8) containing *3d* states brought in by the TM, whereas in the present series, the Samson sample contain no transition metal at all and very little *3d* states.

Once more, numerical experiments would be necessary to assess the origin of those *3d*-like states. Since this is not feasible in the most complex compounds, let us see qualitatively what goes on in the β-Samson compound. An X-ray diffraction pattern is presented in Fig. 33. According to composition and the respective contributions of Al and Mg to the filing of the valence band (e/a=13/5=2.62 e-/at.), the Fermi diameter is expected to fall at $2k_F=31.6$ nm^{-1}. This value is slightly exceeding the position K_P in k-space of the most intense diffraction peaks, which fall in a rather broad range of 25 up to 31.5 nm^{-1}. The classical Hume-Rothery stability mechanism can therefore not be at work. Furthermore, the electron per atom ratio of 13/5 is not the one expected according to the Hume-Rothery rule, which is e/a=21/13. Furthermore, the Fermi sphere is (slightly) larger than the Jones zone and many sharp diffraction peaks are present inside a shell located close to,

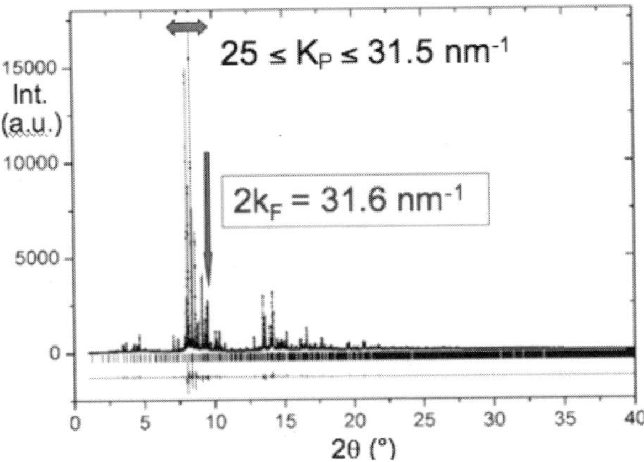

Fig. 33. Diffraction pattern of the β-Samson phase obtained at a synchrotron source (wave length λ=0.035 nm). The position of the Fermi diameter and the range of K_P values that define the Jones zone in the β-Samson compound are indicated by a vertical arrow and a horizontal double arrow, respectively. Observe that the Fermi sphere overlaps the Jones zone. Redrawn from Ref. 8, courtesy of M. Feuerbacher, *CMA*.

but below, $2k_F$. As a consequence, interactions with numerous Bragg planes operate and produce the opening of many van Hove singularities, like in γ-Al$_5$V$_8$, and as a matter of fact, the Bergman phase which present a similar overlapping of Jones zone by the Fermi sphere (not illustrated here). We assign to this mechanism the formation of the *d*-like states pointed out in the previous sub-section. The more Bragg peaks fall immediately below $2k_F$, or equivalently the larger β_c is, the more pronounced will be the stabilization effect of *d*-like states due to orbital hybridization with *s* and *p* states. Because *d*-like states are not brought by a transition metal, but form in a self-consistent way *via* interaction with the crystal itself, it seems appropriate to call this mechanism 'self-hybridization'.

6. Properties of Al-Transition Metal(s) CMAs

6.1. *The essential property of Al-TM CMAs*

The existence of a pseudo-gap at the Fermi energy is probably the most salient feature that characterises aluminium-TM complex metallic alloys.[82] The pseudo-gap is the depletion in the DOS that separates, when it exists, the top of the valence band from bottom of the conduction band (in insulators, it is the gap of forbidden energies which shows up there). The pseudo-gap is intrinsic to Hume-Rothery compounds. We will study in this section a number of Al-based crystals and quasicrystals of the Al-Cu(-Fe) system, which show a more or less pronounced pseudo-gap at E_F. Later in this chapter, we will come to the conclusion that it has deep influence on several properties. It is not always existing however in the total DOS, and is actually associated with the Al $3p$ band in some Al-TM CMAs like the decagonal phase, but not necessarily to the other bands.[83]

A low DOS at E_F was measured by different techniques, consistent with the existence of the pseudo-gap. A typical value of the total DOS at E_F deduced from specific heat measurements is about 1/3 of the free electron value for CMAs based on sp elements and is as low as 1/10 of the free electron value in Al-TM CMAs. Similarly, a clear decrease of the spectral intensity was found in the vicinity of the Fermi cut-off in photoemission spectra of many CMA systems, which can be fitted to retrieve quantitative parameter of the pseudo-gap (width and depth).

This procedure is unfortunately uncertain in systems like Al-Ni-Co where the spectral intensity is dominated by TM d states in the vicinity of E_F. In contrast to photoemission, XES, which has the advantage of both chemical and momentum selectivity associated with dipolar selection rules, has proved very useful to determine the partial density of states around each element in the compound.[62] In Al-TM systems, XES shows that Bragg scattering combined with sp-d hybridization leads to the formation of a wide pseudo-gap in the partial Al $3p$ DOS.

Another method used to probe the DOS at E_F is core-level XPS. Core level lines in X-ray excited photoemission spectra of metallic compounds show an asymmetric tail due to the screening response of the valence electrons to the creation of the core hole.[83] Intrinsic energy losses linked to the excitation of electron-hole pairs that are created simultaneously with the core hole are responsible for the asymmetry of the line shape. The probability of these processes decreases rapidly with the electron-hole pair energy. Therefore, the main contribution comes from low-energy electron excitations across the Fermi level. This implies that the asymmetry of the XPS core level lines depends on the local DOS at the

Fermi level (Fig. 34). As a matter of fact, a systematic loss of the asymmetric tail of TM *2p* core levels was pointed out in highly complex CMAs like i-Al-Pd-Mn, i-Al-Cu-Fe, ζ'-Al-Pd-Mn and o-Al_6Mn, which is consistent with the existence of a deep pseudo-gap at the Fermi energy in these Hume-Rothery alloys.[85]

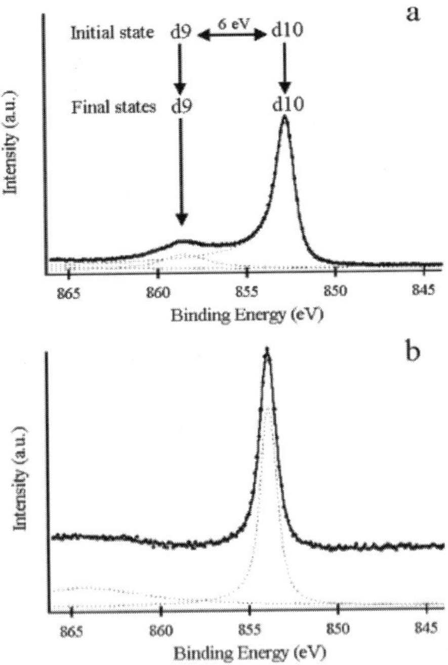

Fig. 34. Ni $2p_{3/2}$ core-levels in pure Ni (a) and in d- $Al_{72}Ni_{11}Co_{17}$ (b) measured by XPS. Courtesy of V. Fournée, *CMA*.

A series of Al-Cu(-Fe) Hume-Rothery compounds of different atomic structures and accordingly different Jones zones were studied using the XES technique in order to investigate the pseudo-gap dependence on crystal composition and structure. The valence band of all these compounds was analyzed with emphasis put on the Al *3p* DOS. In γ-Al_4Cu_9 as well as in all other Al-Cu(-Fe) samples, the Al *3s-d* and Al *3p* sub-bands split in two distinct parts located on each side of the maximum of the Cu *3d* distributions, around 4 eV.[86] Accordingly, the shapes of the Al *3s-d* and Al *3p* distribution curves depart dramatically from that in pure Al, which is a typical parabolic curve distorted by experimental broadening and cut-off effect at E_F (Fig. 29). This result simply means that the free-electron model is no longer valid for Hume-Rothery alloys. The interaction of Cu with Al can be interpreted within the framework of a Fano-like interaction between highly localized states and extended states.[87] The coincidence that the two Al partial spectral curves display both a marked depletion right at the energy of the maximum of the localized Cu $3d$ states proves that Al states still retain an extended-like character in these compounds (see Fig. 28 for an illustration of this effect, but with Al_3Ni, the Ni $3d$ states playing the same role as do Cu $3d$ states in the present context). Using the same methodology, we have analyzed several CMAs of much larger unit cell, including icosahedral (i-) and approximants crystals from the Al-Cu-Fe and Al-Cr-Fe systems. Again, like in B2-cubic β-$Al_{55}Cu_{33}Fe_{12}$, it was observed that the valence band of Al-Cu-Fe samples shows that Fe $3d$ states overlap the Al sub-bands edges near E_F, about 4 eV below E_F, whereas Cu $3d$ states and the Al sub-bands overlap each other over the whole extent of the valence band, namely over about 12 eV.

In all samples, it was observed that the intensity at E_F of the Al sub-bands departs from its value in fcc Al. It is lower compared to the pure metal, and the corresponding valence edge has no longer its half-maximum intensity set at E_F. In contrast, a more or less pronounced depletion appears in the Al DOS at E_F, thus pointing out the formation of the pseudo-gap. This is especially true for Al $3p$ states because (a) in pure fcc aluminium, these states are genuinely of extended character and therefore are more sensitive to changes of the electronic interactions than d states, and (b) they are obtained alone by XES whereas this technique delivers always d and s states together. To quantify the pseudo-gap of the Al $3p$ DOS, we shall restrict ourselves to using the intensity at E_F, labelled $n(E_F)$ hereafter. It is expressed in arbitrary units with the maximum intensity in the curve set to 100, and with a value $n(E_F) = 0.5$ in the pure metal, since the inflexion point of the DOS of a free-electron metal is by definition located at E_F, at half of the maximum intensity on the valence band edge. Meanwhile, the half-width δ (since the unoccupied states are not probed by SXES) broadens simultaneously. It is as small as $\delta=0.2$ eV in Hume-Rothery compounds (and $\delta=0$ in a free-electron system like fcc Al), and reaches values close to $\delta=1$ eV in highly complex compounds. A summary of a large number of numerical data for n(E_F) is given in Fig. 35.

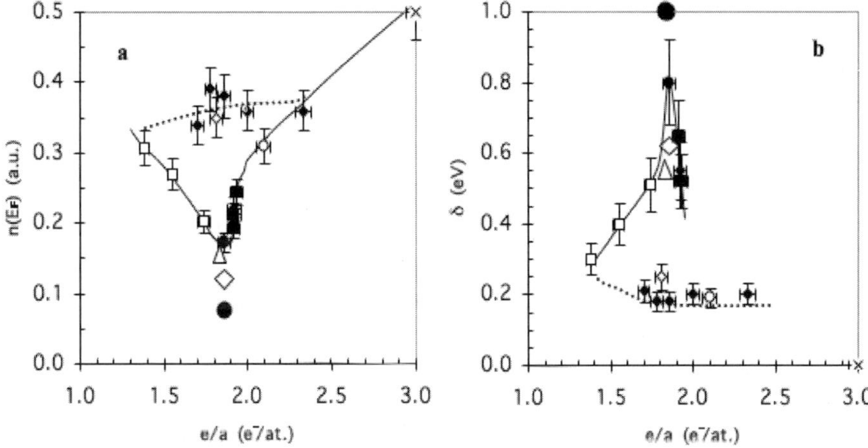

Fig. 35. a) Variation of n(E_F) with e/a ratio from fcc Al (cross at the top right of the figure) to a highly complex icosahedral Al-Pd-Re quasicrystal (large solid dot at the bottom of the figure). Also shown in between are the values of n(E_F) measured for Al-Cu Hume-Rothery compounds (small diamonds, dashed curve), various B2 β-AlCuFe phases of various compositions, icosahedral $Al_{62}Cu_{25.5}Fe_{12.5}$ (large black diamond), and its approximants in the Al-Cu-Fe system (black squares). The open dot is for ω-$Al_{70}Cu_{20}Fe_{10}$ and the small open diamond for φ-$Al_{10}Cu_{10}Fe$. For completeness, icosahedral Al-Cu-Ru (large open diamond) and λ-$Al_{13}Fe_4$ (open triangle) are also given. The solid and dashed lines is only to guide the eye. b) The same symbols are used for the half-width δ of the gap in the same materials.

Figure 35 demonstrates that the minimum of n(E_F) is obtained at e/a = 1.8-1.9 e⁻/at. for quasicrystals with i) a lattice of very high perfection and ii) containing a TM of the mid-series, preferably a *5d* TM alloyed with a TM of the right hand side of the series like Cu or Pd. However, clearly enough, the minimum of n(E_F) (respectively, the maximum of δ), cannot be taken as a unique property because CMAs of very large unit cell and nearly identical chemical composition exhibit almost identical values of n(E_F) as shown in the figure. This conclusion holds true for all CMAs known so far: physical properties vary in (some) inverse proportion to the size of the unit cell and undergo an extreme at the ultimate size of the unit cell, but no real gap is observed when the size of the unit cell

                          Jean-Marie Dubois

approaches its maximum, for instance an infinite size like in an icosahedral quasicrystal. Conversely, simpler compounds like the Al-Cu Hume-Rothery phases shown along the dashed line in Fig. 35 form a distinct branch, which is clearly separated and presents no extremum on this graph. Although the complexity index is much smaller, the comparison of the two branches in Fig. 35 emphasizes the respective influence of the Hume-Rothery mechanism and of the *sp-d* hybridization in the presence of a *3d*, *4d* or *5d* metal.

Actually, a pseudo-gap must be present in all Hume-Rothery compounds, but the surprise in Fig. 35 is to see that it may become very broad in (apparently) simple compounds like the B2 cubic phase (open squares in Fig. 35). This is related to the super-ordering effects described previously in Section 3.4. The depth and width of the pseudo-gap have therefore direct relationship to the complexity of the compound via the number of facets of the Jones zone that touch the Fermi surface. This was dealt with in the previous sub-sections. Before ending this sub-section, it should be kept in mind that we discussed here only the Ap *3p* partial DOS, because this density always shows the pseudo-gap when the Hume-Rothery effect, possibly complemented by orbital hybridization between *p* and *s* or *d* states is at play. The pseudo-gap is however not always visible in the total DOS since it may be hidden by the peak of *3d* states that fall at E_F when the compound contains TM like Cr, or Fe, or even Co, Ni as it is the case, already mentioned, of decagonal phases.

6.2. *Transport properties*

Experimental determinations of transport properties of CMAs are now numerous and supply most of the content of this volume. Measurements of the optical conductivity, thermal conductivity, thermoelectric parameters like the Seebek coefficient were achieved in a rather systematic way and contrasted to theoretical analysis, especially by Macia *et al.*[88] Despite their relevance, we will not report on this aspect of CMAs, due to space limitation, and concentrate only on resistivity data and their connection to $n(E_F)$ as discussed in the previous sub-section.

A blend of several resistivity measurements between liquid helium and room temperature is shown in Fig. 36 for several Al-based CMAs.[89]

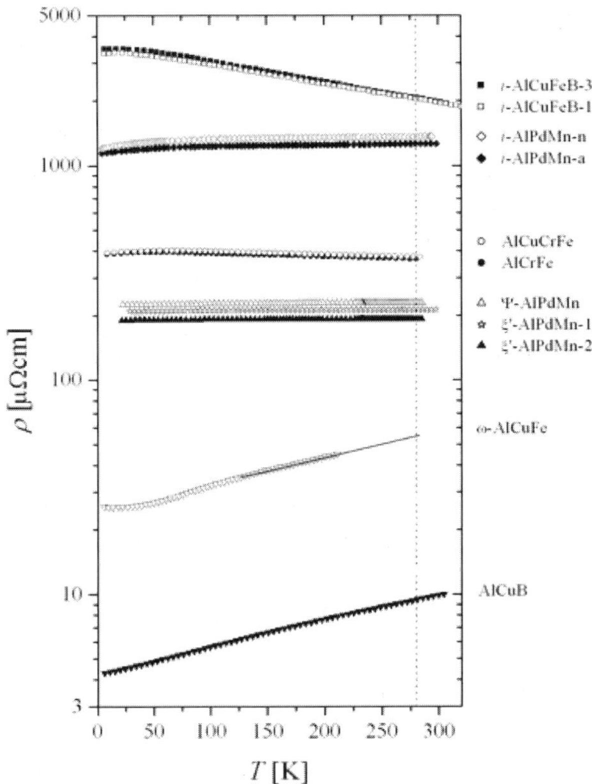

Fig. 36. Resistivity data measured below room temperature is a series of CMAs with complexity index increasing from bottom of the figure towards its top. Observe the sign change of the temperature coefficient of the resistivity, which falls to 0 with the ξ' and Ψ approximants of the AlPdMn decagonal phase. Numbers after the label of the compounds indicate different specimens. The label AlCuB stands for a small unit cell B2-AlCu phase stabilized by addition of a few at.% of boron; i stands for icosahedral, ω for the Al_7Cu_2Fe tetragonal compound. Two large unit cell orthorhombic compounds of the Al-Cr-Fe(-Cu) system are also shown.

The bottom part of the figure plots the data for simple compounds, with a low complexity index, whereas the top part is for CMAs, some of them like the icosahedral phase characterized by large values of β_c. Two different regimes coexist in the figure: low β_c compounds are associated with a positive temperature coefficient of the resistivity, like in normal metals, whereas large values go with a negative temperature coefficient of the resistivity and a large residual resistivity at low temperature. This behaviour is fully atypical for a system made of metals and resembles that of a semiconductor. In between, the ξ' and Ψ approximants (see Section 4.5) of the decagonal phase ($\beta_c = 5.8$ and 7.3, respectively[22]) remarkably exhibit a zero temperature coefficient of the resistivity over the full range from 20K up to 300K. This type of observation has triggered a large research effort to understand better the transition from usual metallic behaviour (small unit cell size) to a behaviour much more resembling that of a semi-conductor (yet, being not a semi-conductor!). In the first regime, the diffusion of charge carriers follows a ballistic law, i.e. is proportional to time t and Einstein's conductivity applies in proportion to $N(E_F)$, the total DOS at E_F. In the second regime, which is called hopping after Mott's theory[90] for disordered systems, conductivity is instead proportional to $[N(E_F)]^2$ and diffusion follows a t^β regime, with $0 \leq \beta \leq 1$.[91] Essentially, it was concluded that the breakdown of Bloch's theorem at large to infinite unit cell size is accompanied by the formation of so-called critical states, neither extended, nor fully localized-like in a totally disordered medium.[92] Further analysis by Mizutani[24,93] and others[94,95] has concluded to a transport mechanism by hopping in highly complex CMAs such as quasicrystals and their approximants, in total contrast to the usual behaviour of alloys.

A major experimental evidence in favour of this model is shown in Fig. 37, using data discussed in the previous section.[81]

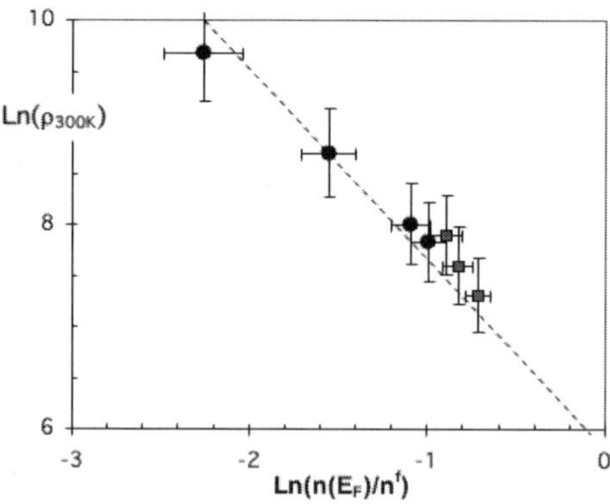

Fig. 37. Plot of Ln ρ at 300K versus $Ln(n(E_F)/n^f)$ showing a slope of -2 (dotted line) as expected by Mizutani from Mott's hopping conduction theory. For calibration purposes, the experimental value of $n(E_F)$ in each sample is referred to that measured in the same conditions with a sample of pure aluminium (n^f).

6.3. *Solid-solid contact*

6.3.1. *Fretting*

Many engineering mechanisms placed on board of spacecrafts exhibit solid-solid surface contacts. These contacts open and close periodically, in some circumstances for several thousands of times. In the mean time, vibrations during launch or during movement of movable accessories can lead to oscillating movements of micrometer amplitude in the contact interval. This type of contact movement is referred to as "fretting". Fretting can eventually degrade the mechanism surface layers. In general, it dramatically increases the tendency of the contacting surfaces to stick together, or "cold-weld", resulting in severe destruction and adhesion forces that jeopardize the whole mission of the satellite.

A special device was developed at the Austrian Research Centre of Seibersdorf (ARCS) in view of studying the effect of fretting onto various contacting bodies. It was used to probe several combinations of CMAs and bulk materials for their tendency towards cold-welding. The test proceeds via repeated closing and opening of a spherical pin to flat disc contact. The pin is brought into contact with the disc for several thousand times under a static load ranging from 20% to 60% of the elastic limit of the weakest material. During the test, the contact is closed without impact and, while being closed during 10 seconds, fretting is applied to the contact. A base pressure of 5.10^{-7} mbar is kept in the machine. The adhesion force, i.e. the force required to re-open the contact, is measured at each opening. Details can be found elsewhere.[96]

Two sintered, bulk icosahedral phases of compositions i-$Al_{62.5}Cu_{25.3}Fe_{12.2}$ and i-$Al_{59.5}Cu_{25.3}Fe_{12.2}B_3$ and one icosahedral single crystal i-$Al_{70}Pd_{22.1}Mn_{7.9}$ as well as five approximants, were analysed[97] against two different hard steels (quoted, respectively, AISI 52100 and SS 316L in the technical literature, see Ref. 96), tungsten carbide (noted WC hereafter), a Ti-based counterpart (Ti_6Al_4V alloy) and an Al-based alloy (noted Al7075 in the technical literature).

First, single crystals of the Al-Pd-Mn system (the icosahedral phase and its ξ'-approximant) were tested in vacuum. Figure 38 presents the adhesion force measured with quasicrystals, approximants and metallic alloys. It is always lower for the quasicrystals and their approximants compared to the metallic alloys versus themselves. Especially, the icosahedral phase of composition i-$Al_{70}Pd_{22.1}Mn_{7.9}$ shows no adhesion against the two steels. The situation however is not always that specific. No adhesion is indeed detected between the three icosahedral specimens in contact with steels AISI 52100, A286 and SS 316L (except for i-$Al_{59.5}Cu_{25.3}Fe_{12.2}B_3$ versus SS316L for which a very low adhesion is noticed). In contrast, high adhesion is found for this icosahedral compound against Ti6Al4V and Al7075 counterparts. Very high adhesion occurs for hard steels against themselves as plot in the figure.

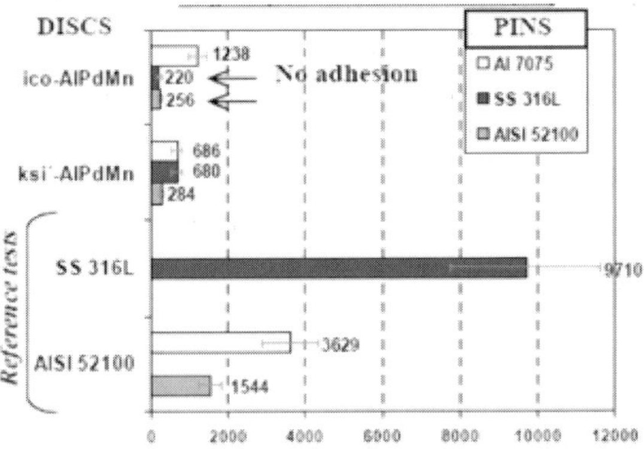

Fig. 38. Summary of fretting tests in high vacuum (10^{-7} mbar) for various disks (left hand side) against pins listed in the right hand side of the figure. Observe the very high adhesion observed for SS316L steel against itself. In comparison, pure quasicrystals against this same steel show no adhesion whatsoever. Redrawn from Ref. 97. Courtesy of M. Sales, ARC-Seibersdorf.

It is remarkable that the adhesion between the two different stainless steels and icosahedral crystals is much lower than for stainless steels in contact with themselves. A similar conclusion is valid for the aluminium alloy Al7075 versus itself. Therefore, the adhesion forces between quasicrystals and steels or titanium (or even the Al-based Al7075 counterpart) are systematically found to be negligible in comparison to typical metal couples used for space applications. The opposite trend is observed when the counterpart is an Al-based alloy, and to a lesser extent, WC.

A tentative explanation of this contrasted behaviour can be suggested here.[98] That the stick force depends on the nature of the metallic counterpart (here, essentially Al or Fe for respectively, the Al-based and hard steel counterparts) strongly suggests that it results from the formation of a band between electronic states in the disk sample and pin. In a very crude assumption, we shall assume the formation of a rigid band with density proportional to the number of *sp*, respectively *d*, states that merge with the Al, respectively Fe, states at the Fermi level (E_F) present in the Al-based or steel pins. The number of *s, p* and *d* states at E_F in our samples is obtained for the very same specimens by X-ray emission spectroscopy as explained earlier in this chapter. The result of this simple computation is shown in Fig. 39. Although far from perfect, matching with the experimental data given above in Fig. 38 or in literature[97,98] it is good enough to support our assumption.

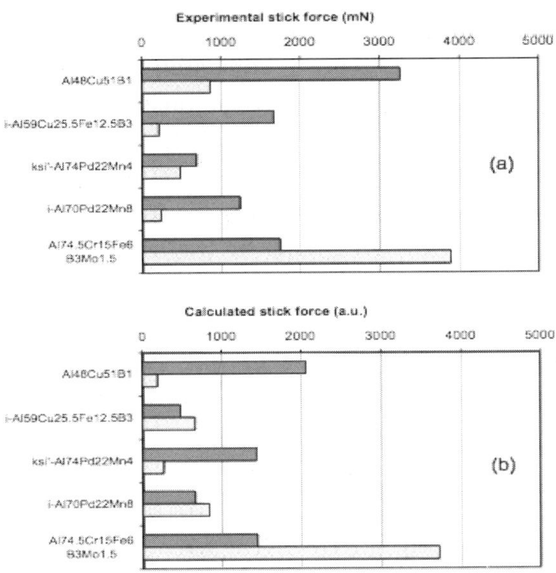

Fig. 39. (a) Experimental stick force (in mN) measured in 10^{-7} mbar vacuum on five compounds as labelled along the vertical axis and against spherical indenters: an Al-based alloy (dark grey) and an Fe-based steel (light grey). (b) Stick force (in arbitrary units), calculated assuming it is proportional to the number of electrons necessary to form a rigid *sp* band with the Al-based indenter (dark grey) or a *d* band with the Fe-based indenter (light grey).

6.3.2. *Friction anisotropy*

Using atomic force microscopy (AFM), Park *et al.* have pointed out that solid-solid contact on the surface of an anisotropic CMA is anisotropic itself. From AFM measurements in ultra-high vacuum on a decagonal single crystal of high lattice quality, they could prove that the directional anisotropy of the friction force is eight times larger along the periodic direction of the surface than when sliding along the aperiodic one.[99,100] The experimental conditions were selected in such a way that the contact (low load) was elastic and that no wear occured. The anisotropy was attributed to the peculiar structural anisotropy of the surface. It disappeared when an amorphous thin oxide film was formed after exposure to air. These observations raise the important question of whether the nature of the quasicrystalline material influences its tribological properties in practical situations and environments, including air and loads large enough to produce wear.

To address these questions, a comparative study of the friction properties of an $Al_{72}Ni_{11}Co_{17}$ decagonal quasicrystal was performed, using the two-fold symmetric surface (parallel to the 10-fold axis) with both AFM and a pin-on-disk tribometer. The specimens for AFM and pin-on-disk experiments were cut from a large single grain of d-$Al_{72}Ni_{11}Co_{17}$ grown at Ames Laboratory, Iowa State University. The pin was a 6 mm diameter hard steel ball (the same high carbon Cr-steel, AISI 52100, as used for the fretting tests described earlier). Due to the fact that the sample surface could not be aligned perfectly horizontal, the pin (when in contact with the sample) oscillated along a vertical direction by a distance of a few micrometers. This was used to follow the angular rotation of the disk with good resolution, although with an arbitrary phase relative to the angular position of the periodic axis of the decagonal crystal. The run distance in these experiments was a fraction of a meter.

The friction force along the two crystallographic directions was measured using AFM as a function of load by scanning the probe tip over a 200 nm line in each direction. A plot of the nanoscale friction force *vs* load (Fig. 40) shows that there is no noticeable difference in friction force along the two directions up to 1.5 μN. Indeed, the amorphous aluminium oxide film present at the surface of the single grain prevents direct contact of the tip with the underlying quasicrystalline material at low loads and explains this isotropic behaviour.

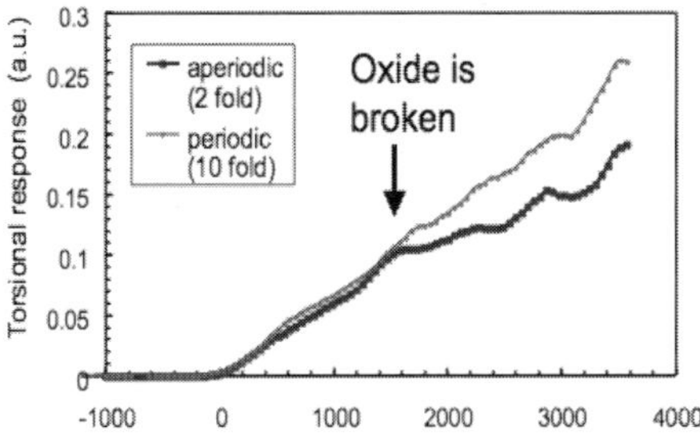

Fig. 40. AFM assessment of the friction force developed in ultra-high vacuum at the surface of a decagonal single grain along periodic and aperiodic directions. Above a certain threshold (here, approx. 1500 mN), the pin pierces the oxide surface layer and comes into contact with the raw material, showing a significant anisotropy in correspondence with the crystal anisotropy (Courtesy: J. Park, Berkeley Nat. Lab., USA).

The two friction force curves diverge above 1.5 μN, the point where the oxide film breaks down. This may be ascertained by imaging the surface before and after the AFM experiment (data not presented in this chapter). The removal of the oxide layer permits contact of the tip with the quasicrystalline metal substrate and explains the appearance of friction anisotropy between periodic and quasiperiodic directions.

The magnitude of the anisotropy, defined as the ratio of friction forces along the 10-fold and 2-fold directions, falls in the range 1.2-1.4. This value is lower than the one found previously on the clean surface at low load and can be explained by the fact that under the present conditions the tip is ploughing through, and removing, the oxide film, which contributes an additional isotropic term to the friction force.

A pin-on-disc apparatus was used on the same quasicrystalline material, again initially covered by an oxide film.[101] On a flat surface perpendicular to the rotation axis the vertical displacement of the pin provides a measure of the wear of both sample and pin. In the present experiment however, the small size of the single grain quasicrystal (less than 1 cm diameter) made it difficult to achieve a perfect perpendicular geometry relative to the pin, so that the displacement due to wear was superimposed on a larger up-and-down oscillation during the rotation of the sample. The small tilt gives rise to a lateral force component that varies with 2π periodicity if the friction coefficient is isotropic. If the friction coefficient was *anisotropic* however, with different values along two directions, then the friction force should exhibit a variation with period π. This is actually what was observed, as shown in Fig. 41. During the initial part of the test (first two periods in the plot) the friction force (top curve) oscillates with 2π periodicity (bottom curve). However, after the pin breaks through the oxide layer the period of the friction force changes to π. The pressure in the chamber is low enough ($\sim 10^{-6}$ mbar total) to prevent growth of a new thick oxide layer between successive rotations that are spaced by a fraction of a second owing to the selected disk rotation speed.

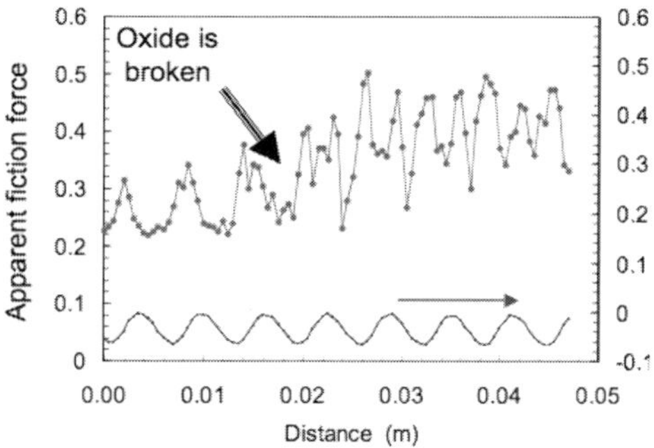

Fig. 41. Friction force signal recorded during a pin-on-disk test performed against hard steel in the conditions described in text. The bottom curve shows the variation of the vertical pin position as a function of run distance of the pin. Observe the change of periodicity of the friction signal after 2-3 cycles (distances above 0.02 m).

In order to estimate correctly the friction coefficient, the effect of the sample misalignment with respect to a horizontal plane must be subtracted from the measured friction force.[102] As mentioned earlier, the periodic variation of the friction force in the isotropic regime is purely geometric and is due to the pin going "up-hill" or "down-hill" as the sample rotates. A schematic diagram of the various forces acting on the pin for a fixed external load is shown in Fig. 42. In this figure, L is the fixed external load, 1N in this experiment. F_L, N, and f are the observed lateral force, the effective normal force, and the actual friction force respectively. It is easy to show, from simple geometrical considerations that $F_L(\alpha, \mu) = L \times \tan(\alpha + \mathrm{atan}(\mu))$, and $N(\alpha, \mu) = L \times \cos(\mathrm{atan}(\mu))/\cos(\alpha + \mathrm{atan}(\mu))$, where α is the angle between the direction of advance of the tip and its projected horizontal direction during the circular motion. Likewise, the actual friction force $f(\alpha, \mu)$ for a given α and μ is: $f(\alpha, \mu) = L \times \sin(\mathrm{atan}(\mu))/\cos(\alpha + \mathrm{atan}(\mu))$. The friction coefficient can then be written as $\mu(F_L/L, \alpha) = \tan(\mathrm{atan}(F_L/L)-\alpha)$. Assuming that α follows a sinusoidal variation, $\alpha = \alpha_0 \sin(2\pi x + \phi)$, μ can be obtained in the isotropic regime by adjusting the parameters α_0 and ϕ so as to compensate the geometric 2π oscillation.

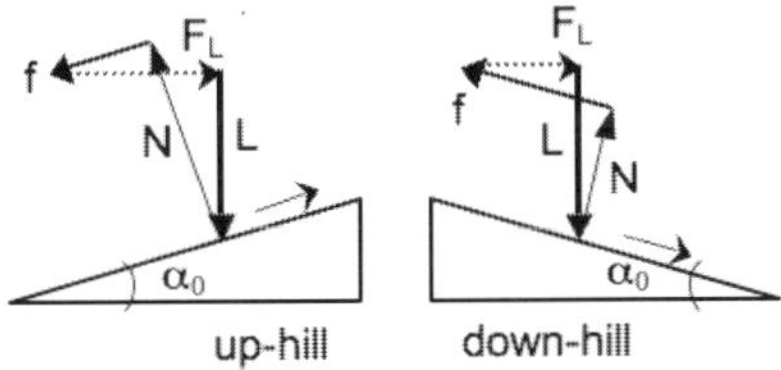

Fig. 42. Schematics of the pin-on-disk test on a flat sample that is slightly tilted from perfect horizontal position by an angle α_0. The load L is applied along the vertical axis whereas the friction force f changes its position and strength when the sample rotates by half a turn.

The bottom curve in Fig. 43 shows the fitting result for N and the top curve the corrected value of the friction coefficient.

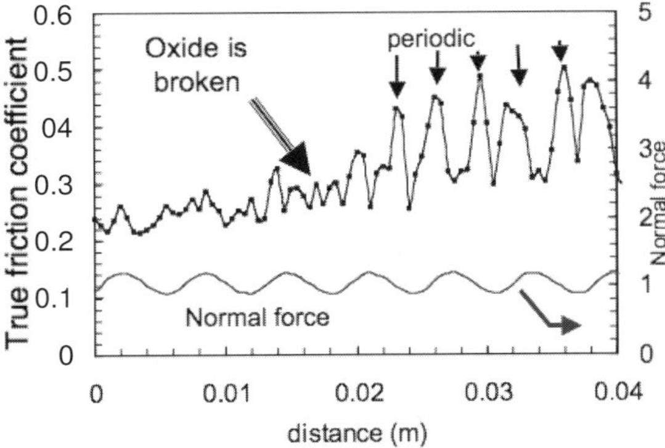

Fig. 43. Fit of the data shown in Fig. 41 as explained in text. The top curve is for the true friction coefficient and the bottom curve for the actual normal force. Observe the change in friction when the oxide layer is pierced by the indenter.

Beyond doubt, the plot reveals the friction anisotropy of the 2-fold decagonal surface in the regime where irreversible removal of the oxide layer occurs. Assuming that the highest friction is along the 10-fold axis, the friction coefficient along this periodic direction is 0.45 ± 0.06, whereas the one along the aperiodic direction is 0.30 ± 0.05. The magnitude of the anisotropy is therefore about 1.5. Note that the two values of friction coefficient are higher than the single value measured on the oxide (0.26 ± 0.05) film before its removal.

Wittman *et al.*[103] also proved quite some time ago that the anisotropy of friction is close to 40% for the decagonal $Al_{62}Co_{15}Cu_{20}Si_{13}$ quasicrystal whereas its hardness anisotropy amounts to at most 5%. Such an anisotropy goes with that of transport properties (electrons, heat) in decagonal CMAs. All friction studies have shown that the unique friction properties of decagonal Al-Ni-Co quasicrystals are an intrinsic property of their peculiar crystallographic structure. The anisotropy, manifested by high and low friction forces along the periodic direction and aperiodic directions, respectively, is present not only on clean surfaces prepared in ultra high vacuum but also in crystals exposed to air, and is manifested after the oxide is broken by wear. Equally important is the finding that the anisotropy disappears when the accumulation of wear destroys the quasicrystalline order. We need to address first the surface energy of CMAs to understand better this ensemble of data.

6.3.3. *Surface energy*

One of the most important properties of a piece of condensed matter is its surface energy. It determines its equilibrium shape. It correlates with its reactivity, melting point, brittleness, etc. It is however hardly accessible to experiment, because it requires equilibrium conditions to be worked out. For instance, an assessment of the surface energy at high temperature may be derived from measurements of the force that is necessary to maintain the shape of a single crystal unchanged when approaching its melting temperature. Although feasible, this method cannot be applied to a whole family of compounds, due to the difficulty to grow many single crystals, and is very sensitive to contamination, surface defects, etc. Alternative approaches were based on observing the equilibrium shape of a small droplet of a known material in its liquid state and deposited in contact with the surface of interest. We will examine such an example in the following section. This approach yields its own difficulties, essentially because the interaction between liquid and solid comprises the so-called liquid-solid interface the energy of which is usually unknown, and because it is also sensitive to surface contaminations which force all experiments to be performed in high vacuum, a condition that is difficult to fulfil at high temperatures, and even more so for many solid samples.

Finally, computer experiments may be used to calculate the surface energy, with limitations again however due to the finite number of atoms that can be handled with nowadays machines and the delicate balance between computations in the bulk and at the surface that must be achieved in the computer. It was nevertheless extensively applied to simple solids, especially metals like in the work published by Vitos *et al.*[104] Unfortunately, once more, surface energy calculations can not be applied regarding CMAs, which most often exceed by far the power of existing computers due to the large number of atoms in the unit cell.

To overcome experimental or computer drawbacks, I have imagined a very simple method that does not deliver the surface energy of a CMA, but only its upper limit, which in the present situation is quite an acceptable piece of data. Let us start with the observation (Fig. 44) that friction is low in the absence of oxygen, which provokes the emission of wear debris and therefore substantial friction when performed in air.[105] In strong contrast, friction decays when oxygen (or air) is pumped out the chamber in which the pin-on-disk apparatus is installed. This behaviour is reversible and may be reproduced at will by pumping out, and reloading, the chamber with air (or oxygen). It is due to the formation of an oxide layer at the surface of the CMA sample, which breaks and forms wear particles in the presence of oxygen. If alternatively the residual pressure of oxygen is low enough in the chamber, friction on the initially oxidized sample decreases rapidly at the beginning of the test when the first worn layer is eliminated from the trace of the indenter.

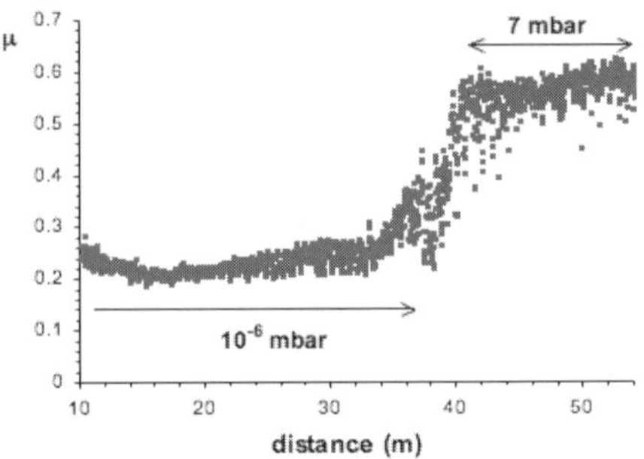

Fig. 44. Friction coefficient measured against hard steel with a sintered specimen of icosahedral Al-Cu-Fe CMA. The figure shows the change in friction when 7 mbar of air is reintroduced in the chamber. This change is reversible if pressure is reduced back to 10^{-6} mbar.

Friction does not increase again if the oxide does not have enough time to grow a new layer during the time elapsed between two successive passages of the indenter. With the apparatus used for the present study, this condition is realised below $10^{-5 \text{ to } -6}$ mbar and a rotation speed of few cycles per second at a trace radius of 3 to 6 mm.

It is interesting to test the method first using specimens of known surface energy, such as e.g. transition metals, or sintered alumina, or window glass. A comparison between an Al-Cu-Fe quasicrystal, a sample of sintered alumina and a piece of hard steel is shown in Fig. 45.

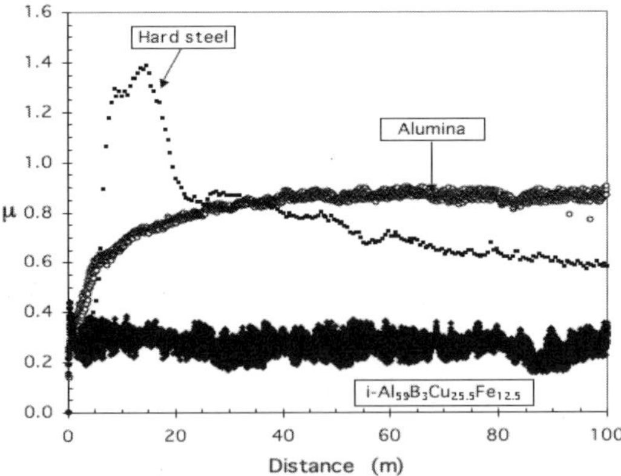

Fig. 45. Friction measured in vacuum (10^{-6} mbar) against hard steel on the same hard steel (solid dots), sintered alumina (open dots), and a sample of sintered icosahedral Al-Cu-Fe-B CMA (bottom curve). Observe the fairly low friction coefficient of Al_2O_3 at the beginning of the test and the plateau reached at the end of the test.

The antagonist was all the same for the three runs, namely a sphere of 6 mm diameter of hard steel. After each run, the ball was examined to assess its wear and replaced by a new one. These conditions apply to all experiments described in this section.

This experiment can be performed using simple metals, or samples of known surface energy. This is the case specifically for the piece of sintered alumina shown in Fig. 45. At the beginning of the test, the hard steel ball comes into contact with the bulk aluminium oxide, whereas at the end of the test, as can be assessed from observation of the trace after withdrawing the sample from the vacuum chamber, the trace is covered by transferred steel to the surface of alumina, and sliding is between the hard steel ball and a film of steel. Meanwhile, the hardness of the solid sample has not changed. We performed such tests on a variety of specimens, beginning with 3d, 4d and 5d metals (Fig. 46).

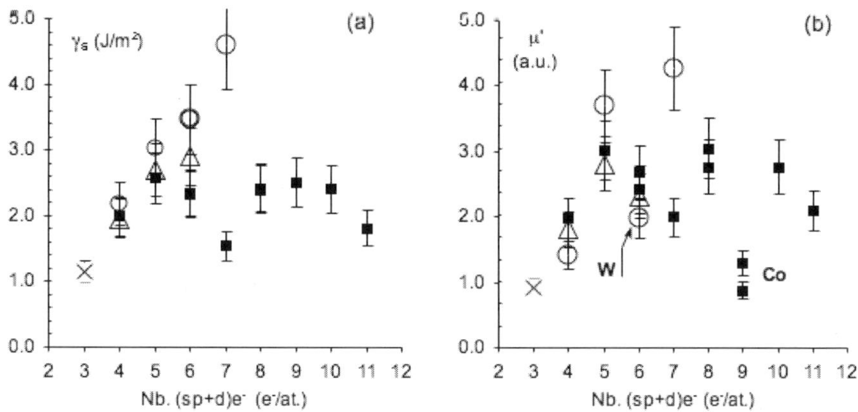

Fig. 46. a) Cohesive energy taken from literature and divided by 2 to give the theoretical surface energy of 3d (solid squares), 4d (triangles), 5d (open dots) metals and aluminium (cross). b) Friction coefficient measured as explained in text and corrected by withdrawing a term inversely proportional to the hardness of the sample as explained hereafter in text. The same symbols are used, respectively, for 3d, 4d, 5d and Al as in a). A minimum is observed at a total number of x=7 (Mn) valence electrons and two maxima at x=5 and x=9, which are typical signatures of the theoretical surface energy shown in a). Exceptions however show up for W (arrow) and Co, as indicated in the figure.

After correcting the friction data for the effect of hardness, which is explained below, the data curves are very reminiscent of the cohesive energy variation with the position of the metal in the d-series. There are only two exceptions, namely Co and W, for which the corrected friction is below the expected level (arrows in the right hand side part of Fig. 46). Most probably, this shift is due to the high stick coefficient of oxygen on those materials, which forbids direct contact of the surface with the indenter and produces lubrication, thus reducing artificially friction.

This correlation leads immediately to a possible relationship between friction measured in such specific conditions (and only in those conditions) and the surface energy of the material of interest. It is shown in the right hand side of Fig. 47, in which the experimental friction coefficient of a large number of calibration samples is presented as a function of the Vickers hardness of those specimens (solid dots). The correlation assumes that the friction coefficient is inversely proportional to the hardness H_V, to account for the ploughing contribution to the hardness, and is proportional to the adhesion energy W_{SP} developed between the solid surface S and the pin P, or:

$$\mu = \alpha / H_V + \beta W_{SP} \qquad (5)$$

where α and β stand for fit parameters.

Considering that no wear occurs during those tests, as can be easily checked after ending the test, we shall assume that W_{SP} represents the reversible adhesion energy between solid and pin: $W_{SP} \equiv \gamma_S + \gamma_P - \gamma_{SP}$. The interfacial energy γ_{SP} is unknown. To get rid of this difficulty, we shall simply assume that the term $\gamma_P - \gamma_{SP}$ cancels, which is a very drastic assumption indeed, but leads us to overestimate γ_S and approach only its upper limit. In the absence of any other means actually to measure or calculate γ_S, the method is valuable. The following equation is therefore used to calibrate the data presented in Fig. 47 and deduce the fit parameters α and β from the set of friction measurements with materials of known γ_S and H_V:

$$\mu \geq \alpha / H_V + \beta \gamma_S \qquad (6)$$

Excellent agreement is obtained between measured and calculated values of the friction measurement, see Fig. 47. The parameter α is equal to 34 and was used in Fig. 46 to correct the data: $\mu' = \mu - 34/H_V$. Observe especially the good agreement found at the highest hardness for sintered alumina before (low friction) and after covering it with transferred steel (high friction) while replacing in the above equation the surface energy of Al_2O_3 by that of iron (standing for steel) at constant hardness. Soft metals, here tin and lead, of course do not fit due to the divergence of the equation when H_V comes too close to 0.

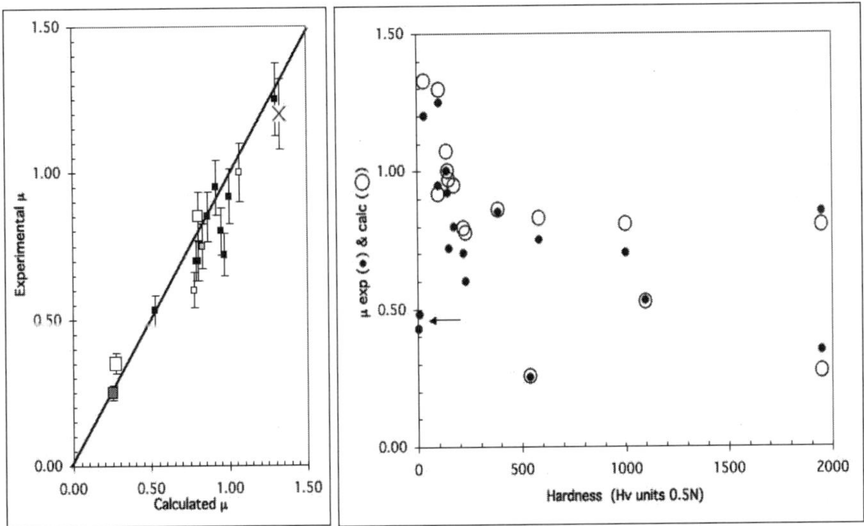

Fig. 47. Left: calculated and experimental friction coefficients for a number of samples of known surface energy and Vickers hardness. The calculation is done according to the equation discussed in text. Right: same data but presented according to the Vickers Hardness of the samples. Solid dots are for experimental data points and open circle for calculated ones. The two data points at the remote right end of the series are for sintered alumina against hard steel, at the beginning of the test and at its end, when steel is transferred to the trace. The arrow shows data for tin and lead, which are too soft to fit with the present model.

We shall thus trust the method and apply it to the estimation of the upper value of the surface energy of CMAs. The experimental data are presented in Fig. 48 as a function of the Vickers hardness of the specimens. Observe first of all that they cover a broad range of values, going from μ as small as 0.2 for some samples up to values close to 0.8.

The distribution of data is not at random: the series at the bottom of the figure stands for alloys containing a significant amount of copper, whereas large values of μ are observed for specimens that contain instead Cr and/or Fe.

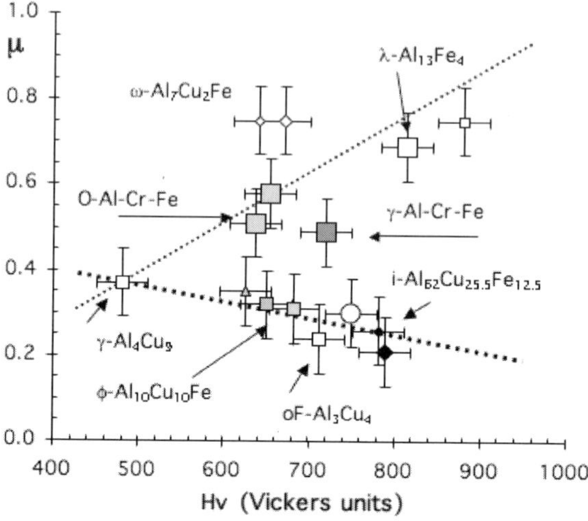

Fig. 48. Friction coefficient measured (in the conditions indicated in text) against a variety of CMA samples. The dotted line guides the eye along the series of Al-Cu(-Fe) samples that show little density of d-states at the Fermi energy. The solid line is for samples with high density of d-states at the Fermi level, due to the presence of Cr or Fe in the composition, instead of Cu. The ω-Al$_7$Cu$_2$Fe compound belongs to the latter family and exhibits a clear metallic behaviour, in strong contrast to the icosahedral Al-Cu-Fe CMA of nearby composition (large and small solid diamonds in the bottom right).

This difference points out a correlation between friction and the existence of d-states at the Fermi level. We will come back to this point later in this section. Using then the fit parameters α and β introduced above and reverting Eq. (6) leads immediately to an estimate of the maximum value of γ_S for those CMA specimens.

The result is shown in Fig. 49 as a function of the experimental friction coefficient used as an order parameter along the x-axis. It is important to notice at this stage that the lowest values of γ_S are observed for the most complex compounds, whereas much simpler ones, especially if they contain Cr or Fe, exhibit larger values of γ_S falling in the range of the surface energy of the constituent metals. To be more specific, $\gamma_S = 0.5\text{-}0.6$ J/m^2 for the Al-Cu-Fe compound and $\gamma_S = 0.8$ J/m^2 for icosahedral Al-Pd-Mn. This is in strong contrast with $\gamma_S = 2$ J/m^2, the

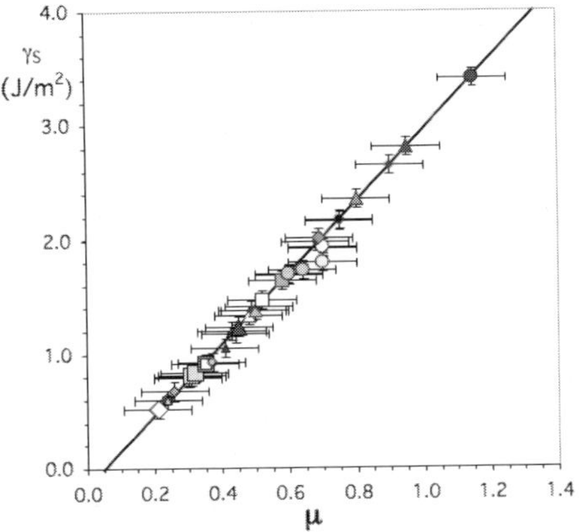

Fig. 49. Representation of the upper limit of γ_S for many CMA compounds, shown as a function of the experimental friction coefficient deduced from the experiments described in text. The open and solid diamonds at the bottom end of the series are for icosahedral compounds of the Al-Cu-Fe system. All complex CMAs fall in the range of γ_S values between 0.5 and 0.8 J/m^2.

value characteristic of the metallic ω-Al$_7$Cu$_2$Fe compound or of orthorhombic Al-Cr-Fe alloys. The largest value of γ_S in this series was found for the Al$_8$V$_5$ compound. Again, a correlation is pointed out in Fig. 50 between the filling of the valence band and γ_S, with a clear tendency to reduce it when the number of s, p and d electrons brought in by the constituents is close enough to 10.

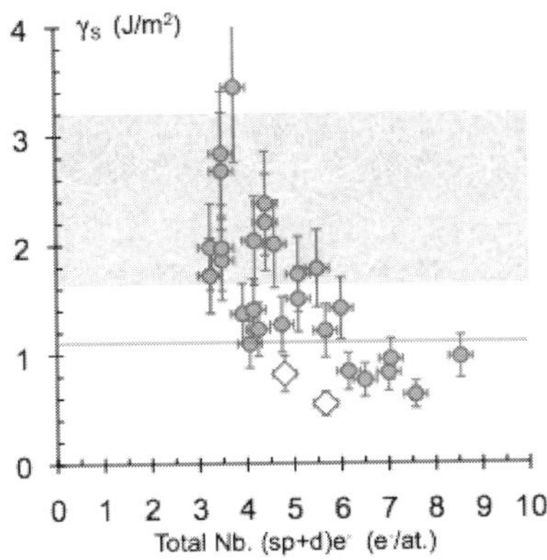

Fig. 50. Correlation observed experimentally between the upper limit γ_S of the surface energy and the total number of s, p and d electrons in the valence band. The two open diamonds stand for the Al-Cu-Fe and Al-Pd-Mn icosahedral compounds.

Once more, icosahedral CMAs (open diamonds in Fig. 50) appear different from CMAs with a regular unit cell and periodicity. Their upper limit of γ_S is small at a same filling of the band. This result is in agreement with other experimental data, especially with attempts to cover a naked quasiperiodic surface with metallic elements like Bi, Al or Ag of reduced surface energy (compared to *3d* transition metals).[106]

6.4. *Wetting against liquid metals*

Several attempts were performed to put this observation on a firmer experimental basis by using a liquid elemental metal deposited at the surface of the solid of interest and measuring its contact angle.[107] An illustration of such an experiment is produced in Fig. 51, with the case of liquid lead on the 5-fold surface of an icosahedral Al-Pd-Mn single grain.

Fig. 51. Small droplets of liquid lead formed in ultra-high vacuum on a 5-fold surface of icosahedral Al-Pd-Mn and frozen-in by cooling the specimen back to room temperature. Observe the pentagonal imprint left at the naked surface by the heat treatment. The contact angle may be measured post-mortem after cross-sectioning a droplet and substrate left underneath. (Courtesy: V. Fournée, IJL, *CMA*).

The experiment is quite tricky and must be performed in ultra-high vacuum to avoid any contamination of the surface that would modify its surface energy. To this end, the surface is first prepared by removing all traces of oxide with the help of an Ar-ion beam and second relaxing it to equilibrium structure by series of subsequent heat treatments and abrasion of the top-most layers. A thin film of lead is then deposited at the surface at low temperature followed by a melting step to produce the droplets. Clearly, Pb "dewets" from the surface, i.e. forms individual droplets the contact angle of which approaches 90°.

This contact angle contains the information on the surface energy according to Young's equation:

$$\gamma_{LV} \cos \theta = \gamma_{SV} - \gamma_{SL} \tag{7}$$

where θ is the contact angle, γ_{LV} the (known) surface energy of the liquid (in the presence of its vapour), γ_{SV}, the surface energy of the solid (also in the presence of the vapour) and γ_{SL} the interfacial energy between liquid and solid. This latter term is unfortunately unknown, which limits drastically the usefulness of the method. Comparison to other substrates is nevertheless meaningful and shows that the value of θ is by far larger than the one observed for lead deposited on other metals (Fig. 52).

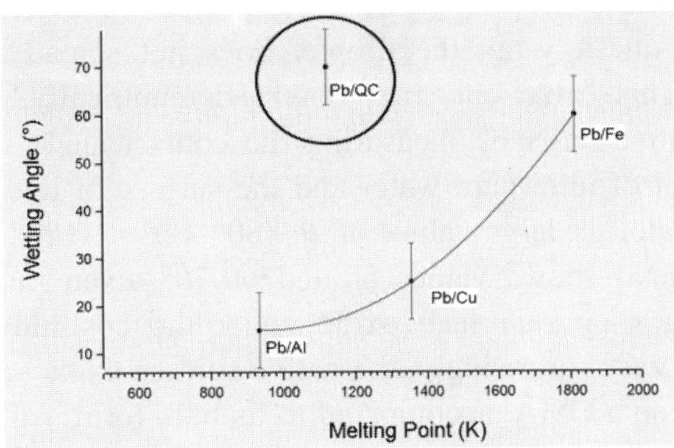

Fig. 52. Empirical correlation observed between the contact angle of liquid lead deposited at the surface of Al, Cu and Fe substrates (the line is only to guide the eye) and the melting temperature of the substrate material. When deposited on an Al-Pd-Mn icosahedral single grain, the value of the contact angle is much larger, see the data point encircled. (Courtesy: V. Fournée, IJL, *CMA*).

Especially, it is about half an order of magnitude larger than when the experiment is performed on pure aluminium, the major constituent of the single grain used for obtaining Fig. 51. Deeper insight may be gained by investigating the deformation of the substrate resulting from the forces exerted by the liquid in the high temperature regime when the quasicrystal is soft enough.[107] Such an experiment has not yet been performed. Conclusively however at this stage, the surface energy of highly complex CMA appears smaller than that of their elemental constituents. It is about 50% that of the most abundant constituent, Al, and it is by far smaller than that of *3d* metallic species.

6.5. *Wetting against polar liquids*

The results presented in the previous sections about the surface energy of highly complex CMAs, and especially quasicrystals, was at the heart of the first potential application of those materials to the production of frying pans.[1,19] The essential quality of a frying pan is de-wetting when one washes it or when food, which is made of about 80% water, comes into contact with the surface. This is the reason why Teflon® is used, a material that has an extremely low surface energy and is non-polar. Similarly, when a droplet of water is placed on a freshly prepared surface of a quasicrystal, the droplet does not spread, but rather, it 'beads up'. This behaviour, first observed empirically,[19] may be put on a quantitative basis by measuring the contact angle, θ, between a minute droplet of ultra-pure water and the surface of interest.[108] CMAs display anomalously large values of θ ($80° < \theta < 110°$), whereas pure aluminium metal shows values around 60-70°, even though both are covered by a similar surface oxide under the conditions of wetting experiments. Very surprisingly, the native surface oxide seems to play a different role on a CMA as compared to its bulk form. An illustration of this effect is shown in Fig. 53. Deriving Young's equation, one comes to:

$$\cos\theta = 2(\gamma_S^{LW} \gamma_L^{LW})^{1/2} / \gamma_L + I_{SL}^{AB} / \gamma_L - 1 \qquad (8)$$

in which γ_L is the surface energy of the liquid, γ_S^{LW} and γ_L^{LW} the Lifshitz-Van der Waals components of the surface energy of the solid and liquid, respectively, and I_{SL}^{AB} stands for all other interfacial interactions between solid and liquid that are not of *LW* type. In this equation, we have assumed that the contact angle θ is well defined, which means that the film pressure $\Pi = \gamma_{SL} - \gamma_{SV}$ is negligible. Indices *S* and *L* denote solid and liquid, respectively, *LW* the Lifshits-Van der Waals forces born from instantaneous fluctuations of the electromagnetic field on both sides of the *S-L* interface, and *AB* any other interaction, for instance dangling bonds, electron donor-acceptor behaviour, etc.

From Eq. 8, it comes immediately that the components of the solid surface energy can be assessed from independent measurements of θ, using various liquids with different γ_L^{LW} and γ_L^{AB}, assuming as in the literature[109] a simple combination rule such as $I_{SL}^{AB} = \sqrt{\gamma_S^{AB} \gamma_L^{AB}}$ and $\gamma_S = \gamma_S^{LW} + \gamma_S^{AB}$. A liquid with $\gamma_L^{AB} = 0$ is called non-polar. With such liquids, varying $\gamma_L^{LW} \neq 0$, a plot of $cos\theta$ as a function of $x = 2\sqrt{\gamma_L^{LW} / \gamma_L}$ produces a linear variation that goes to -1 when x = 0. Similarly, if the solid is non-polar (i.e. $I_{SL}^{AB} = 0$ because $\gamma_S^{AB} = 0$), the plot of $cos\theta$ also is a straight line that goes to -1 for x = 0. This is typically the case of Teflon®, which is a non-polar solid, also of very small surface energy ($\gamma_{Teflon} \approx 18$ mJ/m^2). We used 5 different liquids and measured $cos\theta$ for a flat sample of Teflon according to a methodology that is described elsewhere.[110] The result (Fig. 53) is indeed a straight line that extrapolates to -1 when x = 0.

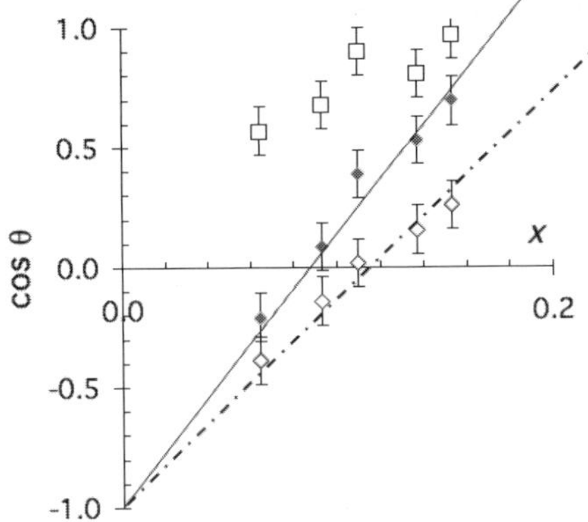

Fig. 53. Plot of $cos\theta$ as a function of $x = 2\sqrt{\gamma_L^{LW} / \gamma_L}$. Five different liquids were used, as shown in the figure, for Teflon® (open diamonds), a single grain of cubic alumina (open squares) and a single crystal of icosahedral Al-Pd-Mn (solid diamonds). Unexpectedly, this later sample exhibits a linear plot of $cos\theta$ vs x, extrapolating to $cos\theta = $ -1 when x=0. Each data point in the figure is an average over at least ten measures of θ. Observe how the two data points at large x align with $cos\theta = $ -1 because the two liquids used are non-polar. The fact that all measures of $cos\theta$ with the icosahedral CMA align on a straight line that goes to $cos\theta = $ -1 for x=0 is a demonstration of its non-polar nature, despite it is covered with the same oxide as the one illustrated in the figure (open squares).

In strong contrast, a single crystal of alumina, which is very much a polar solid due to the formation of dangling bonds at the surface and its electron donor behaviour, exhibits a plot of $cos\theta$ vs x that departs from a straight line as soon as the liquid is polar itself (Fig. 53). Very surprisingly, a polished single grain of icosahedral Al-Pd-Mn CMA shows a straight line of $cos\theta$ vs x. This unexpected result cannot be anticipated from literature, which assigns the wetting behaviour of a solid surface to its extreme surface layers. Here, the native oxide that is always present at the surface of Al-based CMAs studied in ambient atmosphere, seems to play no role at all.

Even more surprising is the fact that the contact angle of ultra-pure water increases when the thickness of the oxide layer is decreased below a threshold value of about 10 nm (Fig. 54). To produce this piece of evidence, we equipped several samples of Al-based CMAs with artificial layers of alumina deposited on top of their native oxide layer. The supplementary layers were prepared by evaporating Al layers at the surface and oxidizing them *in situ* in the evaporator. The final thickness could be assessed by a standard XPS technique.[111]

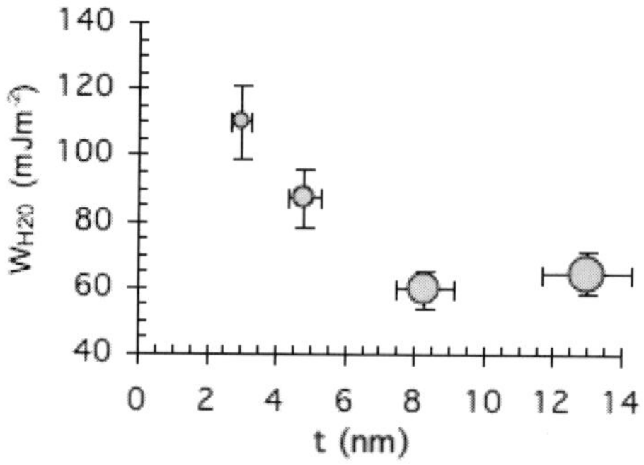

Fig. 54. Reversible adhesion energy of water deposited at the surface of a same CMA sample, but equipped with increasing thicknesses t of the surface oxide layer (see text for details). The size of the symbols increases in proportion to t, to guide the eye. The y-axis is labelled according to the reversible adhesion energy of water $W_{H2O} = (1+cos\theta)\,\gamma_{H2O}$, with $\gamma_{H2O} = 72.8$ mJ/m^2, an equation that is valid only if θ is well defined and close enough to 90°. Except for metallic and oxide samples, it is the case for most specimens used in the present study. Observe how W_{H2O} decreases against normal sense with increasing t values at low enough values of t. A crossover to the normal behaviour is seen around t = 8-10 nm.

Going a step further, we could determine the individual values of γ_S^{LW} and γ_S^{AB} for a full series of CMA samples (Fig. 55). It turns out that the *LW* component is nearly independent of composition and falls within a narrow range of values, slightly below however that of bulk alumina. In contrast, γ_S^{AB} evolves with composition and is found equal to 0 on a highly perfect icosahedral material, whereas it tends towards the value characteristic of (oxidized) aluminium for less complex CMAs.

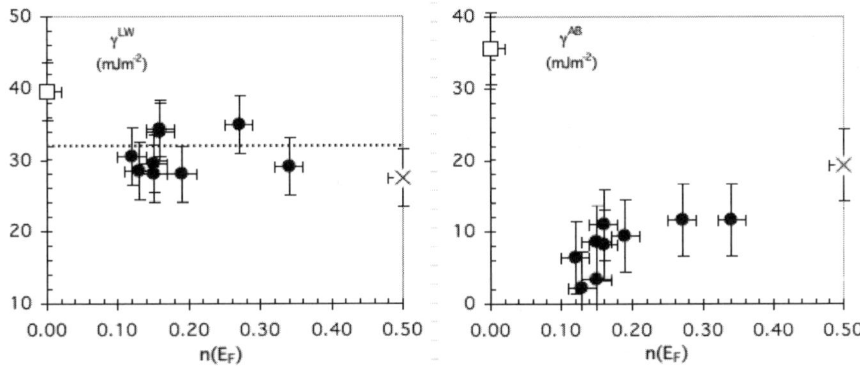

Fig. 55. Evolution of the surface energy components of a series of Al-based CMAs. The left side of the figure shows the *LW* component and the right side the *AB* part of γ_S. The x-axis is drawn according to the Al3*p* density of states at the Fermi energy of the very same specimens, already presented in Fig. 35a earlier in this chapter. Observe the nearly parabolic dependence of γ_S^{AB} with n(E$_F$), whereas little variation of γ_S^{LW} is observed as a function of n(E$_F$). The cross represents pure fcc Al and the open square, bulk alumina. Solid dots are for various CMAs.

Thus, we have arrived at a main result: the degree to which water wets Al-based metals covered by an amorphous oxide depends directly on the density of conduction states in the bulk, and inversely on the oxide layer thickness, for thin enough layers. This result could not have been anticipated from the accepted understanding of the wetting of aluminium oxides by water.[108,109] The existing paradigm would have predicted that the key interactions are very localized *at* the liquid-solid interface (aside from the dispersive forces, which add a constant contribution that depends only on the nature of the covering oxide). Our results, however, imply that the key interactions are the longer-range forces between water molecules in the droplet and the Fermi sea of electron in the CMA substrate beneath the oxide. Based on a careful study of the wetting of a large number of CMA samples, which would take too much space in the present context,[110] we could show that a simple model of the electrostatic interaction of the permanently moving dipoles carried by water molecules and their image in the Fermi sea accounts for the results. It holds true for thin enough layers whereas a cross over towards the ancient paradigm applies when the oxide layer becomes sufficiently thick (t ≥ 8-10 nm).

Basically, the model assigns the sum of all dipole-image interactions to the reaction of the substrate that balances the system of forces involved in Young's equation, see Fig. 56. The reaction of the substrate is usually omitted in literature, but it must exist, at least to take into account gravity. It is represented in Fig. 56 by the dashed vertical arrow.

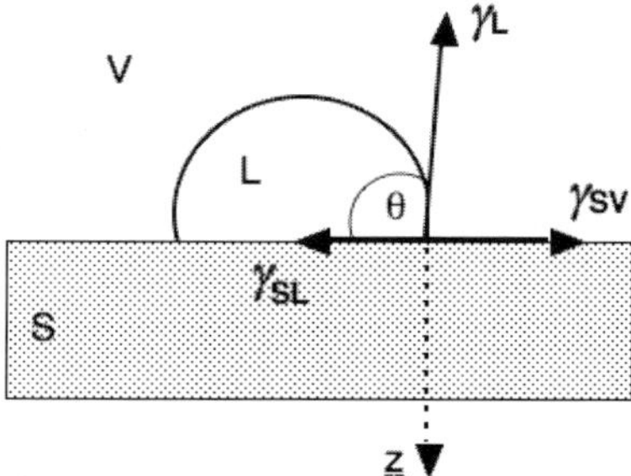

Fig 56. Balance of the surface tensions involved in Young's equation. In the present context, surface energies and surface tensions are equivalent. The contact angle results from equilibrium of the surface energy of the liquid (in the presence of its vapour), the surface energy of the solid (in the presence of the liquid vapour), the interfacial energy and a supplementary term usually not shown in literature which accounts for the reaction of the substrate and is directed along the vertical direction z. In the whole section, we assumed that the contact angle is well defined because the presence of the vapour does not modify the equilibrium conditions.

At the position of the triple line which separates solid, liquid and vapour, a dipole on a given water molecule interacts with its own image and all other images produced by the other nearby molecules. Its own image sits at a distance $d'=d+t/\kappa$ from the oxide-water interface taken as the origin of coordinates along the z-axis. Here, d labels the distance of the dipole above the reference interface, t the oxide thickness and κ is the ratio between the dielectric constants of the oxide and of water, respectively: $\kappa = \varepsilon_o / \varepsilon_w$. Typically, is much smaller than 1: $0.1 \leq \kappa \leq 0.2$. As a consequence, d' is by far larger than t and the image dipole sits within the bulk of the CMA substrates, which explains why the bulk density of states probed by XES is relevant.

Averaging over the effects of (Boltzman) kinetic energy and interaction with all images in the bulk leads to a dependence upon the thickness of the oxide film that varies according to the square of t. The data presented earlier in this section (Figs. 54 and 55) are in agreement with this conclusion. It can also be shown that the strength of the interaction is proportional to the square of $n(E_F)$, essentially because the presence of the oxide film, which is considered in the model as an inert spacer, creates two interfaces, one between oxide and water, the other between oxide and solid. The waves emitted by the dipole partly reflect and partly propagate through those interfaces. The key point is that CMAs of high complexity absorb electromagnetic waves to a very substantial effect. For instance, the reflection coefficient of an Al-Cu-Fe quasicrystal of good lattice perfection is not larger than R=0.6, whereas fcc aluminium is characterised as like any other good metal by R nearly equal to 1.[112] The peculiar wetting properties of Al-based CMAs are therefore a direct consequence of their specific electron transport properties. For thick enough oxide layers, the difference between oxidized aluminium, bulk oxide and CMA cannot be observed. The wetting behaviour is dominated by the presence of the oxide film. At thin enough layer, typically when t is smaller than 8 nm, this unexpected behaviour reveals itself provided $n(E_F)$ in turn is small enough, a condition that is achieved only with highly complex CMAs. Pure aluminium therefore will not depart from wetting properties of bulk alumina because of its marked metallic nature. To close this section, Fig. 57 presents the data gained with many different CMAs on wetting with ultra-pure water, showing excellent agreement with the $n(E_F)^2 / t^2$ expected dependence. The contact angle is related to the reversible wetting energy of water, W_{H2O}, assuming a negligible effect of the vapour film, by the equation $W_{H2O} = \gamma_{H2O}(1 + \cos\theta)$. The data point separate in two branches because we used the Al3p partial density of states, which is available (Fig. 35a), instead of the total density of states that is lacking.

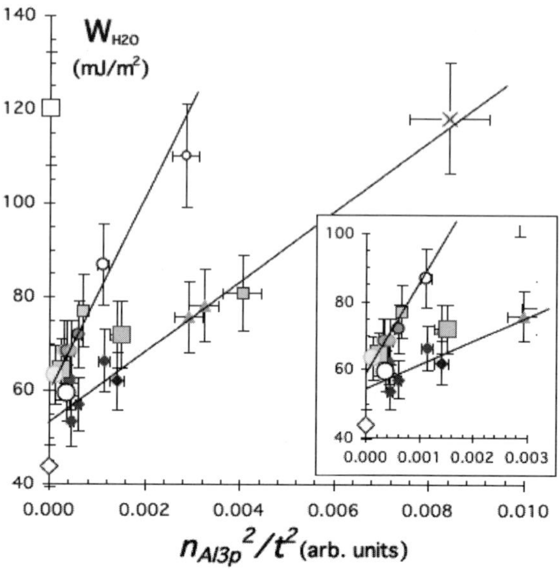

Fig. 57. Reversible adhesion energy of water deduced from contact angle measurements performed on a large variety of specimens and presented as a function of the n^2/t^2 ratio introduced in text. The two open symbols shown on the y-axis are for bulk alumina (square) and Teflon® (diamond). Identical symbols, but of different sizes, are used for a few specimens equipped with artificial oxide layers of varying thickness as explained in text. The data divide in two branches according to the presence or not of *3d* metals (Cr and Fe) that contribute significantly to the density of states at the Fermi energy. The branch at the bottom is for alloys of the Al-Fe-Cu series that show little *d*-states at E_F. Conversely, the *d*-states contribution is much higher in the other Al-Cr-Fe(-Cu) series. Both branches converge (inset) towards a same value at $n^2/t^2 = 0$, from which it is straightforward to deduce that $\gamma_S^{LW} = 32.8 \pm 2\,mJ/m^2$ for all specimens, in accordance with Fig. 55.

7. Inverse Nano-Structuration

It seems useful at this stage of the chapter to emphasize one unique concept that could summarize the major features of CMAs. This concept was already introduced in Vol I of the series of books dedicated to complex intermetallics.[5] It is illustrated in Fig. 58 for the case of mechanical testing in compression. This figure shows, on the left side, stress-strain curves recorded during compression tests of nano- and microcrystalline copper.[113] Remarkably, the nanocrystals exhibit no work hardening, in strong contrast to microcrystalline copper. The absence of work hardening is also a characteristic of complex intermetallics when they become plastic at temperatures approaching the melting point. It is observed with polygrained CMAs as well,[49] as shown in the right hand side part of Fig. 58, but also on single grain specimens.[22]

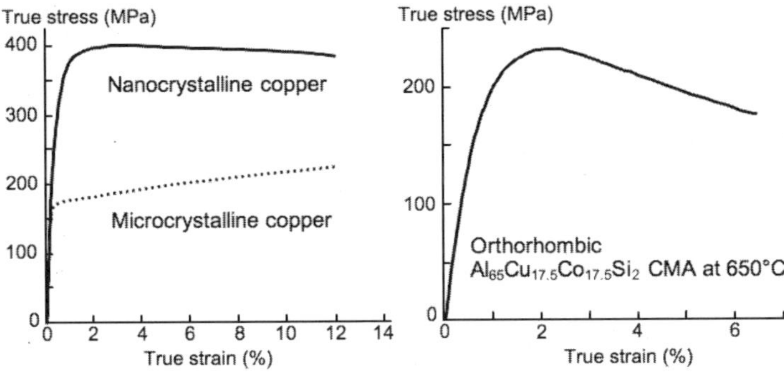

Fig. 58. True stress-true strain curves recorded during compression tests applied to copper[113] (left) and to an orthorhombic CMA[49] (right) the composition of which is indicated in the figure. In contrast to microcrystalline copper, and more generally to metallic alloys, nanocrystalline copper shows no work hardening, a behaviour that is related to the large number of atoms located in grain boundaries and the resulting enhancement of atom mobility. Similarly, CMAs of sufficient complexity are characterised by further degrees of freedom for atoms to move, including in single grained materials, which entails an easier climb of dislocations.

It turns out from comparison of nanograined copper with a bulk CMA that deformation under compression strain of both materials reveals itself in similar ways: enhanced plasticity with no work hardening, large strain at rupture. Many other properties of CMAs follow the same trend, namely they reveal themselves in bulk CMAs, often in single grained CMAs, as they do in nanosized materials. This is especially the case of electron transport properties as well as of phonon mobility. In other words, the effects characteristic of localisation that take place upon reducing the size of individual objects promoted by structuration in the nanoscopic regime are also observed in bulk CMAs, but upon increasing the size of the unit cell. For this reason, I have suggested in Ref. 5 to assemble all such effects pointed out in CMAs under the label "inverse nanostructuration", or INS in short.

8. Conclusion

Many of the tricky properties of CMAs may find technological applications: reduced wetting against polar liquids, reduced friction in solid-solid contact, heat insulation, resistors with zero temperature coefficient, catalysts, thermoelectricity generators, etc. A review was already published by the author.[1] For the time being, this is most probably not the main reason that justifies interest in those materials. The main issue is to understand them, and maybe to find, or prepare, new ones, albeit no fully accepted theory is available to guide the search of such new materials. The discovery and successful preparation[114] of an Al-Cu-Ta compound with a unit cell volume as large as 365 nm^3 is in my opinion the best prove that so much more remains uncovered in the field of metallic alloys, with so far unknown properties. This was "the raison d'être" behind the creation of the *CMA* European Network of Excellence, and therefore the chief motivation to organise yearly a school in Materials Science and edit this series of textbooks.

Acknowledgments

This work was produced in the frame of the CMA network of excellence, contract n°NMP3-CT-2005-500140 between CNRS and the European Commission. The graceful hospitality offered by Prof. An Pang Tsai at NIRIM, Tohoku University, Sendai, Japan, was instrumental in offering the author appropriate conditions for thinking and writing. Grateful thanks are also due to E. Belin-Ferré, P. Brunet, V. Demange, V. Fournée, J. Ledieu, S. Kenzari, M. Sales, and P.A. Thiel for a long lasting collaboration. Thanks are also due to CNRS, Conseil Régional de Lorraine et Communauté Urbaine du Grand Nancy for their funding and support.

References

1. J.M. Dubois, *Useful Quasicrystals* (World Scientific, Singapore, 2005).
2. Proceedings of the 25 Anniversary Conference on Quasicrystals, Eds. R. Lifshitz and D. Shechtman, Phil. Mag. Special Issue (Taylor and Francis, London, 2008).

3. M. Quiquandon, A. Quivy, J. Devaud, F. Faudot, S. Lefebvre, M. Bessière and Y. Calvayrac, *J. Phys.: Cond. Matter* **8** (1996) 2487.

4. S. Bühler-Paschen, this volume, p. 149.

5. *Book Series on Complex Metallic Alloys – Vol. I: Basics of Thermodynamics and Phase Transitions in Complex Intermetallics*, Ed. E. Belin-Ferré (World Scientific, Singapore, 2008). For simplicity, this reference is called Vol. I in this chapter. It was issued from the lectures delivered at the first session of the *CMA* European School on Material Science (EuroSchool). The volumes following in the series will be quoted accordingly: Vol. II is for the book associated with the second session of the EuroSchool (*Vol. II: Properties and Applications of Complex Metallic Alloys*, in preparation, same editor) and Vol. III for the third one (*Vol. III: Surface Science and Engineering of Complex Metallic Alloys*, also in preparation).

6. L. Pauling, *The Nature of the Chemical Bond* (Cornell University Press, New York, 1939).

7. S. Samson, *Acta Cryst.* **19** (1965) 40.

8. M. Feuerbacher *et al.*, *Z. Kristallographie* **222** 257 (2007).

9. J.D. Watson, *La Double Hélice* (Laffont, Paris, 1984).

10. M. Lambert, *Acta Cryst.* **A57** (2001) 1.

11. V. Gerold, *Scripta Met.* **22-7** (1988) 927.

12. *Alliages d'aluminium amorphes ou microcristallins*, J.M. Dubois and G. Le Caër, French Patent n° 2529909 (1982); US Patent n° 4595429 (1986).

13. D. Shechtman, I. Blech, D. Gratias and J.W. Cahn, *Phys. Rev. Lett.* **53** (1984) 1951.

14. C. Janot, J.M. Dubois and J. Pannetier, *Physica B* **146** (1987) 351.

15. L. Pauling, *Nature* **317** (1985) 512.

16. M. Senechal, *Quasicrystals and Geometry* (Cambridge University Press, Cambridge, 1995).

17. A.P. Tsai, A. Inoue and T. Masumoto, *Jpn. J. Appl. Phys.* **26** (1987) L1505. See Ref. 1 for more references.

18. H. Takakura, C. Pay Gomez, A. Yamamoto, M. de Boissieu and A.P. Tsai, *Nature Materials* **6** (2007) 58.

19. *Matériaux de revêtement en alliage d'aluminium*, J.M. Dubois and P. Weinland, French Patent n° 8810559 (1988); US Patent n° 5204191 (1993).

20. *New Horizons in Quasicrystals: Research and Applications*, Eds. A.I. Goldman, D.J. Sordelet, P.A. Thiel and J.M. Dubois (World Scientific, Singapore, 1997).

21. J.M. Dubois, A. Proner, B. Bucaille, P. Cathonnet, C. Dong, V. Richard, A. Pianelli, Y. Massiani, S. Ait-Yaazza and E. Belin-Ferré, *Annales de Chimie*, **19** (1994) 3.

22. K. Urban and M. Feuerbacher, *J. Non-Cryst. Sol.* **334-335** (2004) 143.

23. G. Bergman, J.L.T. Waugh and L. Pauling, *Acta Cryst.* **10** (1957) 254.

24. U. Mizutani, in *The Science of Complex Phases*, Eds. T.B. Massalski and P.E.A. Turchi (The Minerals, Metals and Materials Society, Warrendale, 2005), pp 1-42.

25. F. Laves and H. Wittle, *Metallwirt.* **14** (1935) 545.

26. J.B. Friauf, *J. Am. Chem. Soc.* **49** (1927) 507.

27. Ch. Janot, *Quasicrystals, a Primer* (Clarendon Press, Oxford, 1992) and references therein.

28. S. Kenzari *et al.*, J. *Phys.: Cond. Matter* **20** (2008) 09518 (7 pp.).

29. S. Lyonnard, G. Coddens, Y. Calvayrac and D. Gratias, *Phys. Rev. B* **53** (1996) 3150.

30. R. Blüher, P. Scharwaechter, W. Frank and H. Kronmüller, *Phys. Rev. Lett.* **80-5** (1998) 1014.

31. P.A. Kalugin and A. Katz, *Europhys. Lett.* **21(9)** (1993) 921.

32. M. Gil-Gavatz, D. Rouxel, P. Pigeat, B. Weber and J.M. Dubois, in *Quasicrystals, Preparation, Properties and Applications*, Eds. J.M. Dubois, P.A. Thiel, A.P. Tsai and K. Urban (Materials Research Society, Warrendale, 1999), p. 75.

33. K. Wang and P. Garoche, *Ann. Chim. Fr.* **18** (1993) 475.

34. E. Abe, S.J. Pennycook and A.P. Tsai, *Nature* **42** (2003) 347.

35. M. Feuerbacher, Ref. 5, p. 377.

36. W. Steurer *et al.*, to be published. Paper presented at the 16th Intern. Conf. on Solid Compounds of Transition Elements, Dresden July 2008).

37. C. Dong, Q.H. Zhang, D.H. Wang and Y.M. Wang, *Euro Phys. J. B* **6,** 25-32 (1998).

38. P. Häussler, J. Barzola-Quiquia, D. Hauschild, J. Rauchhaupt, M. Stiehler and M. Hackert, in *The Science of Complex Alloy Phases*, Eds. T.B. Massalski and P.E.A. Turchi (The Minerals, Metals and Materials Society, Warrendale, **43-86**, 2005).

39. P. Gille, in *Vol. I*, p. 73.

40. J.M. Dubois, P. Brunet and E. Belin-Ferré, in *Quasicrystals, Current Topics,* Eds. E. Belin-Ferré, C. Berger, M. Quiquandon and A. Sadoc (World Scientific, Singapore, 2000).

41. P. Brunet, L. Zhang, D.J. Sordelet, M.F. Besser and J.M. Dubois, *Mater. Sci. and Eng.* **A294-296** (2000) 74.

42. P. Weisbecker, G. Bonhomme, G. Bott and J.M. Dubois, *J. Non-Cryst. Sol.* **351** (2005) 1630.

43. V. Demange, J. Ghanbaja, C. Beeli, F. Machizaud and J.M. Dubois, *J. Mater. Res.* **19(8)** (2004) 2285.

44. C.J. Jenks and P.A. Thiel, *J. Molec. Catal. A: Chemical* **131** (1998) 301.

45. S. Kenzari, D. Bonina, J.M. Dubois and V. Fournée, *Scripta Mat.* **59** (2008) 583.

46. H. Mehrer and R. Galler, in *Quasicrystals, Preparation, Properties and Applications*, Eds. J.M. Dubois, P.A. Thiel, A.P. Tsai and K. Urban (Materials Research Society, Warrendale, 1999), p. 67.

47. M. Gil-Gavatz, D. Rouxel, P. Pigeat, B. Weber and J.M. Dubois, in *Quasicrystals, Preparation, Properties and Applications*, Eds. J.M. Dubois, P.A. Thiel, A.P. Tsai and K. Urban (Materials Research Society, Warrendale, 1999), p. 75.

48. P.A. Thiel and J.M. Dubois, *Materials Today* **2-3** (1999) 3.

49. S.S. Kang and J.M. Dubois, *Phil. Mag. A* **66-1** (1992) 151.

50. M. Feuerbacher, in *Vol I*, pp. 377.

51. A.M. Watanabe, A. Kato and A.P. Tsai, *Mater. Sci. Engg.* A385 (2004) 382.

52. M. Wollgarten, M. Bartsch, U. Messerschmidt, M. Feuerbacher, R. Rosenfeld, M. Beyss and K. Urban, *Phil. Mag. Lett.* **71-2** (1995) 99.

53. M. Feuerbacher, H. Klein, P. Schall, M. Bartsch, U. Messerschmidt and K. Urban, in *Quasicrystals, Preparation, Properties and Applications*, Eds. J.M. Dubois, P.A. Thiel, A.P. Tsai and K. Urban (Materials Research Society, Warrendale, 1999), p. 307.

54. H. Klein, M. Feuerbacher, P. Schall and K. Urban, *Phys. Rev. Lett.* **82** (1999) 3468.

55. M. Boudard, H. Klein, M. de Boissieu, M. Audier and H. Vincent, *Phil. Mag. A* **74** (1996) 939.

56. L. Beraha, M. Duneau, H. Klein and M. Audier, *Phil. Mag. A* **76** (1997) 587.

57. W. Hume-Rothery, *J. Inst. Metals* **35** (1926) 295.

58. H. Jones, *Proc. Phys. Soc.* **A49** (1937) 250; A. P. Blandin, in *Phase stability in metals and alloys*, Eds. P. S. Rudman, J. Stringer and R. I. Jaffee (McGraw Hill, New York, 1965), p. 115; T. B. Massalski and U. Mizutani, *Progress in Materials Science* **22**, 151 (1978).

59. J. Friedel and F. Denoyer, *C. R. Acad. Sci., Ser. 2* **305**, 171 (1987).

60. R. Asahi, H. Sato, T. Takeuchi and U. Mizutani, *Phys. Rev. B* **71** (2005) 165103.

61. B.K. Agarwal, *X-ray Spectroscopy, Optical Series S987* (Springer Verlag, Berlin, 1979).

62. E. Belin-Ferré, *J. Phys.: Cond. Matter* **14** (2002) R789.

63. H. Jones, *Proc. Roy. Soc. A* **144** (1934) 225.

64. G. Trambly de Laissardière, D. Nguyen Manh, L. Magaud, J.P. Julien, F. Cyrot-Lackmann and D. Mayou, *Phys. Rev. B,* **52** (1995) 7920; G. Trambly de Laissardière, D. Nguyen Manh and D. Mayou, *Prog. Mat. Sci.,* **50** (2005) 679.

65. G.V. Raynor, *Prog. Met. Phys.,* **1** (1949) 1.

66. A.P. Tsai, in *New Horizons in Quasicrystals: Research and Applications*, Eds. A.I. Goldman, D.J. Sordelet, P.A. Thiel and J.M. Dubois, (Singapore, World Scientific, 1997), p. 1.

67. A.P. Tsai, J.Q. Guo, E. Abe, H. Takakura and T.J. Sato, *Nature,* **408** (2000) 537; H. Takakura, C. P. Gomez, A. Yamamoto, M. de Boissieu and A. P. Tsai, *Nature Materials,* **6** (2007) 58.

68. A.P. Tsai, *J. Non Cryst. Solids*, 334-335 (2004) 317.

69. D. Gratias, Y. Calvayrac, J. Devautd-Rzepski, F. Faudot, M. Harmelin, A. Quivy and P.A. Bancel, *J. Non Crystal. Solids* **153-154** (1993) 482.

70. A.J. Bradley and H.J. Goldschmidt, *J. Inst. Metals* **65** (1939) 389.

71. F. Faudot, A. Quivy, Y. Calvayrac, D. Gratias and M. Harmelin, *Mater. Sci. Eng.,* **A 133** (1991) 383.

72. C. Dong, J.M. Dubois, M. de Boissieu, C. Janot and J. Pannetier, *J. Phys.: Cond. Matter,* **2** (1990) 6339.

73. U. Mizutani, in *Vol. I, Chapter 11*, p. 319, and *Vol III*, to appear in 2009, and references therein.

74. U. Mizutani, R. Asahi, H. Sato and T. Takeuchi, *Phys. Rev. B* **74** (2006) 235119.

75. U. Mizutani, R. Asahi, T. Noritake and T. Takeuchi, *J. Phys.: Condens. Matter* **20** (2008) 275228.

76. G. Trambly de Laissardière, Z. Dankhazi, E. Belin, A. Sadoc, D. Nguyen Manh, D. Mayou, D.A. Papaconstantopoulos and M.A. Keegan, *Phys. Rev. B*, **51** 4035 (1995).

77. V. Fournée, I. Mazin, D.A. Papaconstantopoulos and E. Belin-Ferré, *Phil. Mag. B.*, **79** (1999) 205; G. Trambly de Laissardière and T. Fujiwara, *Phys. Rev. B*, **50** (1994) 5999; E. Belin-Ferré, G. Trambly de Laissardière, P. Pêcheur, A. Sadoc and J.M. Dubois, *J. Phys.: Cond. Matter*, **9** (1997) 9585.

78. D.A. Papaconstantopoulos, *Handbook of the band structure of elemental solids* (Plenum Press, New York, 1986).

79. P. Nozières and C. T. de Dominicis, *Phys. Rev.*, **178** (1969) 6105; G. A. Rooke, *J. Phys. C*, **1** (1968) 767 and 776.

80. P. Donnadieu, H.L. Su, A. Proult, M. Harmelin, G. Effenberg and F. Aldinger, *J. Physique (France)*, **6** (1996) 1153.

81. E. Belin-Ferré, V. Fournée and J.M. Dubois, *Mat. Trans. JIM*, **42-6** (2001) 911.

82. A. Traverse, L. Dumoulin, E. Belin and C. Sénémaud, in *Quasicrystalline Materials*, Eds. C. Janot and J.M. Dubois (World Scientific, Singapore, 1988), p. 399.

83. E. Belin-Ferré, Z. Dankazi, V. Fournée, A. Sadoc, C. Berger, H. Müller and H. Kirchmayr, *J. Phys.: Cond. Matter* **8** (1996) 6213.

84. S. Sunjic and A. Doniach, *J. Phys. C: Solid State Phys.*, **3** (1970) 285.

85. V. Fournée, J.W. Anderegg, A.R. Ross, T.A. Lograsso and P.A. Thiel, J. Phys.: Condens. Matter, **14** (2002) 2691

86. V. Fournée, E. Belin-Ferré and J.M. Dubois, *J. Phys.: Cond. Matter* **10**, 4231-4244 (1998).

87. K. Terakura, *J. Phys F: Met. Phys.*, **3** 1773 (1977).

88. E. Macia, *Phys. Rev. B*, **66** (2002) 174203; C.V. Landauro, E. Macia and H. Solbrig, *Phys. Rev. B*, **67** (2003) 184206.

89. E. Belin-Ferré, M. Klansek, Z. Jaclic, J. Dolinsek and J. M. Dubois, *J. Phys.: Condens. Matter*, **17** (2005) 6911.

90. N.F. Mott, *J. Non-Cryst. Solids*, **1** (1968) 1; *Phil. Mag.*, **19** (1969) 835.

91. D. Mayou, in *Quasicrystals, Current Topics*, Eds. E. Belin-Ferré *et al.* (World Scientific, Singapore, 2000).

92. C. Sire, in *Lectures on Quasicrystals*, Eds. F. Hippert and D. Gratias (Les Editions de Physique, Les Ulis, 1994).

93. U. Mizutani, *J. Phys.: Condens. Matter*, **10** 4609 (1998).

94. C. Janot, *Phys. Rev. B*, **53-1** (1996) 181.

95. V. Demange, A. Milandri, M.C. de Weerd, F. Machizaud, G. Jeandel and J.M. Dubois, *Phys. Rev. B*, **65** (2002) 144205.

96. A. Merstallinger and E. Semerad, in *Test Method to Evaluate Cold Welding under Static and Impact Loading*, In-house-Standard by Austrian Research Centre Seibersdorf, Vol. **1** (1995) and Vol. **2** (1998).

97. M. Sales, A. Merstallinger, P. Brunet, M.C. de Weerd, V. Khare, G. Traxler and J.M. Dubois, *Phil. Mag.*, **86 (6-8)** (2006) 965.

98. J.M. Dubois, M.C. de Weerd, J. Brenner, M. Sales, G. Modzen, A. Merstallinger and E. Belin-Ferré, *Phil. Mag.*, **86 (6-8)** (2006) 797.

99. J.Y. Park, D.F. Ogletree, M. Salmeron, R.A. Ribeiro, P.C. Canfield, C.J. Jenks and P.A. Thiel, *Science*, **309** (2005) 1354.

100. J.Y. Park, D.F. Ogletree, M. Salmeron, C.J. Jenks and P.A. Thiel, *Tribology Lett.*, **17** (2004) 629.

101. J.M. Dubois, in Ref. 5, Vol. I., p. 1.

102. J.Y. Park, D.F. Ogletree, M. Salmeron, C.J. Jenks, P.A. Thiel, J. Brenner and J.M. Dubois, *J. Mat. Research*, **23-5** (2008) 1488.

103. R. Wittmann, K. Urban, M. Schandl, M. and E. Hornbogen, *J. Mat. Research*, **6** (1991) 1165-1168.

104. L. Vitos, A.V. Ruhan, H.L. Skriver and J. Kollar, *Surface Science*, **411** (1998) 186.

105. I.L. Singer, J.M. Dubois, J.M. Soro, D. Rouxel, J. Von Stebut, in *Quasicrystals, Eds. S. Takeuchi et T. Fujiwara (World Scientific, Singapore, 1998)*, p769-772.

106. V. Fournée, H.R. Sharma, M. Shimoda, A.P. Tsai, B. Unal, A.R. Ross, T.A. Lograsso and P.A. Thiel, *Phys. Rev. Lett.* 95 (2005) 155504; H.R. Sharma, V. Fournée, M. Shimoda, A.R. Ross, T.A. Lograsso, P. Gille and P.A. Thiel, *Phys. Rev. B* **78** (2008) 155416.

107. C. Bergman, C. Girardeaux, C. Perrin-Pellegrino, P. Gas, J.M. Dubois and N. Rivier, *J. Phys.: Cond. Matter* **20** (2008) 314010 (7pp); V. Fournée, Vol. III and references therein.

108. W.R. Tyson, Canadian Metall. Quarterly 14-4 (1975) 307; A. Carré, in *Phénomènes d'Interface*, Ed. J. Briant (Technip, Paris, 1989).

109. R.J. Good, *J. Adhesion Sci. Technol.* **6-12** (1992) 1269. See this article for references to Bronsted, Lewis and others regarding the theory of contact angles and wetting by liquids.

110. E. Belin-Ferré and J.M. Dubois, *Int. J. Mat. Res.* **97-7** (2006) 985.

111. B.R. Strohmeier, *Surf. Interf. Anal.* **15** (1990) 51.

112. V. Demange, A. Milandri, M.C. De Weerd, F. Machizaud, G. Jeandel and J.M. Dubois, *Phys. Rev. B* **6-14** (2002) 144205 1.

113. Y. Champion, C. Langlois, S. Guérin-Mailly, P. Langlois, J.L. Bonnentie, and M. Hÿtch, *Science* **300** (2003) 310.

114. M. Conrad, B. Harbrecht, T. Weber, D. Y. Jung and W. Steurer, Larger, larger, largest – a family of cluster-based tantalum-copper-aluminides with giant unit cells. Part B: the cluster structure, Preprint (2009).

CHAPTER 2

SOLUTION GROWTH OF INTERMETALLIC SINGLE CRYSTALS: A BEGINNER'S GUIDE

Paul C. Canfield

Ames Laboratory, Department of Physics and Astronomy
Iowa State University, Ames, Iowa 50011, U.S.A.
E-mail: canfield@ameslab.gov

Progress in experimental solid state physics, as well as in related disciplines, is often hampered by limited access to single crystals of novel materials. This lack of single crystalline samples constrains researchers to study or measure materials that are available rather than the materials that are most relevant to the questions at hand. The extreme example of this is the allegorical Ph.D. student who has to measure the one crystal that the advisor has on hand, regardless of the physical question to be probed. As part of the second Euroschool on complex metals I presented lectures on the design, discovery, growth and characterization of novel intermetallic compounds. In this paper I will focus on the basics of solution growth of novel compounds.

1. Introduction

The 2nd Annual European School in Material Science was held in Ljubljana, Slovenia from May 21 to 26, 2007. The focus of this summer school was the synthesis and properties of novel, bulk, intermetallic compounds. With this in mind I presented a series of four lectures on growth (first lecture) and characterization (second lecture) of novel compounds with an emphasis on the tuning of magnetism (the third lecture) and superconductivity (the forth lecture). These lectures were based on my past twenty (or so) years of experience in the broad field of design, discovery, growth, characterization and understanding of novel materials. Whereas we have published literally hundreds of papers on the

93

characterization (and hopefully added to the understanding) of a wide number of materials ranging from local moment magnetism,[1-4] to quasicrystals[5-10] and spinglasses[11-14] and from superconductors[15-22] to heavy fermions,[23-28] we have only published a small number of papers that explicitly outline how we synthesize the vast majority of these compounds: by solution growth.[29-32] This paper is an attempt to augment and compliment these technical, solution growth papers. I will try to outline, in a step by step fashion, how to plan a solution growth, how to assemble a growth and, near the end of the growth, how to decant the remaining solution from the crystals that (hopefully) have grown. This paper is meant to provide a novice with some concrete instructions and examples.

Before beginning with the details it is vitally important to make one point clear. Solution growth of intermetallic compounds is not a sterile, theoretical exercise; it involves the use of element and compounds with varying degrees of toxicity; it involves the use of sharp and very hot tools; it involves the use of high temperature furnaces. Before starting a program of research that involves the solution growth of intermetallics you should consult your local safety office and make sure that they advise and approve of your plans. Even in my laboratory, a lab that has over two decades of collective experience with solution growth, when we try a new variant or a new flux, we examine the procedure, discuss possible problems and make sure that we are preceding in as safe a manner as possible. The growth of materials in this manner is not difficult, but it does involve steps that demand respect and careful attention.

2. What Do You Need?

Solution growth of intermetallic compounds requires a solution, something to hold it in, something to provide a protective environment, and something to provide a controllable temperature. In addition, a decanting device is often very useful too. Let's review each of these "needs" in turn.

If you want to grow CeSb out of a Sn flux, then you clearly need some Ce, Sb and Sn.[29] Often growth labs will have a collection of various high purity (and even lesser purity) elements and compounds on hand for use. Like a well stocked liquor cabinet, you save the "good stuff" (in this case the really high purity stuff) for special occasions, such as when you have already optimized a growth using less pure (cheaper) stuff.

A comment on purity is appropriate at this point. Be very careful to understand what the purity of your starting materials is. It is important to know what you are using and what different purity numbers mean. For example, many chemical retailers will provide purities on a "metals basis". This is (unfortunately) often a very deceptive number and provides a classic example of "caveat emptor". Metals basis purity only considers other metals as impurities. Nickel with 5% iron would have a metals basis purity of 95% (so far so good). On the other hand nickel with 5% oxygen in it (e.g. in the form of a nickel oxide) and 0.5% iron would have a metals basis purity of 99.5%. The oxygen is not counted at all. Worse yet is the case of oxides. I have purchased VO_2 that was sold as 5-9's (i.e. 99.999%) purity that contained 10% V_2O_5. The chemical retailer was not lying, but this is an ugly way of making a mixed phase material appear to be purer than it is.

Once the composition of the solution is know the next step is to decide how to hold it. Generally this is done in a crucible. The most common crucible for intermetallic growth is high density, sintered Al_2O_3. (In a similar manner the most common crucible for oxide growths is pure Pt.) The choice of crucible is generally one determined by cost and stability. Al_2O_3 is stable against attack from many of the low melting elements that are used as solutions: Al, Cu, Zn, Ga, Ge, In, Sn, Sb, Pb, and Bi. On the other hand thermite-like reactions are a problem with solutions that are sufficiently concentrated in rare earth metals, Mg, or some transition metals (such as Ti, Hf and Zr). These metals can form a more stable oxide and sometimes release Al into the melt. In these cases other crucibles such as yttria, zirconia, or elemental Ta have to be used.[30-32] For this article only the use of Al_2O_3 will be discussed. As a rule of thumb, about ten atomic percent of rare earth can be added to one of the low melting elements listed above and held in an Al_2O_3 crucible without attack (up to ~1200°C). Lesser amounts of other reactive elements can also be held in Al_2O_3, sometimes at lower temperatures.

The solution, sitting in the crucible usually needs to be protected from the air; from possible reaction with oxygen, nitrogen, etc. This is most easily done by sealing the crucible into an amorphous silica ampoule. This requires a hydrogen-oxygen blow torch, a well ventilated glass bench, and a pumping station for evacuating and back filling the ampoule before sealing.

The growth then needs to be heated and cooled in a controlled manner. This is usually done with a furnace that has a programmable temperature controller. Generally a furnace that can go as high as (or higher than) 1200°C is needed. If the furnace available only reaches 1000°C, then this precludes growths that require higher temperatures. Generally heating is done rather fast (100°C/hour) and cooling is done in steps and rather slowly (1 – 10°C/hour or slower). A temperature controller with multiple steps is convenient, but not required. Generally the furnace should be one that can be opened rapidly and allow for the removal of the hot ampoule at the end of the growth.

Once the growth cycle is completed it is often very convenient to decant the remaining solution from the crystals that have (hopefully) grown. If the excess liquid is not removed, it will solidify around the crystals and will have to be removed by chemical and/or mechanical means once the growth is at room temperature. Decanting of the still hot growth requires a lab centrifuge with metal cups to hold the ampoule.

So then, to answer the question, "what is needed?": the elements or compounds that will comprise the initial solution; a crucible to hold the melt; a quartz bench to form and evacuate the ampoule that protects the melt from air; a programmable furnace; and a centrifuge for decanting off the excess solution at the end of the growth.

3.   Planning the Growth

The first step in planning a growth is to have an idea. This is often the hardest step in the process. The best way around this difficulty is to follow the advice of Linus Pauling, who said, "The way to get good ideas is to get lots of ideas and throw the bad ones away."

Once a specific compound has been determined to be of interest, or a specific compositional phase space has been identified as promising, then it is useful to see if any information exists about it in the form of phase diagrams. For simplicity we can examine the schematic phase diagram shown in Fig. 1a. If the compound AB_2 is to be grown, then, given that it is a paratectically melting compound, methods that utilize cooling of a stoichiometric (or near stoichiometric) melt of AB_2 will not give single phase AB_2. This means that methods such as arc melting, Bridgeman growth, or zone refining can yield a mixture of AB, AB_2, and B. AB_2 can, though, be readily grown via the cooling of a solution rich in B. As long as the initial melt composition is more B-rich than that of the composition of the liquid at the paratectic temperature (P on diagram), then the primary solidification (on cooling) will be AB_2.

A potential cooling curve is shown as a dotted line in Fig. 1a. This line represents the temperature dependent composition of the liquid phase as the growth cools. Between temperatures T_1 and T_2 the liquid stoichiometry does not change. Once $T < T_2$, though, crystalline AB_2 starts to form and the remaining solution becomes increasingly B-rich. In general it is desirable to stop cooling the growth at some temperature T_3 that is greater than the eutectic temperature so that the excess solution can be decanted off. In general a binary phase diagram is not known precisely and reported versions of any given binary phase diagrams can differ from each other (sometimes significantly). Since rapid cooling through T_2 can cause multiple nucleation and dendritic growth (neither of which are conducive to the growth of large, well formed single crystals) slow cooling of the melt should begin at some temperature above T_2. For the growth of AB_2 a possible growth cycle could consist of heating to T_1, dwelling at T_1 for several hours to allow for homogenization of the melt, dropping to $T_2 + 50°C$ in an hour or two and then cooling from $T_2 + 50°C$ down to T_3 over 50-100 hours. At T_3 the growth could then be taken out of the furnace and decanted.

The compound AB can be grown from a stoichiometric melt, at least in theory, but in many cased the temperatures associated with the primary solidification (the temperatures above the paratectic temperature on the B-rich side and the eutectic temperature on the A-rich side) can be too high. Given that the growths that will be discussed in this introductory paper will be sealed in amorphous silica (see[30,33] for higher temperature growths and how to handle them) it should be noted that the maximum temperature that can be used during the growth procedure is very close to 1200°C. Amorphous silica softens and looses it structural integrity for temperatures significantly above 1200°C and either collapses or explodes (depending on the internal pressure at that temperature). In cases like this it can be useful to find a third element (or combination of elements) to act as a flux: a solution that has a lower temperature surface of primary solidification of the compound AB. This is shown as a schematic, pseudo-binary, phase diagram in Fig. 1b. In such a case a small amount of elements A and B can be added to a lot of element C. The cooling curve, with temperatures T_1, T_2, and T_3 is operationally the same as that discussed above for Fig. 1a. In this case, finding a viable element C can be the tricky part since data on ternary (or higher) phase diagrams can be very limited or simply non-existent. A real example of growths using phase diagrams like those show in Fig. 1 is the growth of $CeSb_2$ and Ce_2Sb out of excess Sb and Ce respectively and the growth of CeSb out of Sn as a third element.[29]

Whether based on consultations of phase diagrams, previous growths, experience and/or intuition, a growth starts with the determination of an

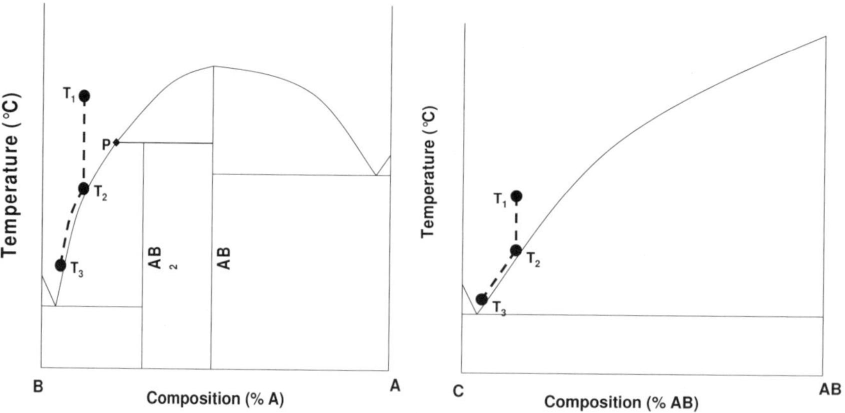

Fig. 1a (left panel). Schematic binary phase diagram for the A – B system. The dotted curve represents a possible cooling curve and the associated composition of the solution. Point P indicates the composition of the liquid/solid line at the paratectic temperature.
Fig. 1b (right panel). Schematic pseudo-binary phase diagram for the C – AB system. The dotted curve represents a possible cooling curve and the associated composition of the solution.

initial composition of the melt and a heating and cooling profile. These can (and usually are) be adjusted in subsequent growths based on information extracted from earlier attempts. Ultimately, if a growth is successful, a growth can be optimized and a standard composition and temperature profile can be used to grow it repeatedly and reliably.

4. Assembling the Growth

Once the growth has been planned it is time to get into the lab and actually assemble it. At this point we will start using the materials described in passing above. Figure 2 is a picture of a length of amorphous silica tubing that has been sealed and flattened on one end and two 2 ml Al_2O_3 crucibles. These will be used to provide a growth environment.

Fig. 2. Silica tube (sealed and flattened on bottom) and two 2ml Al_2O_3 crucibles.

Figure 3 shows the crucibles with some of the elemental metals placed inside.

                                    Paul C. Canfield

Fig. 3. Al_2O_3 crucibles being packed with elemental metals for growth of RAl_3 compound (elemental Al shown).

Generally it is best to have higher melting elements or alloys surrounded by lower melting ones, as shown schematically in Fig. 4a. Once the growth crucible is assembled, the "catch" crucible can be packed with silica wool (Fig. 4a, b). The purpose of this crucible is to immobilize the excess liquid at the end of the growth the silica wool helps to separate the crystals from this excess fluid. If there were no silica wool, or if there is an inadequate amount, then some of the crystals can move over to the catch side and still be embedded in the ultimately solidified, excess liquid. In some cases the growth can be assembled without a catch crucible and the silica wool can simply be placed on top of the growth crucible. In such cases, though, there can be chemical attack between the solution and the silica tube or, if the solution expands upon cooling, the tip of the ampoule will crack and shatter as the plug of decanted solution solidifies. Both of these cases can lead to a loss of inert environment while the crystals are still hot and often leads to surface oxidations (or worse).

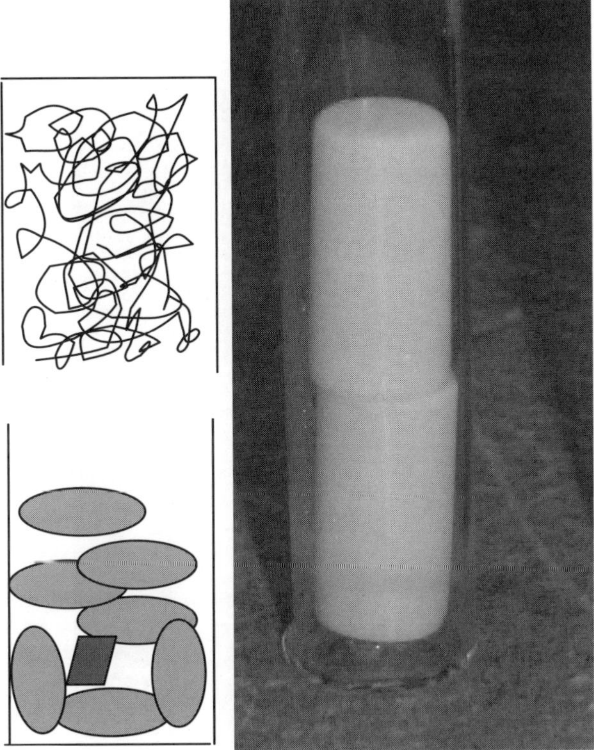

Fig. 4. (a-left panel) Schematic diagram of growth and catch crucibles. The smaller, parallelogram, shape indicates a higher melting element and larger, elliptical, shapes represent the lower melting element. The upper crucible is packed with silica wool to provide a filtering and immobilization of the excess liquid at the end of growth cycle, during the decanting step. (b- right panel) Picture of growth and catch crucibles in silica tube.

Figure 5 shows the silica tube after necking. The growth and catch crucibles are easily seen as well as a plug of silica wool on top of the catch crucible. This wool acts as a cushioning device that prevents the crucibles from slamming into the sealed tip of the tube and shattering it during the decanting step. It is important that during the necking step the silica tube is softened (by the hydrogen-oxygen blow torch) and allowed to collapse with minimal pulling. This allows for necks that are thick walled. If the tubing is pulled and thinned, then the seal can be weak or

                              Paul C. Canfield

the silica can suddenly bubble inward and shatter as the neck is heated up for the seal.

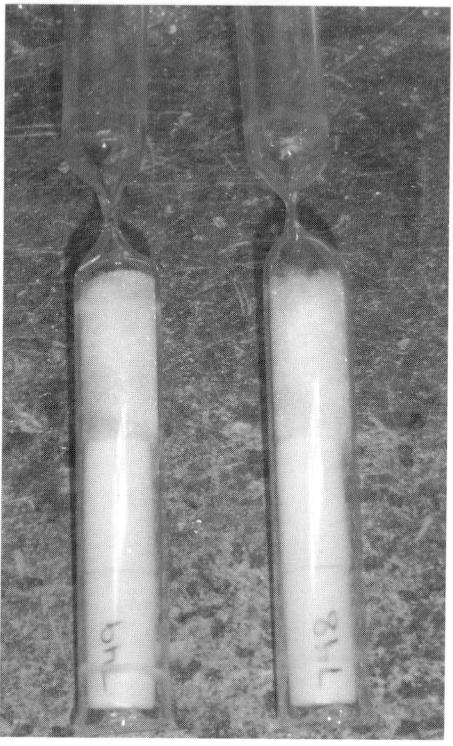

Fig. 5. Growth ampoules after necking, but before sealing. Note the silica wool above the catch crucibles.

The necked tubes are attached to a pumping system, evacuated and back filled with an inert gas (argon is often used). Depending upon the growth, the ampoules can be sealed in partial vacuums ranging from less than a torr up to ¼ atmosphere or higher. Often partial pressures of Ar that will result in near ambient pressure at the maximum growth temperature are used when the melt contains elements that have vapor pressures in excess of 0.1 atmosphere or when the solution contains elements that are both volatile and reactive with the silica tubing (such as Al). Once the ampoule has been pumped and flushed (usually three or more times) and back filled to the desired pressure of Ar, it is sealed off with the hydrogen/oxygen torch. The sealed ampoule is then place in a larger Al_2O_3 crucible for mechanical support, as shown in Fig. 6. The ampoule should be cleaned with a mild organic solvent, such as ethanol, to remove grease/finger prints, which can, at elevated temperatures, lead to attack on the silica.

Fig. 6. Sealed growth ampoules in 50 ml Al_2O_3 crucibles for mechanical support. Note: in this image, the growth and catch crucibles are hidden by the 50 ml, support crucibles.

5. Running the Growth

At this point the sealed ampoules can be placed into the furnace (Figs. 7 and 8). This is done by donning the thermally insulating gloves and using the tongs shown in Fig. 8. In addition a face visor is always used to protect the face and neck. These furnaces are heated by two banks of SiC heaters, one on top and one underneath the hearth plate that the ampoules are supported by. If the solution contains extremely volatile elements (e.g. Zn) then the ampoules can be raised up toward the upper bank of heating elements. This helps to prevent re-condensation of the solvent on top of the silica wool by making the top of the ampoule slightly hotter than the bottom.

Fig. 7. Two growth ampoules in box furnace. Furnaces such as this can easily hold up to 10 growth ampoules and allow for rapid sampling of phase space.

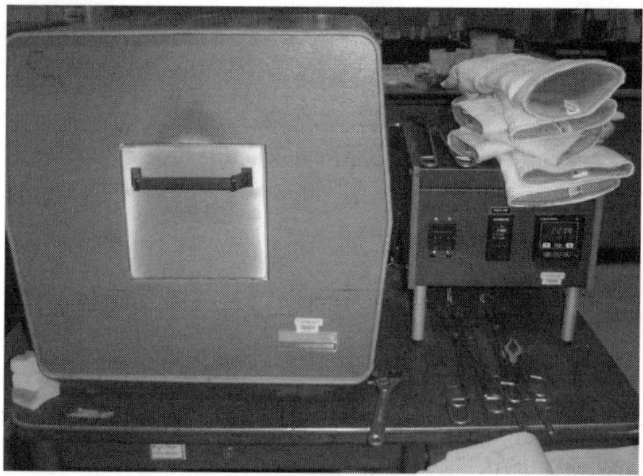

Fig. 8. 1500°C, SiC heater element, box furnace used for solution growth. Power supply/ temperature controller shown on right. High temperature gloves (on top of temperature controller) as well as tongs (on top and underneath of temperature controller) also shown.

Once the ampoules are in the furnace the temperature profile can be programmed into the temperature controller and recorded into the lab notebook. In general heating to the maximum temperature can be achieved within a few hours. A dwell, of several hours, at maximum temperature is often used so as to encourage complete dissolution and mixing of the solution. Cooling of the solution can take from as little as 5 hours to as long as several hundreds of hours, depending upon the specific growth. Exploratory growths are generally fast whereas highly optimized growths are, in general much slower. A temperature controller that allows for the input of hours between temperature set points, rather than cooling rate, is preferable since this, de facto, can allow for almost any arbitrary and small cooling rate. For example, if the melt is to be cooled from 575°C to 510°C in 100 hours this is trivial to program if hours to cool is input, but awkward if the rate has to be input. With the usual 0.1°C/hour resolution this requires a choice between 0.6°C/hour (108.3 hours) or 0.7°C/hour (92.8 hours). After the slow cooling is complete the furnace should be programmed to dwell at the final temperature for a long time so that there is no possibility of the furnace shutting off before the growth is removed.

6. Decanting

Decanting is often the step that many researchers over look when thinking about solution growth, especially decanting in a centrifuge. This, it turns out, is a very useful and, in many cases, vital step. Just inverting the growth ampoule in the furnace, and letting gravitational acceleration (9.8 m/s^2) pull at the remaining solution, is not enough to lead to significant separation of crystals from the solution. In many cases the metallic solution is quite viscous and, in addition there can often be a thin layer of crystals or slag on the meniscus that prevents or hinders fluid flow. Figure 9 shows the solution to these problems: a centrifuge equipped with a metal rotor and cups. A back of the envelope calculation shows that an ampoule spinning at 1000 rpm (slow for a lab centrifuge) with the sample volume about 10 cm from the axis of rotation provides a factor of 100 increase in the acceleration felt by the solution. This is more than enough to remove the excess solution, and in many cases actually produce mirrored growth facets on the exposed crystals (see Fig. 10).

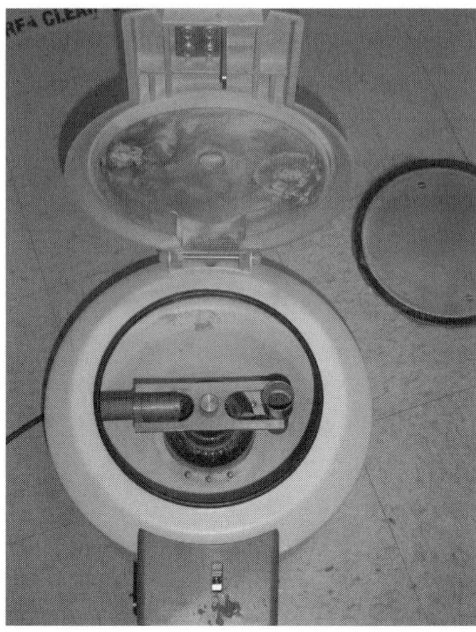

Fig. 9. Centrifuge with brass rotor and cups. Refractory fiber is used at the bottom of the sample cup to provide cushioning. Small coins of the world (as part of our international growth effort) are used as counter weights.

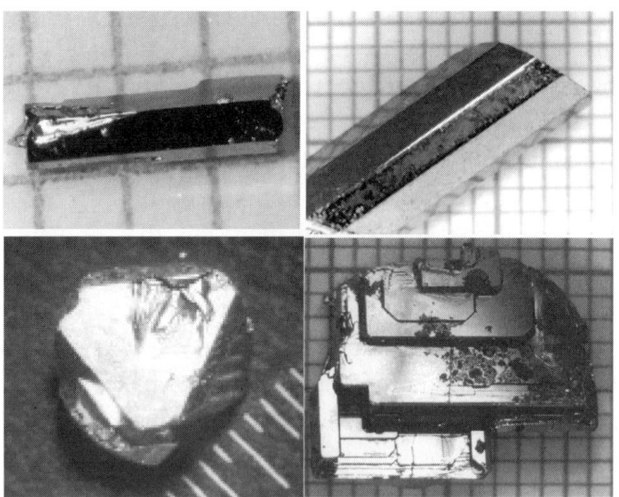

Fig. 10. Examples of crystals grown via solution growth as described in text. Upper left: YbAgGe (trigonal unit cell)[25]; upper right: decagonal phase AlNiCo[7]; bottom left: YFe_2Zn_{20} (cubic)[23]; lower right: $LaAgSb_2$ (tetragonal).[2]

The decanting process is a dynamic one. The centrifuge, rotor and cups are at room temperature and the growth can be at temperatures as high as 1200°C. This process works because the growth is inside of a crucible that is itself inside of a silica ampoule (in many cases and evacuated ampoule). If the ampoule is removed from the furnace in a prompt manner and quickly tipped into the cup of the centrifuge, then the excess liquid can be decanted off before it drops in temperature by more than 10 or 20°C. To decant thermal gloves and a visor need to be worn and tongs are used to grab the 50 ml alumina crucible. The ampoule is then tipped into the centrifuge cup, the lid of the centrifuge is slammed shut, and the centrifuge is started. This whole process (from opening the furnace door to starting the centrifuge) should take no more than 5 seconds. The centrifuge need to run for less than 10 seconds since the decanting takes place in the first second or two of acceleration. Too much time with the centrifuge on actually increases the risk of breaking the ampoule and even deforming or breaking the crystals. Once the centrifuge rotors come to rest the ampoule can be removed and placed in a safe place to cool.

7. Opening the Growth and Planning the Next One

When the ampoule has cooled to close to room temperature it can be opened. Care should be used when opening it since it contains fibrous glass and once opened will also have shards of glass and, depending on the elements used, hazardous waste. Opening the ampoule in a well ventilated space similar to that was used for the glass work is a good way to minimize these risks. Often the growth ampoule will contain some, or all, of the well formed single crystals, with the remaining ones being caught on the top of the silica wool in the catch crucible. In many cases it is necessary to break open the growth crucible so as to extract the crystals that have nucleated off of the inner walls. In general the Al_2O_3 crucibles are not re-usable, so breaking them is not an unforgivable waste. (This is very different from the case of oxide growth out of Pt crucibles that are cleaned and reused for tens of growths.)

One of the bonuses of solution growth is that the crystals often manifest well developed (and even mirrored) facets. A few examples are shown in Fig. 10. These are clearly examples of successful growths. Clearly, not every attempt is so successful. Below are a few "trouble shooting" comments. Much more detail and theoretical background can be found in texts on crystal growth.[34,35]

Total spin: Called this because all of the material in the growth crucible is still liquid at the decanting temperature and therefore is "spun" to the catch crucible side. In terms of Fig. 1, this means that the spin temperature was above T_2. Although this is irritating (the growth produced no crystals) it can actually be a very useful data point. It means that the starting temperature of the next growth can be reduced to a temperature close to the spin temperature that resulted in the total spin. This can lead to much lower cooling rates, since time in not being wasted in cooling through large expanses of the homogeneous liquid phase.

No spin: As suggested by the name, nothing leaves the growth crucible during the decanting (spinning) step. This usually means that the decanting temperature was too close to or below the local eutectic point. The simplest thing to do is repeat the growth and decant at a higher temperature. Often the growth and catch crucibles can simply be resealed into a new ampoule and used again. Note: if there has been some spin or if there has been some attack on the growth crucible, then this is not advisable since the stoichiometry of the initial melt has been changed in an unquantifiable manner.

Growth crucible cracked/ruptured: The growth crucible can be damaged due to a number of reasons. The most common is a thermite-type reaction, as described above. If this is suspected then either the amount of reactive metal in the melt needs to be reduced, or a different type of crucible needs to be used. Sometimes the elements placed in the growth crucible can react exothermally when they melt and mix. When this is the case, the energy released can shatter the crucible during the initial heating of the growth. If this is suspected, then a pre-melting of these elements (e.g. in an arcmelter) will allow for the release of energy outside of the crucible. One example of this is Al and Pd. When they are melted together they release considerable energy (glow brightly) and shatter Al_2O_3 crucibles. On rare occasions the growing crystals become so large that they actually break the crucibles as the span the inner diameter. This is a "good" problem since it means that the crystals are large. Often in this case the cracks are not large enough to cause any leaking of the melt.

Small and/or poorly formed crystals: This can happen for many different reasons. The most common one is associated with cooling too fast. It is important that the melt does not cool to rapidly through T_2 so raising the temperature at which slow cooling starts sometimes helps. On the other hand, reducing the cooling rate often is important as well. Essentially more time (i.e. a lower cooling rate) is the easiest, first adjustment to a growth when there is a need to improve crystal size or quality

Growth of a second phase: In some growths there can be well formed, larger crystals with smaller crystals of a different morphology on top. For example, crystals with a cubic morphology and linear dimension of 5 mm with little rods growing off of them at skew angles. This is a clear hint that during the cooling process a second phase (rods) started to grow at lower temperatures than the primary phase (cubes). The simplest way to try to avoid this is to repeat the growth with a higher decanting temperature (increase T_3).

8. Final Remarks

This paper is meant to provide a detailed introduction to the mechanics of solution growth. It is meant to augment earlier papers on this technique[29-33] while presenting new details that are often presumed to be "obvious" or "known". Solution growth of novel materials is not limited to intermetallic compounds. Many of the same ideas are applicable to oxides, salts, and organic compounds.[34,35] This paper, as well as these references will hopefully provide the interested researcher with the information necessary to start using this powerful technique. In a century that will be faced with a myriad of technical challenges that will have to be addressed by the discovery and design of new materials with specific, complex properties, the more scientists that are engaged in the search for such compounds, the better humanities chances are of surviving to see the dawn of the 22nd century.

 Paul C. Canfield

Acknowledgments

The design, growth and discovery of novel materials requires long term support of scientific research of the most fundamental and basic type. My group has benefited greatly from the exactly such support from the U.S. Department of Energy, Basic Energy Sciences. Great thanks, and credit, are due to its program managers (both past and present). Work at the Ames Laboratory was supported by the Department of Energy, Basic Energy Sciences under Contract No. DE-AC02-07CH11358.

References

1. P. C. Canfield, S. L. Bud'ko, B. K. Cho, A. Lacerda, D. Farrell, E. Johnston-Halperin, V. A. Kalatsky and V. L. Pokrovsky, *Phys. Rev.* **B55**, 970 (1997).
2. K. D. Myers, S. L. Bud'ko, I. R., Fisher, Z. Islam, H. Kleinke, A. H. Lacerda and P. C. Canfield, *Journal of Magnetism and Magnetic Materials* **205**, 27 (1999).
3. S. L. Bud'ko, P. C. Canfield, C. H. Mielke, A. H. Lacerda, *Phys. Rev.* **B57**, 13624 (1998).
4. S. L. Bud'ko, Z. Islam, T. A. Wiener, I. R. Fisher, A. H. Lacerda, and P. C. Canfield, *Journal of Magnetism and Magnetic Materials* **205**, 53 (1999).
5. I. R. Fisher, Z. Islam, A. F. Panchula, K. O. Cheon, M. J. Kramer, P. C. Canfield and A. I. Goldman, *Phil. Mag.* **B77**, 1601 (1998).
6. Z. Islam, I. R. Fisher, J. Zarestky, P. C. Canfield, C. Stassis and A. I. Goldman, *Phys. Rev.* **B57**, R11047 (1998).
7. I. R. Fisher, M. J. Kramer, Z. Islam, A. R. Ross, A. Kracher, T. Wiener, M. J. Sailer, A. I. Goldman and P. C. Canfield, *Phil. Mag.* **B79**, 425 (1999).
8. I. R. Fisher, M. J. Kramer, Jt. T. A. Wiener, Z. Islam, A. R. Ross, T. A. Lograsso, A. Kracher, A. I. Goldman and P. C. Canfield, *Phil. Mag.* **B79**, 1673 (1999).
9. I. R. Fisher, M. J. Kramer, Z. Islam, T. A. Wiener, A. Kracher, A. R. Ross, T. A. Lograsso, A. I. Goldman and P. C. Canfield, *Materials Science & Engineering*, **A294-296**, 10 (2000).
10. I. R. Fisher, X. P. Xie, L. Tudosa, C. W. Gao, C. Song, P. C. Canfield, A. Kracher, K. Dennis, D. Abanoz and M. J. Kramer, *Phil. Mag.* **B82**, 1089 (2002).
11. P. C. Canfield and I. R. Fisher, *Journal of Alloys and Compounds* **317-318**, 443 (2001).
12. I. R. Fisher, Z. Islam, J. Zarestky, C. Stassis, M. J. Kramer, A. I. Goldman and P. C. Canfield, *Journal of Alloys and Compounds* **303-304**, 223 (2000).
13. I. R. Fisher, K. O. Cheon, A. F. Panchula, P. C. Canfield, M. Chernikov, H. R. Ott and K. Dennis, *Phys. Rev.* **B59**, 308 (1999).

14. T. A. Wiener, I. R. Fisher, S. L. Bud'ko, A. Kracher and P. C. Canfield, *Phys. Rev.* **B62**, 15056 (2000).

15. S. L. Bud'ko, G. Lapertot, C. Petrovic, C. E. Cunningham, N. Anderson and P. C. Canfield, *Phys. Rev. Lett.* **86**, 1877 (2001).

16. P. C. Canfield and S. L. Bud'ko *Scientific American* **292**, 62 (2005).

17. P. C. Canfield and G. W. Crabtree, *Physics Today* **56**, 34 (2003).

18. P. C. Canfield, P. L. Gammel and D. J. Bishop, *Physics Today* **516**, 40 (1998).

19. B. K. Cho, P. C. Canfield and D. C. Johnston, *Phys. Rev. Lett.* **77**, 163 (1996).

20. P. C. Canfield, S. L. Bud'ko and B. K. Cho, *Physica C*: **262**, 249 (1996).

21. B. K. Cho, P. C. Canfield and D. C. Johnson, *Phys. Rev.* **B52**, R3844 (1995).

22. H. Suderow, S. Vieira, J. D. Strand, S. Bud'ko and P. C. Canfield, *Phys. Rev.* **B69**, 060504/1 (2004).

23. M. S. Torikachvili, S. Jia, E. D. Mun, S. T. Hannahs, R. C. Black, W. K. Neils, Martien, Dinesh, S. L. Bud'ko and P. C. Canfield, *Proceedings of the National Academy of Sciences of the United States of America,* **104**, 9960 (2007).

24. A. Hiess, J. X. Boucherle, F. Givord, J. Schweizer, E. Lelievrc-Bcrna, F. Tasset, B. Gillon and P. C. Canfield, *J. of Physics: Conden. Matter* **12**, 829 (2000).

25. S. L. Bud'ko, E. Morosan and P. C. Canfield, *Phys. Rev.* **B69**, 014415/1 (2004)

26. Z. Fisk, P. C. Canfield, W. P. Beyermann, J. D. Thompson, M. F. Hundley, H. R. Ott, E. Felder, M. B. Maple., M. A. Lopez de la Torre et al., *Phys. Rev. Lett.* **67**, 3310 (1991).

27. A. Yatskar, N. K. Budraa, W. P. Beyermann, P.C. Canfield and S. L. Bud'ko, *Phys. Rev.* **B54**, R3772 (1996).

28. A. Yatskar, W. P. Beyermann, R. Movshovich and P. C. Canfield, *Phys. Rev. Lett.* **77**, 3637 (1996).

29. P. C. Canfield and Z. Fisk, *Phil. Mag.* **B65**, 1117 (1992).

30. P. C. Canfield and I. R. Fisher, *Journal of Crystal Growth,* **225**, 155 (2001).

31. Y. Janssen, M. Angst, K. W. Dennis, P. C. Canfield and R. W. McCallum, *Review of Scientific Instruments,* **77**, 056104/1 (2006).

32. Y. Janssen, M. Angst, K. W. Dennis, R. W. McCallum and P. C. Canfield, *Journal of Crystal Growth,* **285**, 670 (2005).

33. Z. Fisk and J. P. Remeika, in *Handbook on the Physics and Chemistry of Rare Earths*, Eds. K. Gschneidner, Jr. and L. Eyring (Amsterdam, Elsevier, 1989) Vol. 12.

34. B. R. Pampkin, *Crystal Growth* (International series of monographs on the science of the solid state, Pergamon Press, 1975.) Vol. 6.

35. D. Elwell and H. J. Scheel, *Crystal Growth from High-Temperature Solutions* (Academic Press, 1975).

CHAPTER 3

THERMAL CONDUCTIVITY OF COMPLEX METALLIC ALLOYS

Ana Smontara, Ante Bilušić, Željko Bihar and Igor Smiljanić

Laboratory for the Study of Transport Problems, Institute of Physics
P.O. Box 304, 10000 Zagreb, Croatia
E-mail: ana@ifs.hr

The term "complex metallic compounds" comprises a great number of compounds characterized by their large unit cells that can contain up to a thousand atoms. The paper views the heat and electric transport properties of the class of these compounds known also as quasi-crystalline approximants, with the unit cells formed of polytetrahedral structures also present in quasicrystals. The simultaneous long-range periodicity on the physical-length scale and the presence of non-periodicity of polytetrahedral structures on the short-range length scale causes a crossover of space localization and delocalization of atomic vibrations and current carriers. The electrical conductivity in the temperature range 1.5 K- R.T. is weak temperature dependent that is explained as an interplay between the delocalization (due to the long-range periodicity) and the localization (due to the short range polytetrahedral order) of charge carriers, while the thermal conductivity vs. temperature behavior is similar to those for thermal insulators, with the remark that half of the net heat transport is governed by the charge carriers.

1. Introduction

The aim of this chapter is to explain the main mechanisms by which heat is conducted in complex metallic alloys. In general, there are several mechanisms by which heat can be transmitted through a solid and many processes which limit the effectiveness of each mechanism. In a non-metal, heat is conducted by means of the thermal vibrations of the atoms. In a simple metal this mode of heat transport makes some contribution, but the observed thermal conductivity is almost entirely due to the electrons. It is not automatic that the thermal conductivities of the two types of solid must be very different in magnitude, as they are in the case of electrical conductivity, but their dependences on temperature and on the imperfections in individual specimens are rather different. In many solids, such as alloys, both transport mechanisms can make comparable contributions to the observed conductivity, and the relative proportions vary with composition and temperature. Complex metallic alloys (CMAs) belong to these solids. They exhibit complex structures that contain some hundred up to several thousand atoms in the unit cell.[1] Examples are the 'Bergman phase' $Mg_{32}(Al,Zn)_{49}$ with 162 atoms in the unit cell[2], orthorhombic ξ'-$Al_{74}Pd_{22}Mn_4$ (258 atoms/unit cell)[3,4,5] and the related Ψ phase (about 1500 atoms/u.c.)[5], λ-Al_4Mn (586 atoms/unit cell)[6], cubic β-Al_3Mg_2 (1168 atoms/unit cell)[7], and the heavy-fermion compound $YbCu_{4.5}$, comprising as many as 7448 atoms in the supercell.[8] These giant unit cells contrast with elementary metals and simple intermetallics whose unit cells in general comprise only from single up to a few tens of atoms. The giant unit cells with lattice parameters of several nanometers provide translational periodicity of the CMA crystalline lattice on the scale of many interatomic distances, whereas on the atomic scale the atoms are arranged in clusters with polytetrahedral order, where icosahedrally coordinated environments play a prominent role. The clustering feature of CMA structures together with the two competing physical length scales the quasiperiodic short-range atomic order and the periodic long-range order may have a significant impact on the physical properties of the material, such as the electronic structure and the lattice dynamics. On this basis, CMA materials are expected to exhibit novel transport properties, like a combination of metallic electrical conductivity with low thermal conductivity, and electrical and thermal resistances tunable by varying the composition.

2. Basics of the Thermal Conductivity Measurements

Thermal conductivity is a physical property important from both scientific and technological points of view. For a scientist, thermal conductivity offers valuable information about heat carrying channels present in a system (e.g., electrons, phonons, magnetic excitations, etc.), important scattering mechanisms and the presence of eventual phase transition. For an engineer thermal conductivity is crucial for many fields of technology where good thermal insulation or greasing is important.

Many of the methods for measuring thermal conductivity in the temperature region below room temperature are similar, the differences being mainly in thermometry. Here, we present thermal conductivity apparatus[9,10], developed in our laboratory, suitable for use in the temperature range 1.7 to 300 K.

Thermal conductivity is an intrinsic property of a material that relates an existing temperature gradient, $\vec{\nabla}T$, that causes a heat flux, $\vec{Q}$. Generally, thermal conductivity is a tensor[11]:

$$Q = \hat{\kappa} \cdot \nabla \cdot T. \tag{1}$$

In the one-dimensional approximation, thermal conductivity tensor collapses to a thermal conductivity coefficient, κ, whereas the problem of the measurement of thermal conductivity becomes a problem of determination of thermal flux through a sample and measurement of the temperature drop across it:

$$\frac{P_Q}{A} = \kappa \cdot \frac{\Delta T}{l}. \tag{2}$$

In Eq. (2), P_Q is a heat power sent through a sample of the cross section A, while ΔT is a temperature difference between two points across the sample, separated by l.

There are two ways how the heat power P_Q can be determined: a relative and an absolute one. In a relative method, a reference sample, with a known thermal conductivity, is placed between a heater and the sample. P_Q is calculated using Eq. (2), with ΔT determined across the reference sample via a differential thermocouple. The sample thermal conductivity is then determined by independent measurement of ΔT across the sample. In an absolute method, the sample heater is glued directly to the sample holder whose other end is thermally anchored to a heat sink, i.e. the sample-holder body (see Fig. 1).

                    Ana Smontara et al.

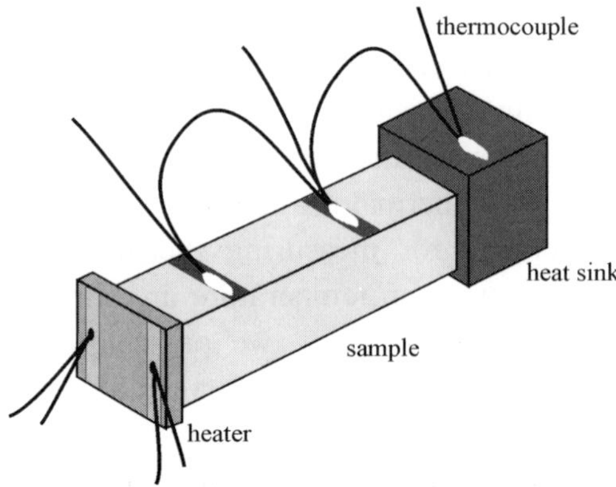

Fig. 1. Measurement scheme of absolute, steady state method of the thermal conductivity measurement. White spots represent IMI 7031 varnish, used to thermally anchor thermocouple to sample, and sample to heater and heat sink as well.

As a heater, we use a RuO_2 chip-resistor, due to its superior stability after several coolings to the liquid helium temperature. Heat power through the sample P_Q is supposed to be equal to the electrical power P_E of the chip-resistor ($P_E = U \cdot I$, where U is a voltage drop, and I a current). Sample and chip-resistor are glued via IMI 7031 (former GE 7031) varnish, in order to provide the best possible thermal contact between each other. It is also used for thermally anchoring the sample to a heat sink. The temperature drop across the sample, ΔT, is monitored via differential thermocouple chromel-gold with 0.07 at.% of iron. To avoid the presence of other metals on joints due to soldering, wires are welded, that makes spatially small contact. The temperature drop is reached in a steady-state mode, i.e. when a time variation of $\Delta T(t)$ is within user-defined stability conditions. For more precise determination of the mean temperature of the sample, another differential thermocouple shows the temperature difference between the sample cold end and the heat sink. Thermocouple joints are glued by IMI 7031 to the sample and heat sink, with special attention in order to avoid electrical contact between the sample and the thermocouple. Otherwise, extra thermoelectric voltage would be added to the thermocouple voltage.

The sample holder contains two vacuum spaces: one with the sample that has to be in high vacuum conditions (see Sec. 2. 1), and the auxiliary one that serves as a heat-link controller with surrounding liquid helium. In order to cool the system down to the liquid helium temperature, a small amount of gaseous helium is put into the auxiliary sample space providing the heat link. To heat up the system to higher temperatures, heat link has to be broken, what is done by pumping out the auxiliary vacuum space to a rotary vacuum.

2.1. *Heat losses in thermal conductivity measurements*

The main demand for any apparatus for thermal conductivity measurements is to minimize losses of heat generated by the sample heater. Generally, there are three ways how heat can be taken away from the sample: convection, conduction, and radiation.

Gas that surrounds the sample is the main source of heat losses by convection. Consequently, the sample has to be in high-vacuum conditions. Figure 2 shows the pressure dependence of the thermal conductivity, given in arbitrary units (a.u.), at 300 K.

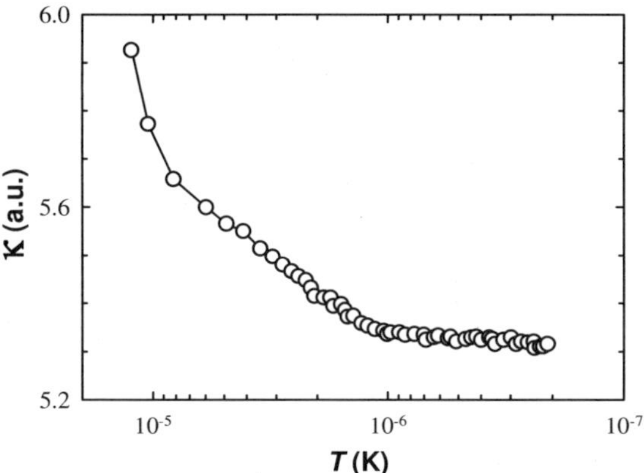

Fig. 2. Pressure dependence of the thermal conductivity, given in arbitrary units at $T = 300$ K, as a function of the pressure.

One can see a saturation of the thermal conductivity coefficient below the pressure of approximately 5×10^{-7} mbar. This is the pressure measured on the top of the turbo molecular pump, and is not a real pressure, which can be even one order of magnitude lower, in the sample space, due to long pumping time and relatively big number of vacuum-line size reductions and pumping-direction changes.

Another way how heat can be taken away from the sample is heat conduction through thermocouple and sample-heater wires, thermally connected with sample and sample heater, respectively. The solution of this problem is to make the thermal resistance of these wires as high as possible, so that the heat transferred by them can be neglected. The most critical parts of the thermocouple are gold–iron wires whose thermal conductivity coefficient is exceptionally high. In order to increase their thermal resistance, they should be as thin and as long as possible. In that sense, we have chosen thermocouple wires of 25 μm in diameter (for studying the thermal conductivity of complex metallic alloys which cross section usually is order 1 mm^2) and 10 cm of length. In order to determine an electrical power developed on the sample heater, one needs to measure precisely heater resistance by four-terminal method. Again, wires are as thin and as long as possible, with the choice of material for voltage reading wires to be as pure thermal conductor as possible (manganese is used). Current supplying leads are copper ones; otherwise the heat developed on highly resistive manganese wire would heat up the entire system. We have chosen both copper and manganese wires to be 15 cm long, and 50 μm in diameter. Table 1 shows a comparison between heat losses at different temperatures T on a gold–iron (thermocouple) and copper (heater current) wires with thermal conductivity κ, calculated for the above mentioned wire lengths and diameters. ΔT is a typical temperature difference developed across the sample (it is usually kept to be around 1% of the surrounding temperature). Columns entitled Q_{wire} and Q_{heater} give calculated heat losses and heat developed at RuO_2 chip-resistor sample heater, respectively. It can be seen that heat losses are negligible.

At high temperature an important contribution to heat losses is radiation. According to Stefan-Boltzmann's law, irradiated heat power is

$$Q_{irr} = \sigma \cdot E \cdot S \cdot (T^4 - T_0^4) \tag{3}$$

Table 1. Comparison of heat losses (column Q_{wire}) in thermocouple (Au) and heater current-supply wires (Cu) with a heat generated by the heater (column Q_{heater}). T is the temperature, κ is the thermal conductivity, and ΔT is a typical temperature drop across the sample.

T(K)	κ(W/mK)	ΔT(K)	$Q_{wire}(\mu W)$	Q_{heater}(mW)
300	300 (Au)	3	~ 5	1-10
	400 (Cu)		~ 7	
100	300 (Au)	1	~ 2	1-5
	400 (Cu)		~ 3	
10	3.000 (Au)	0.1	~ 1.5	0.1-1
	3.000 (Cu)		~ 1.5	

where σ is a Stefan-Boltzmann's constant ($\sigma = 5.67 \times 10^{-8}$ W/m^{2}K^4), E is effective emissivity factor, S irradiating area, T sample temperature, and T_0 temperature of the surrounding (i.e., the sample holder). Under consumption that the sample is surrounded by a thermal shield, and that both of them are of a cylindrical shape, the effective emissivity factor is[12]

$$E = \frac{\varepsilon_{sh} \cdot \varepsilon_s}{\varepsilon + (1 - \varepsilon_s) \cdot \varepsilon_{sh} \cdot S_{sh}/S_s}, \tag{4}$$

where ε_s (ε_{sh}) is the emissivity of the sample (shield), and S_s, and S_{sh} are areas of the sample and shield, respectively. As a shield we can take a housing of the sample space. According to Drude, emissivity is given by

$$\varepsilon = 365 \cdot \sqrt{\rho/\lambda_m}, \tag{5}$$

where ρ is the electrical resistance, and λ_m a radiation wave-length with the highest intensity, calculated by Wien's law $\lambda_m T = 2.9 \times 10^{-3}$ mK. At $T = 300$ K, $\lambda_m = 9.7$ μm. For $\rho = 100$ μΩm, emissivity ε is of the order of 1, whereas for brass (a material the sample holder is made of) it is between 0.1 and 1[a]. For some typical values of sample and shield areas ($S_s = 10$ mm^2, $S_{sh} = 1000$ mm^2), the effective emissivity factor is, according to Eq. (3), of the order of 10^{-2}. At $T_0 = 300$ K, typical value of

---

[a]See web-page http://www.newportus.com/Products/Techncal/MetlEmty.htm.

$(T-T_0)$ is 0.5 K, that, according to Eq. (2), gives $Q_{irr} \sim 10$ μW. It is acceptably low, since it is about 1% of the heater power (see Table 1).

In the first order approximation, Eq. (2) can be rewritten as $Q_{irr} \propto T^3 \cdot \Delta T$, that gives a characteristic $\kappa(T) \sim T^3$ upturn due to radiation effects.

2.2. *Example – thermal conductivity of magnetite Fe_3O_4*

Thermal conductivity curves measured on a sample of magnetite Fe_3O_4, which undergoes the first order a metal-insulator phase transition[13] at $T = 120$ K and electrical resistivity jumps for two orders of magnitude,[14] are compared with three other magnetite samples, taken from literature,[15] and are shown in Fig. 3. The main heat carriers in magnetite are phonons, with a rather strong magneto-elastic interaction, that makes the thermal conductivity coefficient very sensitive to small changes in magnetic sub lattice. From Fig. 3 one can see that a small amount of Co, with different effective Bohr magneton numbers,[16] changes the thermal conductivity drastically. The difference between the other three samples of Fe_3O_4 (the one we were measuring and the other two from the

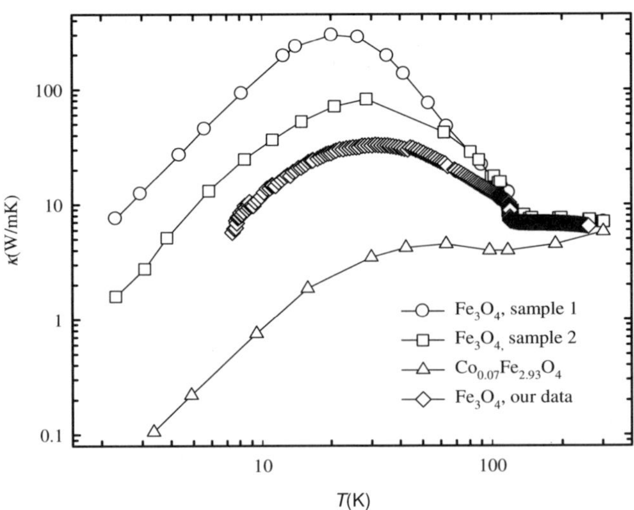

Fig. 3. Thermal conductivities of different samples of magnetite Fe_3O_4, taken from literature[15] (denoted as 'sample 1', 'sample 2', and $Co_{0.07}Fe_{2.93}O_4$), and measured on our apparatus (denoted as 'our data').

reference 15 can be explained by small sample-dependent changes in magnetic substructure. This graph confirms that experimental data of the thermal conductivity taken on our apparatus is competitive with others. *Note* the absence of $\kappa(T) \sim T^3$ upturn, that confirms that radiation effects are negligible.

3. The Analysis of Experimental Thermal Conductivity Data

The theory of thermal conductivity, of intrinsic intellectual interest, must be primarily considered as a tool for the interpretation of experimental results and, if possible, their prediction.[17] In this section we will restrict ourselves to metallic alloys which are relatively well behaved. They show a situation where there is both lattice conduction, κ_l, and electronic conduction, κ_e; however, the behavior of the electrons is simplified by their degeneracy. The approach to electronic thermal conductivity and that to lattice thermal conductivity are quite different. In the case of electrons the problem is normally associated to the determination of the Lorenz number, $L = \kappa_e/\sigma(T)\cdot T$, assuming that the electrical conductivity, $\sigma(T) = \rho(T)^{-1}$, is known. This is generally a reasonable assumption since it is easier to measure the electrical conductivity than the thermal conductivity.

If the electrical conductivity is known, say, at room temperature, then it is relatively easy to predict its value at high temperatures, but at low temperatures the effects of an unknown residual resistance may become important, since the number of electrons may vary from one sample of the same material to another, or from one temperature to another. Thus information on the electrical conductivity is essential if the thermal conductivity is to be successfully analyzed.

The lattice heat conduction, on the other hand, has to be treated as a complete entity. One normally considers it is better to calculate the electronic thermal conductivity as accurately as possible, so that the lattice conductivity is obtained by subtracting the electronic conductivity from the total conductivity. The existence of certain methods of analyzing thermal conductivity data should influence the design of experiments whose purpose is not only to measure but also to understand the thermal conductivity of a particular material. So, if the material is an electrical conductor, it is highly recommended to determine the electrical conductivity over the same temperature range as the thermal conductivity. In the case of metals and alloys, as well as complex metallic alloys this is really essential.

3.1. *Thermal conductivity of metals and alloys*

The thermal conductivity of pure metals is almost entirely electronic, therefore it is extremely important to have measurements of the electrical conductivity so as to test the Wiedemann-Franz law ($\kappa_e = L\sigma(T){\cdot}T$). Furthermore it is important, particularly at low temperatures, that the thermal and electrical conductivity are determined on the very same specimen, since the effects of impurities and lattice defects will, vary from one sample of the same metal to another. This is not as necessary at high temperatures, where lattice scattering predominates. The standard pattern of the temperature dependence of the electronic thermal conductivity κ_e in pure aluminum and copper metals, can be seen from Fig 4. It is characterized by an independence of temperature at temperatures higher than about $(1/2)\theta_D$, θ_D is Debye temperature,, at lower temperatures κ_e begins to increase owing to the decreasing strength of the lattice scattering. Eventually κ_e through a maximum and begins to decrease in accordance with the behavior to be expected when static imperfection scattering predominates.

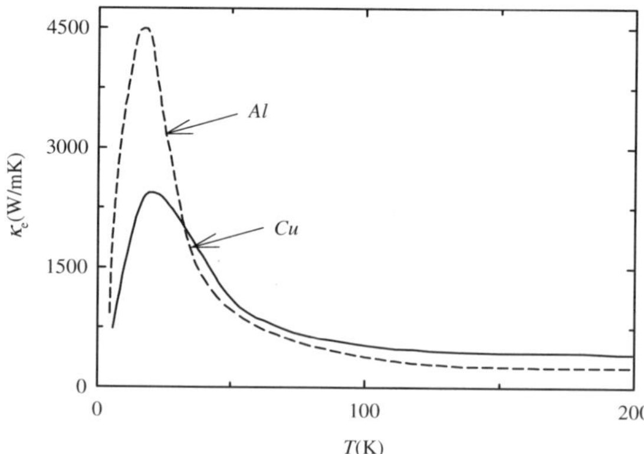

Fig. 4. Standard pattern of the temperature dependence of the electronic thermal conductivity of pure aluminum and copper metals.

There have been a large number of investigations of the thermal conductivity of metallic alloys, almost exclusively below room temperature. The scientific interest arises from the possibility of determining the lattice thermal conductivity in a metallic conductor to a degree of accuracy not possible in the case of pure metals. The basic question concerns the separation of the thermal conductivity into electronic and lattice (phonon) contributions,

$$\kappa = \kappa_e + \kappa_l \tag{6}$$

At high temperatures ($T > (1/2)\theta_D$), for those alloys where the solvent metal obeys the Wiedemann-Franz law, it will be safe to assume the same for the alloy. However, as the temperature increases, the accuracy with which κ_l can be estimated decreases, because, while κ_e is independent of temperature, κ_l is proportional to T^{-1}. At low temperatures there is a difficulty due to the fact the Wiedemann-Franz law breaks down badly for the intrinsic resistance. In the alloys the static imperfection scattering is greater, so errors in estimating intrinsic resistance are less significant, for dilute solid solutions, it will be possible to use the value of intrinsic resistance obtained for pure solvent metal. For these reasons it is usually possible to determine κ_l quite accurately at liquid helium temperature, $T \approx 4$ K. The electronic thermal conductivity itself is not of a great interest in the study of alloys. One point that has been investigated concerns the applicability of Matthiessen's rule that states resistivity is additive ($\rho = \Sigma\rho_i$). For example, in the case of nickel alloys the deviation from that rule is larger for thermal resistance than for electrical resistance. A large part of the interest in the lattice thermal conductivity of alloys arises from the fact that it is possible to study electron scattering of phonons in these materials. Moreover, alloys contain a considerable amount of lattice disorder so it is not surprising that scattering by point defects, stacking faults and dislocations can become manifest at higher temperature.

3.2. *Thermal conductivity of complex metallic alloys*

Among the still relatively small number of known complex metallic materials, large, high-quality single crystals of the orthorhombic

 Ana Smontara et al.

ξ'-AlPdMn and its Ψ-modification; monocrystalline $Mg_{32}(Al,Zn)_{49}$; the cubic β-Al_3Mg_2 polycrystalline and monocrystalline samples as well as the polycrystalline samples of the ε-phases in the AlPd(Fe,Co,Rh) systems (see Table 2), were recently grown.[1-7] In following, we present a study of their thermal conductivity-combined with the electrical resistivity data obtained.[18-25]

A semi quantitative model of thermal conductivity appropriate to complex metallic alloys with the large-scale periodicity of the lattice and the small-scale atomic clustering structure has been used previously in the investigations of the approximant and quasicrystal phases in the giant unit-cell orthorhombic phases in the AlCrFe systems.[26-27] According to that model, thermal conductivity parameter $\kappa(T)$ is divided into three terms

$$\kappa(T) = \kappa_e(T) + \kappa_D(T) + \kappa_H(T) \tag{7}$$

Table 2. Sample characterization data.

Number	Composition	Abbreviation
1.	ξ-$Al_{73}Pd_{22.9}Mn_{4.1}$-1	ξ-AlPdMn-1
2.	ξ-$Al_{72.7}Pd_{23.2}Mn_{4.1}$-2	ξ-AlPdMn-2
3.	Ψ-$Al_{72.9}Pd_{22.9}Mn_{4.2}$	Ψ-AlPdMn
4.	$Mg_{29.4}(Al,Zn)_{51.6}$	$Mg_{32}(Al,Zn)_{49}$
5.	β-β' $Al_{3.26}Mg_2$-mocrystal	β-AlMg (m)
6.	β-β'- $Al_{3.26}Mg_2$-polycrystal	β-AlMg (p)
7.	$Al_{73}Pd_{25}Fe_2$	APF2
8.	$Al_{73}Pd_{23}Fe_4$	APF4
9.	$Al_{71}Pd_{24}Fe_5$	APF5-1
10.	$Al_{75}Pd_{205}Fe_5$	APF5-2
11.	$Al_{73}Pd_{20}Fe_7$	APF7
12.	$Al_{75.5}Pd_{14.5}Fe_{10}$	APF10
13.	$Al_{74}Pd_{12.5}Co_{13.5}$	APC13.5
14.	$Al_{73.5}Pd_{11}Co_{15.5}$	APC15.5
15.	$Al_{74}Pd_{15}Rh_{11}$	APR11
16.	$Al_{74.5}Pd_9Rh_{16.5}$	APR16.5
17.	$Al_{74}Rh_{24}$	AR24

Note: The samples from No. 1.-3. and 7. to 17. belong to a family of related structure so-called ε-AlPd(Mn,Fe,Co,Rh) phases.[28]

The electronic contribution to the thermal conductivity κ_e can be calculated using the Wiedemann-Franz law

$$\kappa_e = L_0 T / \rho \tag{8a}$$

and the electrical resistivity $\rho(T)$ data, measured on the same specimens. (*Note*: In general, the Lorenz number has not a constant value $L_0 = 2.44 \cdot 10^{-8}$ $W\Omega K^{-2}$). Calculation of the Lorenz number $L(T)$ for the *i*-AlCuFe quasicrystals[29] has shown that its "relaxation-time-approximation" value L_0 may have a correction of the order of 10%. For that reason it is reasonable to assume a slight modification of the Wiedemann-Franz law

$$\kappa_e = (1 + c) L_0 T / \rho \tag{8b}$$

where the correction factor c (of the order 0.1) should be determined from the fit. c may also be considered as a compensating factor for the experimental error in the resistivity $\rho(T)$, which is of the same order of magnitude. The lattice contribution $\kappa_l(T) = \kappa(T) - \kappa_e(T)$, could be analyzed by considering (*i*) the propagation of long-wavelength acoustic phonons (for which the complex metallic alloys structure is an elastic continuum) within the Debye model and (*ii*) hopping of localized vibrations within the icosahedral cluster substructure, which participate in the heat transfer via thermally activated hopping. In the simplest model, hopping of localized vibrations is described by single activation energy E_a, yielding a contribution to the thermal conductivity

$$\kappa_H = \kappa_H^0 \cdot \exp(-E_a / k_B T) \tag{9}$$

where κ_H^0 is a constant. The Debye thermal conductivity is written as[17]

$$\kappa_D = C_D \cdot T^3 \cdot \int_0^{\theta_D/T} \tau(x) \cdot \frac{x^4 \cdot e^x}{(e^x - 1)^2} \, dx \tag{10}$$

where $C_D = (4\pi k_B^4 / v_{mean} h^3)$, v_{mean} is the average sound velocity, θ_D the Debye temperature, τ the phonon relaxation time and $x = (h\omega/2\pi k_B T)$, where $(h/2\pi)\omega$ is the phonon energy. The different phonon-scattering processes are incorporated into the relaxation time $\tau(x)$ and we assume that Matthiessen's rule is valid, $\tau^{-1} = \Sigma \tau_j^{-1}$, where τ_j^{-1} is a scattering rate related to the j-th scattering channel. In analogy to the AlPdMn approximant and quasicrystals,[20,26] it is appropriate to consider two dominant scattering processes in the low temperature range (from 8 to 300 K): (1) scattering of phonons on structural defects of stacking-fault type with the scattering rate $\tau_{sf}^{-1} = Ax^2 \cdot T^2$ (note that, since $x^2 \propto \omega^2 \cdot T^{-2}$, τ_{sf}^{-1} does not show an explicit temperature-dependence, but frequency-dependence $\tau_{sf}^{-1}(7a^2/10v_{mean})\gamma^2\omega^2 N_s$, a is a lattice parameter, γ is the Grüneisen parameter and N_s is the linear density of states) and (2) *umklapp* processes with the phenomenological form of the scattering rate pertinent to quasicrystals,[20,25-26] $\tau_{um}^{-1}=Bx^\alpha \cdot T^\beta$ (yielding frequency and temperature dependence $\tau_{um}^{-1} \propto \omega^\alpha \cdot T^{\beta-\alpha}$, where the exponents α and β should be determined from the fit to experimental data), so that the total scattering rate is $\tau^{-1}=\tau_{sf}^{-1}+\tau_{um}^{-1}$.

Note: Debye temperatures of the complex metallic alloys are not known as yet from complementary measurements. For that reason, the used θ_D value is that for the related icosahedral i-AlPdMn quasicrystals, commonly found to be close to 500 K.[30] Since the $\kappa(T)$ data of the complex metallic alloys are available only up to 300 K, it turned out that the fit was insensitive to a slight change of this θ_D value, so that a fixed θ_D=500 K was used. Hence, the Debye constant C_D was not taken as a free parameter, it was instead calculated by using $v_{mean} \approx 4000$ ms[-1], a value determined for the i-AlPdMn from ultrasonic data.[31]

3.2.1. ξ' and Ψ -phases in the AlPdMn complex metallic system

The ξ'-AlPdMn family is representative of complex metallic alloys. These are characterized by two substantially different physical length scales, one defined by the quasiperiodic short-range atomic order within the cluster substructure and the other by the periodic long-range order, so that the competition between these two scales influences transport properties of the material. The study of three ξ'-AlPdMn high-quality single crystals (one in a Ψ-superstructure state with 1500 atoms in the unit cell, the other two in a mixed basic ξ' phase (360 atoms/unit cell) and the Ψ phase) has shown that the electrical resistivity measured along the pseudo fivefold [010] direction of the samples between 300 and 4 K exhibits temperature variation of less than 2% (see Fig. 5). The room-temperature values amount 213 $\mu\Omega$cm for the ξ'-AlPdMn-1 sample,

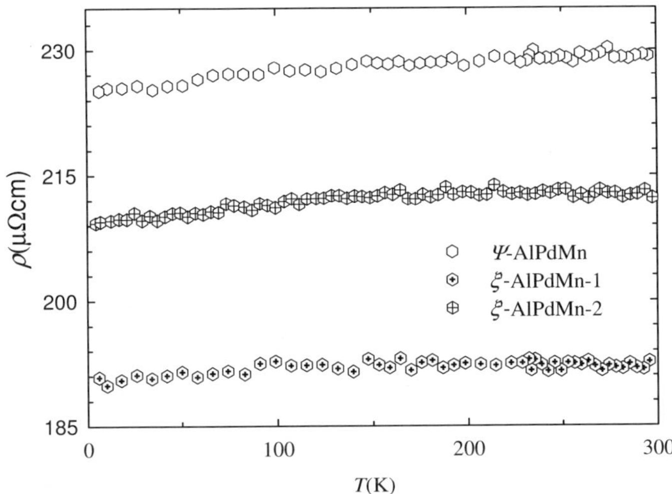

Fig. 5. Temperature-dependent electrical resitivities $\rho(T)$ of the ξ' and Ψ-AlPdMn samples between 4 and 300 K.

192$\mu\Omega$cm for the ξ'-AlPdMn-2 and 229 $\mu\Omega$cm for the Ψ-AlPdMn, and the $\rho(T)$ decrease from 300 K to 4 K amounts $R = (\rho_{300K}-\rho_{4K})/\rho_{300K} = 1.4\%$ for the ξ'-AlPdMn-1 sample, $R = 0.5\%$ for the ξ'-AlPdMn-2, and $R = 1.7\%$ for the Ψ-AlPdMn (Table 3). The qualitative explanation of the temperature-compensated $\rho(T)$ is based on the competition between two different physical length scales. The large-scale periodicity promotes the regular metallic (positive-temperature-coefficient resistivity, whereas the short-scale quasiperiodicity favors the negative-temperature-coefficient resistivity, and the competition of both results in a nearly temperature-independent compensated $\rho(T)$. A similar situation of a nearly temperature-compensated $\rho(T)$ was also found in the large unit-cell quasicrystal approximants of the AlCrFe family.[26]

The differences in the absolute resistivity values of the three ξ'-AlPdMn samples are small, being scattered at e.g. room-temperature in the interval ±8% around the average value 211 $\mu\Omega$cm. While the measurement error in the temperature dependence of $\rho(T)$ of each sample is small, of the order 1%, there are larger errors present relating to the sample geometry determination and to the homogeneity of the current through the samples that affect the absolute value of the resistivity. Due to these effects, the absolute resistivity values shown in Fig. 5 may present an up to 20% error, which is larger than the scattered range of the experimental ρ values. From this point of view, the absolute resistivity values of all three samples are within the estimated error range, showing no significant differences.

Table 3. The room-temperature (R.T.) electrical resistivity ρ and thermal conductivity data κ of the investigated samples, where $\Delta\rho=\rho(300\ \mathrm{K})-\rho(4\mathrm{K})$ and $\Delta\kappa=\kappa(T)-\kappa_e(T)$ where electronic thermal conductivity is estimated using Wiedemann-Franz law ($\kappa_e=L_o\sigma(T)\cdot T$).

Sample	$\rho^{R.T.}$ ($\mu\Omega$cm)	$(\Delta\rho/\rho^{R.T.})$ (%)	$\kappa^{R.T.}$ W/mK	$\kappa_e^{R.T.}$ W/mK	$(\Delta\kappa/\kappa)^{R.T.}$ (%)
ξ-AlPdMn-1	213	1.4	8.6	3.3	62
ξ-AlPdMn-2	192	0.5	7.3	3.6	51
Ψ-AlPdMn	229	1.7	5.0	2.0	60
$Mg_{32}(Al,Zn)_{49}$	46.8	15	24	17	29
β-AlMg (m)	35.5	7	34	20	41
β-AlMg (p)	41.3	5	42	20	52
APF2	112	50.5	18.25	6.56	64
APF4	386	3.8	5.08	3.19	37
APF5-1	270	15.3	2.86	1.85	35
APF5-2	306	1.0	4.51	2.46	45
APF7	586	2.9	2.90	1.27	56
APF10	376	2.0	4.38	1.36	67
APC13.5	260	34.2	8.57	3.65	57
APC15.5	438	3.8	3.59	0.68	79
APR11	579	2.0	1.92	1.29	33
APR16.5	308	16.0	5.02	2.52	50
AR24	1212	3.1	2.20	0.57	74

The temperature dependence of the thermal conductivity parameter $\kappa(T)$ of all three samples is displayed in Fig. 6. The $\kappa(T)$ values at R.T. are 8.6 *W/mK* for the ξ-AlPdMn-1 sample, 7.3 *W/mK* for the ξ-AlPdMn-2 and 5.0 *W/mK* for the Ψ-AlPdMn. These low $\kappa(T)$ values are of the same order as those of *i*-AlPdMn quasicrystals[20,32], where they are considered to be a consequence of both the low electronic density of states at E_F and the non periodicity of the lattice, making the electronic and lattice contributions to the heat transport small.

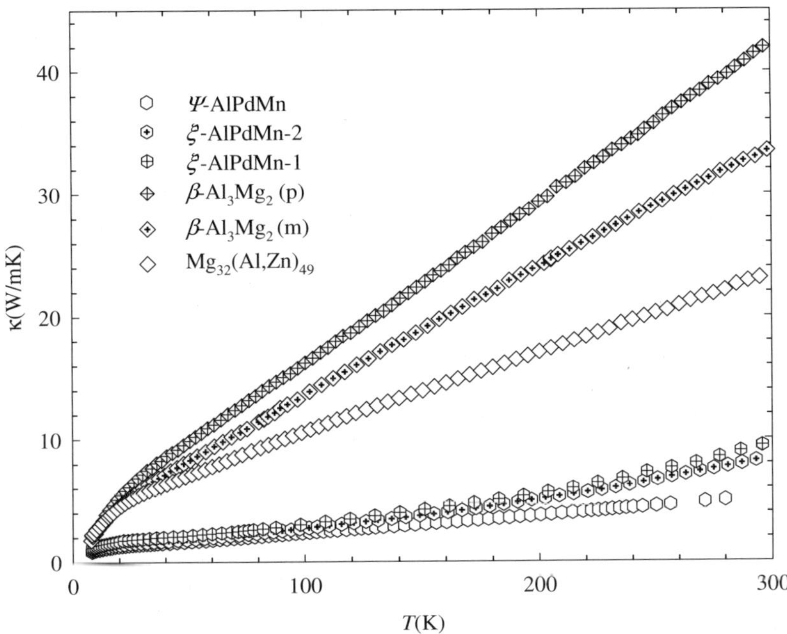

Fig. 6. Temperature-dependent thermal conductivities $\kappa(T)$ of the: single grain ξ' and Ψ-AlPdMn samples; β-Al$_3$Mg$_2$ monocrystaline and polycrystalline samples; and Mg$_{32}$(Al,Zn)$_{49}$ monocrystalline sample between 8 and 300 K.

The electronic contribution to the thermal conductivity κ_e can be calculated using the Eq. (8b) and the measured electrical resistivity $\rho(T)$ data, presented in Fig. 5.

The lattice contribution to the thermal conductivity $\kappa_l = \kappa - \kappa_e$ is analyzed, in analogy to quasicrystals and amorphous solids, by considering the propagation of long-wave length phonons within the ξ'-AlPdMn structure and, at elevated temperatures, the hopping of localized vibrations (Fig. 7).

While the long-wavelength phonons may be analyzed within the Debye model, hopping of localized vibrations represents thermally activated motion, which we assume can be described by an activation energy E_a. The theoretical fits of $\kappa(T)$ for the ξ'-AlPdMn-1, ξ'-AlPdMn-2 and Ψ-AlPdMn samples (Fig. 8) using Eq. (7) show very good agreement with the experimental data in all the investigated temperature range and the fit parameters are given in Table 4.

Table 4. Fit parameters of the thermal conductivities $\kappa(T)$ from Figs. 8, 10, 11 and 15.

Sample	E_a(meV)	$A(10^6 s^{-1}K^{-2})$	α	β
ξ'-AlPdMn-1	57	5.9	3.9	4
ξ'-AlPdMn-2	53	6.4	3.8	4
Ψ-AlPdMn	21	9.6	3.5	4
$Mg_{32}(Al,Zn)_{49}$	15.7	5.8	2.6	4
β-AlMg (m)	16.9	7.1	2.0	4
β-AlMg (p)p	20.8	5.8	2.0	4
APF2	32.2	4.2	2.0	3.6
APF4	15.4	22.5	2.0	4.3
APF5-1	18.6	11.3	2.2	3.5
APF5-2	18.8	10.1	2.4	3.4
APF7	15.0	12.5	1.3	3.9
APF10	14.3	6.3	1.6	4.6
APC13.5	22.3	6.1	2.5	3.6
APC15.5	15.8	9.4	2.0	3.8
APR11	15.2	24.6	2.4	4.0
APR16.5	20.5	12.0	2.1	3.4
AR24	14.8	18.4	2.0	3.7

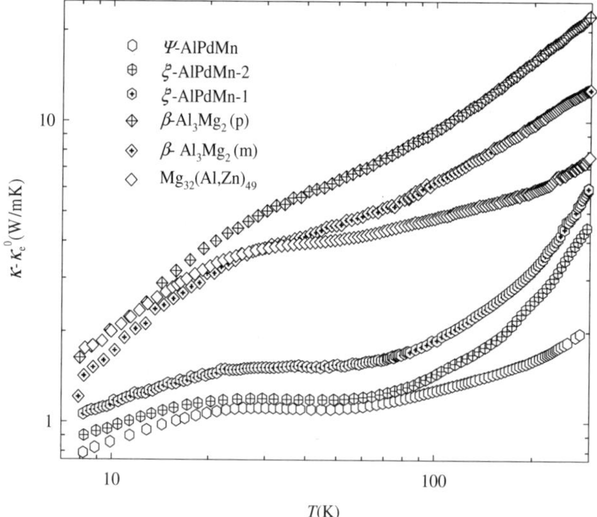

Fig. 7. Temperature-dependent lattice thermal conductivities $\kappa_l = \kappa(T)-L_0\sigma(T)\cdot T$ of the ξ' and Ψ-AlPdMn single grain samples, β-Al$_3$Mg$_2$ monocrystaline and polycrystalline samples and Mg$_{32}$(Al,Zn)$_{49}$ monocrystalline sample between 300 and 8 K.

On all graphs in Fig. 8, the individual contributions κ_e, κ_D and κ_H to the total $\kappa(T)$ are also displayed.

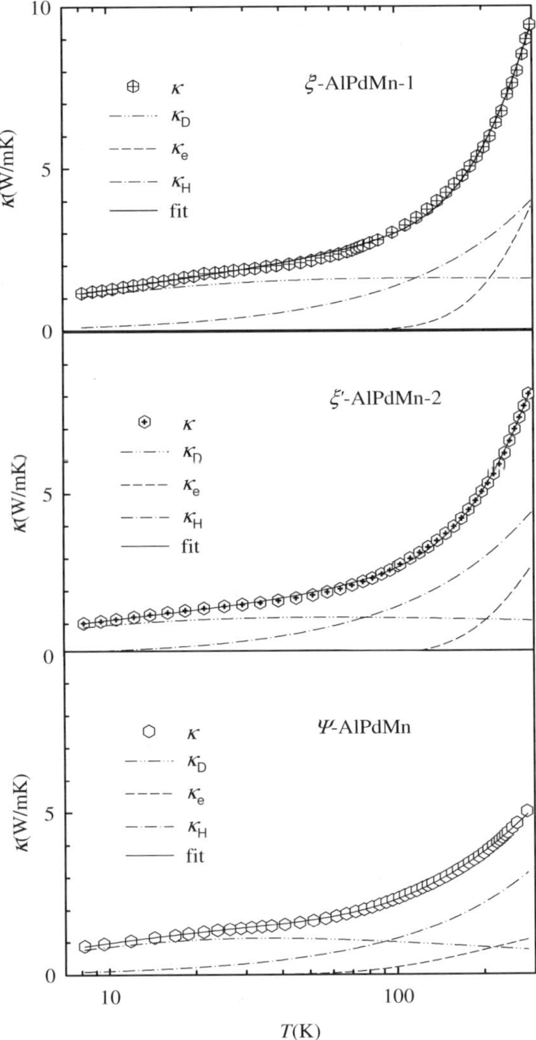

Fig. 8. Temperature-dependent thermal conductivities $\kappa(T)$ of the ξ' and Ψ-AlPdMn samples. Dash-dot lines are the fits with Eq. (7) and the fit parameters are given in Table 4. The three contributions to the total $\kappa(T)$ are shown separately: the electronic contribution κ_e-short-dash line, the Debye contribution κ_D-dash-dot-dot line and the hopping contribution κ_H-dash-dot line.

Due to the very small temperature dependence of $\rho(T)$ that enters the Wiedemann-Franz law, the electronic contribution $\kappa_e(T)$ exhibits practically linear temperature dependence and the correction factor c was found to be either small (amounting 0.16 for the ξ'-AlPdMn-1 and ξ'-AlPdMn-2 samples) or negligible (0.03 for the Ψ-AlPdMn). The Debye contribution $\kappa_D(T)$ exhibits a maximum at about 40 K and a decrease at higher temperatures. A similar behavior is commonly found in periodic solids, where it is attributed to *umklapp* processes and phonon-phonon scattering.[33] For all three samples, the hopping contribution κ_H becomes significant above 100 K. The activation energies for hopping were found similar for the ξ'-AlPdMn-1 and ξ'-AlPdMn-2 samples $E_a \approx 55$ meV, whereas $E_a \approx 21$ meV of the Ψ-AlPdMn is about twice smaller. This smaller E_a value reflects the considerably less steep $\kappa(T)$ increase at temperatures above 100 K of the Ψ-AlPdMn sample, as compared to the other two (see Fig. 7).

The origin of this difference could be due to a slightly different degree of structural order in the samples. The E_a values correlate with the existence of dispersionless vibrational states for energies higher than 12 *meV* in *i*-AlPdMn quasicrystals. Dispersionless states indicate localized vibrations and are considered to be a consequence of a dense distribution of energy gaps in the phonon excitation spectrum, which prevents extended phonons from propagating through the lattice, whereas localized vibrations may still be excited. Therefore, localized vibrations also appear to be present in the giant-unit-cell ξ'-Al-Pd-Mn material, where their origin may be attributed to the icosahedral cluster substructure. The parameter A, which is for all samples close to $10^7\text{s}^{-1}\text{K}^{-2}$ makes it possible to estimate the linear density of stacking faults N_s. We take typical values for the lattice parameter $a \approx 1$nm and the Grüneisen parameter $\gamma \approx 2$ and get $N_s = 0.8$ μm^{-1}. This micrometer-scale N_s value is comparable to those reported for quasicrystals.[34-36] The stacking-fault-like structural defects may thus be considered as the source of phonon scattering at low temperatures in the ξ'-AlPdMn systems. The parameter α define phonon scattering by *umklapp* processes in a phenomenological way. The α values of the three samples are between 3.5 and 3.9 (see Table 4). At R.T. for the ξ'-AlPdMn-1, $\kappa_e = 3.3$ W/mK and $\kappa_l = \kappa - \kappa_e = 5.3$ W/mK and with the ratio $\kappa_e/\kappa_l = 0.62$, while for the

ξ'-AlPdMn-2 κ_e = 3.6 W/mK, κ_l = 3.7 W/mK and κ_e/κ_l = 0.97, and for the Ψ-AlPdMn κ_e = 2.0 W/mK, κ_l = 3.0 W/mK and the ratio κ_e/κ_l = 0.67. For all samples, κ_e and κ_l are thus of comparable size at room temperature. This is different from both simple metals, where the electronic contribution is usually 1–2 orders of magnitude larger than the lattice contribution, and Al-based quasicrystals, where electrons carry less than 1% of the heat. From this point of view, the investigated ξ'-AlPdMn giant unit-cell materials lie somewhere in between regular metals and quasicrystals.

3.2.2. β-Al$_3$Mg$_2$ *complex metallic alloy*

β-Al$_3$Mg$_2$ is a lightweight material (density 2.2 *g/cm^3*) with an extraordinarily large face-centered cubic unit cell and a lattice constant a_0 = 2.82 *nm* containing approximately 1168 atoms in the unit cell.

Two different samples shaped in the form of rectangular prisms with usual dimensions 2x2x7 mm^3, were investigated under nominally the same experimental conditions. The first was a monocrystal with the long axis along the [100] direction, whereas the second was a polycrystal. The polycrystalline sample was included in the study for comparison with the monocrystal in order to check for the influence of grain boundaries on the transport properties. The samples consisted of a single phase with no inclusion of secondary phases with a rather high density of stacking-fault-like defects, as a consequence of thermal stresses during the cooling stage after growth.

The electrical resistivity data are displayed in Fig. 9. Both samples exhibit a positive-temperature-coefficient resistivity, indicating that lattice vibrations play an important role in the temperature dependence of resistivity. The monocrystal exhibits residual resistivity $\rho_{4\ K}$ = 33.2 $\mu\Omega$cm and room temperature resistivity $\rho_{300\ K}$ = 35.5 $\mu\Omega$cm, so that the total increase from 4 to 300 K is by a factor R = $(\rho_{300\ K} - \rho_{4\ K})/\rho_{4\ K}$ = 7%. The polycrystal exhibits slightly higher values, $\rho_{4\ K}$=39.2 $\mu\Omega$cm, $\rho_{300\ K}$ = 41.4 $\mu\Omega$cm and R = 5%, which may be attributed to the grain boundaries in the sample that act as additional scattering centers. The temperature dependence of the thermal conductivity parameter $\kappa(T)$ of both samples β-Al$_3$Mg$_2$ is displayed in Fig. 6. The $\kappa(T)$ values at 300 K

                    Ana Smontara et al.

are $\kappa_{300\ K} = 34$ *W/mK* for the monocrystal and $\kappa_{300\ K} = 42$ *W/mK* for the polycrystal (see Table 3).

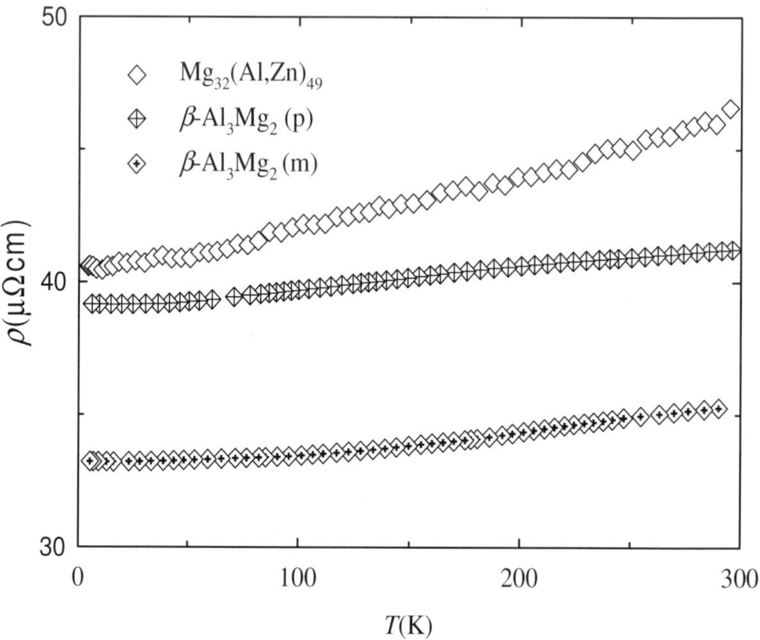

Fig. 9. Temperature-dependent electrical resitivities $\rho(T)$ of the monocrystaline and polycrystalline β-Al$_3$Mg$_2$ samples and of the Mg$_{32}$(Al,Zn)$_{49}$ Bergman phase monocrystalline sample between 4 and 300 K.

The electronic contribution to the thermal conductivity κ_e is calculated using the Eq. (8b) and the measured electrical resistivity $\rho(T)$ data, presented in Fig. 9. A correction factor c is small, of the order 0.1 or less that may be considered as a compensating factor for the errors in the sample-geometry estimation and in the homogeneity of the current in both the resistivity and the thermal conductivity experiments.

The lattice contribution to the thermal conductivity is analyzed by considering the propagation of long-wavelength phonons within the β-Al$_3$Mg$_2$ structure and, at elevated temperatures, the hopping of localized vibrations. While the long-wavelength phonons may be analyzed within the Debye model, the hopping of localized vibrations represents thermally activated motion described by activation energy E_a. Model of *quasiumklapp* phonon scattering in quasicrystal structure,[11] show that the acoustic vibrational spectrum contains many gaps (recall that the ideal crystal has only one gap, at the Brillouin zone boundary).

The dense energy-gap distribution implies that *umklapp*-type phonon scattering processes are much more frequent in quasicrystalline than in crystal structure. Calculation of the scattering rate leads to the power-law-type frequency and temperature dependence of the scattering rate[31] $\tau_{um}^{-1} \propto \omega^2 \cdot T^4$. One can expect that the local quasicrystal-like order of β-Al$_3$Mg$_2$ should result in a power-law *umklapp* scattering rate. It was assumed phenomenologically *umklapp* $\tau_{um}^{-1} = B\omega^{\alpha} \cdot T^4$, where the parameter B and exponent α should be determined from the fit. The total scattering rate is then a sum of $\tau^{-1} = \tau_{sf}^{-1} + \tau_{um}^{-1}$. The Debye temperature of β-Al$_3$Mg$_2$ is not as yet known from complementary measurements, so that a trial value fits of $\kappa(T)$ for the monocrystal and polycrystal (see Fig. 10) using Eq. (7) show excellent agreement with the experimental data over the whole investigated temperature range, and the fit parameters are given in Table 4.

The correction factor c to the Wiedemann-Franz law was found to be small ($c = 0.1$) for the polycrystal and negligible ($c = 0.008$) for the monocrystal. The Debye contribution $\kappa_D(T)$ exhibits a maximum at about 50 K for both samples. Similar behaviour is commonly found in periodic solids, where the maximum is attributed to phonon-phonon *umklapp* processes.[33] The hopping contribution κ_H becomes significant above 60 K. The activation energies for hopping were found $E_a = 16.9$ meV for the monocrystal and $E_a = 20.8$ meV for the polycrystal. These E_a values are similar to those determined in ξ and Ψ-AlPdMn alloys (see Sec. 3.2.1 and Table 4). Therefore, localized vibrations appear to be present in the giant-unit-cell β-Al$_3$Mg$_2$ material as well, where their origin may be attributed to the icosahedral cluster substructure, a feature that is characteristic of the icosahedral qusicrystals as well.

The parameter A, which is for both samples close to 10^7 s^{-1}K^{-2}, makes it possible to estimate the linear density of stacking faults N_s. The obtained value $N_s = 0.8$ μm^{-1} correlates well to the TEM contrast images of the β-Al$_3$Mg$_2$ samples (see Fig. 1a. of ref. 37), where stacking faults on the 0.2 μm scale were directly observed. The stacking-fault structural defects are thus an important source of phonon scattering at low temperatures in the β-Al$_3$Mg$_2$ material. The parameters B and α define phonon scattering by *umklapp* processes in a phenomenological way. A power-law dependence of the *umklapp* scattering rate provides evidence that dense distribution of energy gaps exists in the acoustic phonon branch of β-Al$_3$Mg$_2$. For both samples the fits yielded $\alpha=2.0$ (see Table 4), which yields the frequency and temperature dependence of the *umklapp* rate $\tau_{\mathrm{um}}^{-1} = \omega^2 \cdot T^2$.

Comparing the electronic and lattice contributions to the thermal conductivity of the β-Al$_3$Mg$_2$ samples, we find at 300 K $\kappa_e = 20$ W/mK and $\kappa_l = \kappa - \kappa_e = 14$ W/mK with the ratio $\kappa_e / \kappa_l = 1.4$ for the monocrystal and $\kappa_e = 20$ W/mK, $\kappa_l = 22$ W/mK and $\kappa_e / \kappa_l = 0.9$ for the polycrystal. For both samples, κ_e and κ_l are thus of comparable size at room temperature, and the investigated β-Al$_3$Mg$_2$ giant-unit-cell material, similar to ξ-AlPdMn giant unit-cell materials, also lies somewhere between regular metals and quasicrystals.

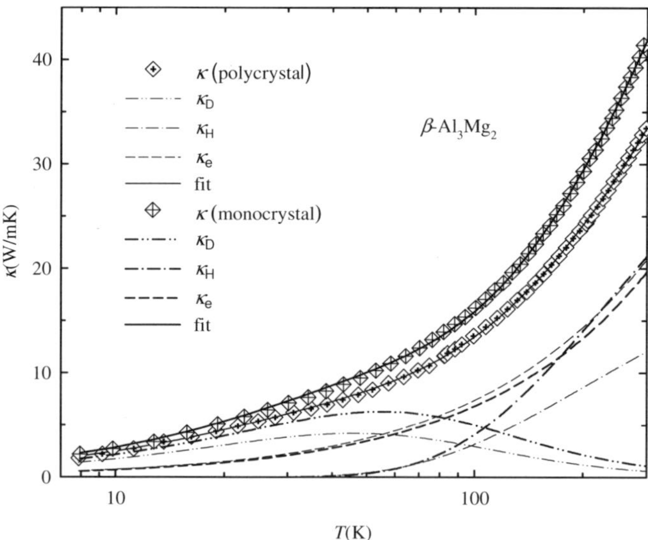

Fig. 10. Temperature-dependent thermal conductivities $\kappa(T)$ of the β-Al$_3$Mg$_2$ monocrystaline and polycrystalline samples between 8 and 300 K. Dash-dot lines are the fits with Eq. (7) and the fit parameters are given in Table 4. The three contributions to the total $\kappa(T)$ are shown separately: the electronic contribution κ_e-short-dash line, the Debye contribution κ_D-dash-dot-dot line and the hopping contribution κ_H-dash-dot line.

3.2.3. $Mg_{32}(Al,Zn)_{49}$ complex metallic alloy

The structurally homogeneous monocrystalline material (free of secondary phases) of the $Mg_{32}(Al,Zn)_{49}$ Bergman phase comprising 162 atoms in the unit cell, was studied, in order to investigate how the exceptional structural complexity of the giant unit cell and the coexistence of two competing length scales, one defined by the unit-cell parameters and the other by the cluster substructure, affect the physical properties of a metallic material.

The electrical resistivity of the material between 4 K and R.T. is in the range $\rho \approx$ 40-47 $\mu\Omega$cm and exhibits a positive increase to higher temperatures, where this type of $\rho(T)$ can be attributed to the inelastic electron-phonon interaction.

The thermal conductivity of the Bergman phase contains an electronic contribution κ_e that was estimated by the Wiedemann-Franz law, and a lattice contribution κ_l that can be reproduced by the sum of a Debye term (long-wavelength phonons) and a term due to hopping of localized vibrations. While hopping becomes the dominant lattice heat-carrying channel at temperatures above 70 K, the Debye term dominates at low temperatures and exhibits a maximum at about 40 K. At the lowest measured temperature (8 K), the scattering of phonons on stacking-fault-like defects limits the heat transport, whereas at higher temperatures, *umklapp* processes become excited with a $\tau_{um}^{-1} \propto \omega^{2.6}\cdot T^{1.4}$ dependence of the *umklapp* rate.

The room-temperature value of the electronic conductivity is 2.4 times larger than the lattice conductivity, but κ_e and κ_e are still of comparable size. This feature is also intermediate between simple metals and alloys on the one hand and Al-based quasicrystals on the other hand, and it seems to be a general feature of the Al-based giant-unit-cell CMAs.

Though the Bergman-phase CMA compound appears to be an excellent candidate to investigate the influence of high structural complexity and the competition of two different length scales on the physical properties of a metallic solid, no spectacular differences with respect to simple metals and alloys were found. The dominant icosahedral local atomic coordination in the giant unit cell makes the

Bergman phase resemble amorphous metals on the local scale, which is evident with the temperature-dependent electrical resistivity. This also shows that the electronic transport properties of the Bergman phase are predominantly determined by the small-scale atomic clustering, whereas the large-scale periodicity of the CMA lattice is of more or less marginal importance (Fig. 11).

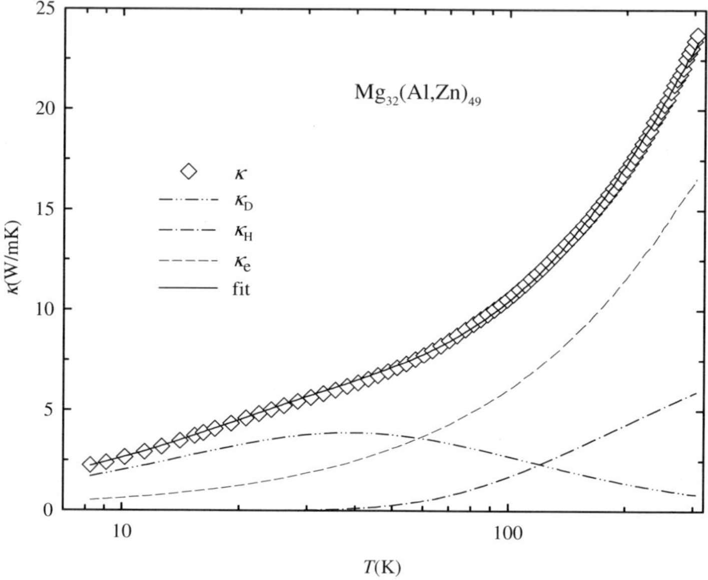

Fig. 11. Temperature-dependent thermal conductivities $\kappa(T)$ of the and Mg$_{32}$(Al,Zn)$_{49}$ sample between 8 and 300 K. Solid line is the fits with Eq. (7) and the fit parameters are given in Table 4. The three contributions to the total κ(T) are shown separately: the electronic contribution κ_e-short-dash line, the Debye contribution κ_D-dash-dot-dot and the hopping contribution κ_H-dash-dot-line.

3.2.4. ε-phase in the AlPd (Fe,Co,Rh) complex metallic system

An interesting family of related complex metallic alloys structures, was revealed in the Al-Pd alloys system and its ternaries with transition metals (Mn,Fe,Co,Rh). In these systems, complex metallic phases were observed in specific compositional ranges and, depending on composition, exhibit a row of giant unit-cell orthorhombic structures designated as ε-phases, and those aperiodic in one dimension.[38] The investigated ε-AlPd(Fe,Co,Rh) samples were pure polycrystalline materials, not polluted by secondary phases. The nominal compositions and abbreviations of their compositions are given in Table 2.

The resistivities of the ε-AlPd(Fe,Co,Rh) samples are shown in Fig. 12, whereas their room-temperature (RT) values $\rho_{300\,K}$ and the ratios $\rho_{300\,K}/\rho_{4\,K}$ (representing a rough measure of the temperature dependence of the resistivity) are given in Table 3.

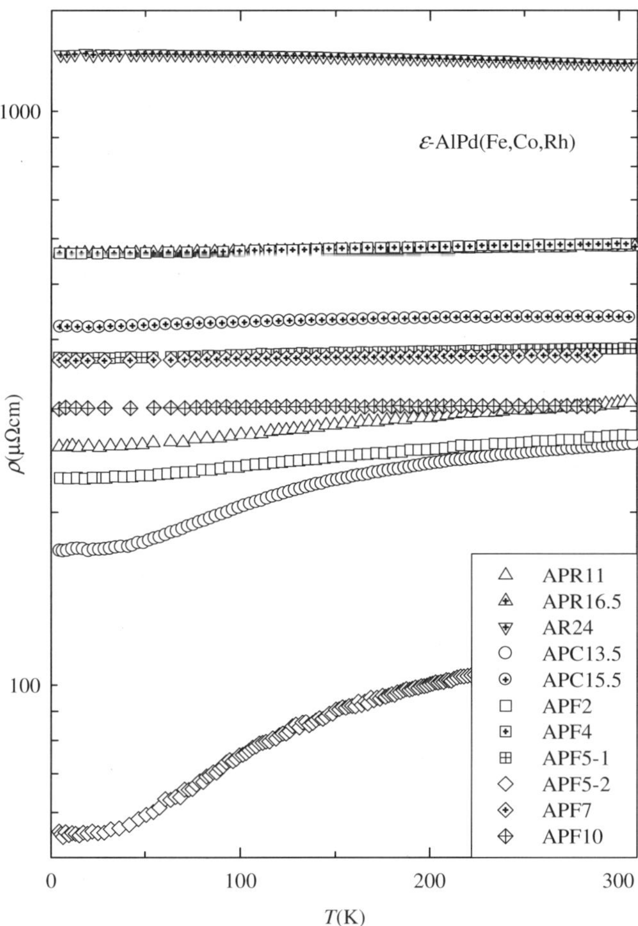

Fig. 12. Temperature-dependent electrical resitivities $\rho(T)$ of the ε-AlPd(Fe,Co,Rh) phases.

The R.T. resistivities of most samples are in the range 200-600 $\mu\Omega$cm and exhibit small temperature variations of a small % between 300 and 4 K. Exceptions are the samples APF2 with lower resistivity ($\rho_{300\,K}$ = 112 $\mu\Omega$cm and ($\rho_{300\,K}/\rho_{4\,K}$) = 2.02) and AR24 with the highest resistivity ($\rho_{300\,K}$ = 1212 $\mu\Omega$cm). The AR24 sample also exhibits a small negative temperature coefficient ($\rho_{300\,K}/\rho_{4\,K}$) = 0.97), in contrast to all the other samples, which show positive temperature coefficients. The temperature-dependent thermal conductivities $\kappa(T)$ of all samples are displayed in Fig. 13. The amazing result is the low $\kappa(T)$ values, which are at R.T. in the range $\kappa_{300K} \approx$ 2-10 W/mK.

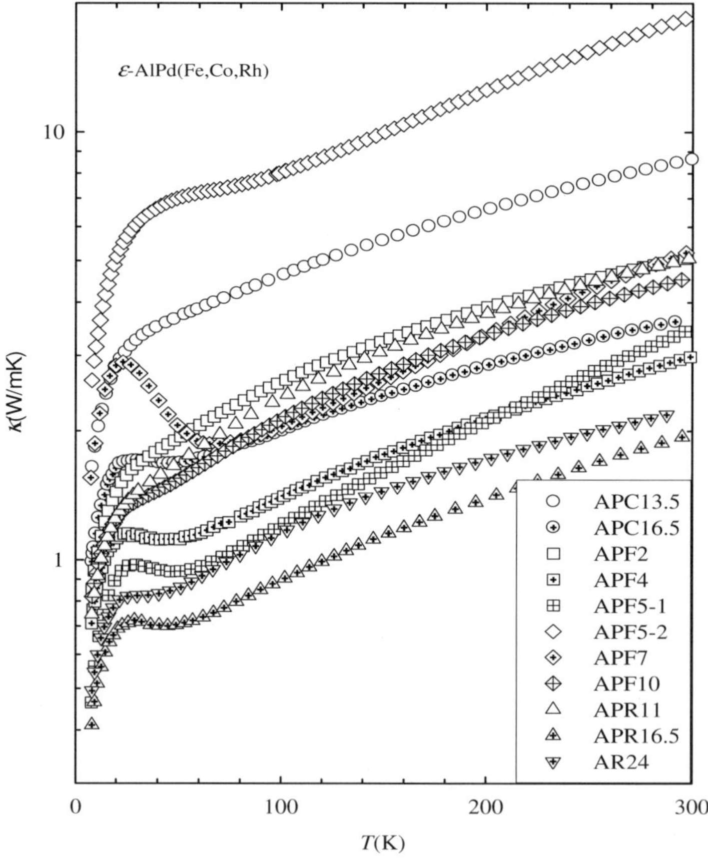

Fig. 13. Temperature–dependent thermal conductivities $\kappa(T)$ of the investigated ε-AlPd(Fe,Co,Rh) phases between 8 and 300 K.

The conductivities of the AR24 and APR11 samples are even as low as κ_{300K} = 2 W/mK. These κ_{300K} values, surprisingly low for alloys of regular metals, are comparable to the thermal conductivity of amorphous SiO_2 shown in Fig. 14 with κ_{300K} = 2.8 W/mK, a known electrical and thermal insulator, as well as to the technologically widespread thermally insulating material yttrium-doped zirconia ceramics $Zr_{1-x}Y_xO_{2-x/2}$ (x<0.2), where κ_{300K} = 2 W/mK.[39] (Here it is important to note that $\kappa(T)$ of Zr/YO_2 ceramics remains constant at this value up to 1200°C, while it continuously increases due to the temperature-dependent electronic and hopping contributions in the investigated ε-phases).

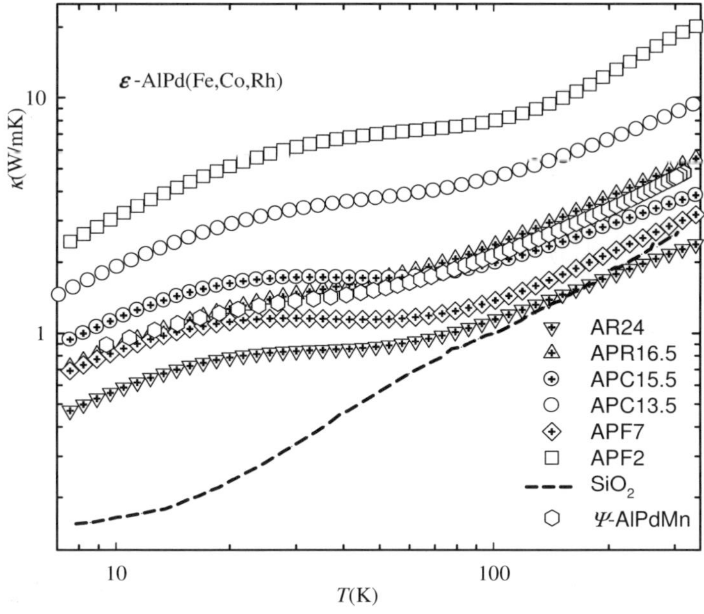

Fig. 14. Temperature-dependent thermal conductivities $\kappa(T)$ of a selected ε-AlPd(Fe,Co, Rh) phases between 8 and 300 K. Thermal conductivity of the thermal insulator amorphous SiO_2 (dashed line) is shown for comparison. $\kappa(T)$ of a monocrystalline Ψ (equivalent to ε_{28})-AlPdMn is shown by open hexagon.

The $\kappa(T)$ curves of the ε-phase samples were analyzed by means of Eq. (7). The fits to $\kappa(T)$'s of ε-AlPd(Fe,Co,Rh) samples are shown in Fig. 15, whereas the fit parameters are shown in Table 4. There is no fit parameter involved in κ_e, so that the total lattice contribution $\kappa_l = \kappa - \kappa_e$ is well defined, whereas its subdivision into $\kappa_D + \kappa_H$ has some freedom and should be considered as qualitative. As the Debye contribution exhibits a maximum at temperatures between 20 and 40 K and declines above, whereas hopping contribution becomes significant above about 70 K, showing an exponential increase, the temperature range where κ_D and κ_H are of comparable magnitude is small and the two contributions do not affect each other much in the fitting procedure. A common observation to all ε-phase samples is the small Debye significant hopping contribution at elevated temperatures.

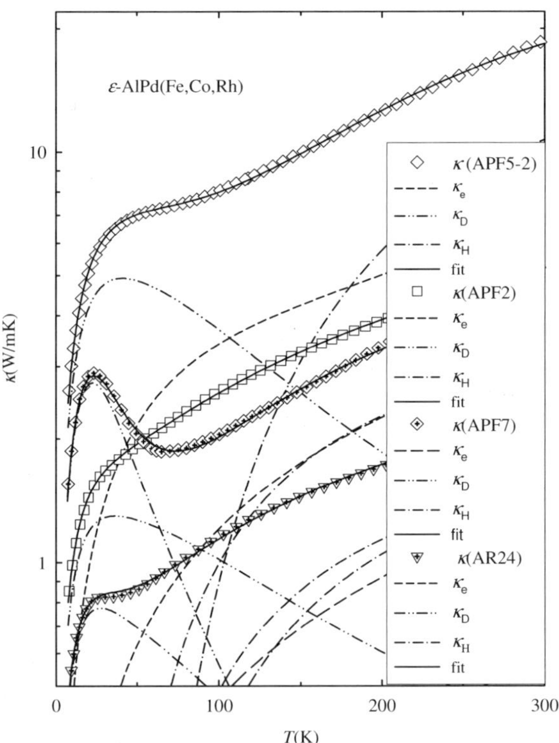

Fig. 15. Temperature-dependent thermal conductivities $\kappa(T)$ of the ε-phases. Solid lines are the fits with Eq. (7) and the fit parameters are given in Table 4. The three contributions to the total $\kappa(T)$ are shown separately: the electronic contribution κ_e–dashed line, the Debye contribution κ_D–dash-dot-dot line and the hopping contribution κ_H–dash-dot line.

A qualitative explanation of this result is that large atomic clusters of icosahedral symmetry suppress the heat transfer by phonons, whereas localized cluster vibrations still provide a weak lattice heat conduction channel. In addition, conduction electrons contribute to the usual channel to the heat conductivity, but since the electrical resistivities of the ε-phases are moderate (of few 100 $\mu\Omega$cm), electronic thermal conductivity contribution is also not large. Consequently, the total thermal conductivities of the ε-phase materials are about the same as that of amorphous SiO_2.

Here it is important to *note* that whereas SiO_2 is both a thermal and an electrical insulator, the ε-phase CMA materials offer a combination of a thermal insulator with an electrical conductor, which can be considered as a "smart" property of the material, of to date unrealized conventional solid-state materials. It is also important to stress that $\kappa(T)$ values, obtained on polygrain materials, are practically the same as those obtained on ξ and Ψ-phases of the AlPdMn single grain samples (see Fig. 6). This can be seen from Fig. 14, where $\kappa(T)$ of a monocrystalline AlPdMn of predominant Ψ phase along the [010] crystal direction is also shown.

This demonstrates that polycrystallinity does not have a significant influence on thermal response of the ε-phase material. Very likely the cluster substructure already provides enough phonon scattering centers to suppress phonon propagation, so that additional scattering at the grain boundaries does not add much to this effect. Similar consideration seems to be appropriate for the electron scattering in the electronic transport (recall that electrical resistivities of monocrystalline ξ and Ψ phase in the AlPdMn fall within the range of the resistivities of polygrain samples displayed in Fig. 5). The thermal conductivity fit parameters in Table 4 and allow us to derive some average properties of the ε-phases. The average hopping activation energy of the localized vibrations is $E_a=18$ meV, which is similar to the values observed in ξ and Ψ phases in the AlPdMn. The average values of the exponents α and β that govern the frequency and temperature dependence of the *umklapp* rate are $\alpha = 2.0$ and $\beta = 3.8$ and, yielding the average dependence $\tau_{um}^{-1} = \omega^2 \cdot T^{1.8}$. This is close to the phenomenological power-law $\omega^2 \cdot T^2$, sometimes observed in systems with a dense distribution of energy gaps in the vibrational spectrum, like in β-Al$_3$Mg$_2$ (see Sec. 3.3.2). As a comparison, the *umklapp* rate in regular crystals with the gap only at the Brillouin zone boundary is of the form $\tau_{um}^{-1} = \omega^2 \cdot T \cdot \exp(-\theta_D/\delta T)$, where δ is a dimensionless parameter of the order one.

Very small thermal conductivities, $\kappa < 4.5$ W/mK in the temperature interval 2-300 K, were reported for a series of related CMA materials, the Al-based Mackay-type 1/1 cubic approximants.[40] In contrast to a phenomenological analysis of the $\kappa(T)$ dependence, the explanation of the anomalously low thermal conductivity in[40] was derived from first-principle calculations of the phonon dispersion, using crystallographic structure data of the investigated materials. It was found that the combination of small group velocity and the large number of optical phonon branches caused by the large number of atoms in the unit cell greatly reduces the lattice thermal conductivity by increasing probability of the *umklapp* processes in the phonon scattering. The presence of vacancies in the structure further contribute to this tendency.

We expect that a similar microscopic picture also explains the anomalously low thermal conductivity of the ε-phases. There is one marked difference in the temperature-dependent $\kappa(T)$'s of the ε-phases and these approximants. The approximants exhibit very weak temperature dependence $\kappa(T)$ above 100 K (see Fig. 2 of ref. 40), while the ε-phases exhibit an exponential increase above about 70 K (Figs. 13-15) that we model phenomenological by a thermally-activated behavior with an activation energy E_a and attribute to hopping of localized vibrations (i.e., a locally given -vibrating atomic cluster "hits" its neighbor). Presently the structures of the ε-phases are only approximately known, therefore a first-principle analysis along the steps outlined in[40] should still be performed after the structures are known in more detail.

4. Conclusions

An interesting feature of the complex metallic alloys is their low thermal conductivity, which is at room temperature about the same as that of known thermal insulators amorphous SiO_2 and Zr/YO_2 ceramics. While SiO_2 and Zr/YO_2 are also electrical insulators, the complex metallic alloys exhibit electrical conductivity typical of metallic alloys, so that this material offers interesting combination of an electrical conductor with a thermal insulator. The reason for the weak heat conductivity of the complex metallic alloys appears to be structural: large and heavy atomic clusters of icosahedral symmetry in the giant unit cell that suppress the propagation of extended phonons, so that the lattice can no longer efficiently participate in the heat transport (apart from the remaining weak channel of hopping of localized complex metallic alloys, low thermal conductivity at a moderate electrical conductivity can be expected to be a general property of this class of materials. Complex metallic alloys can thus be considered as a promising candidate towards development of a technologically interesting "smart" material that joins the properties of an electrical conductor with a thermal insulator, a feature that is mutually exclusive in conventional metals.

Acknowledgments

This work was done within the activities of the 6th Framework EU Network of Excellence "Complex Metallic Alloys" (Contract No. NMP3-CT-2005-500140, and has been partially supported by the Ministry of Science, Education and Sports of Republic of Croatia through the Research Projects Nos. 035-0352826-2848 "Thermal and charge transport in highly frustrated magnets and related materials" and 177-0352828-0478 "Transport and magnetic properties of nanostructured complex metallic compounds". The authors thank Pr. Dr. J. Dolinšek for very useful information.

References

1. See, for a recent review, K. Urban and M. Feuerbacher, *J. Non-Cryst. Solids* **334&335**, 143 (2004).
2. G. Bergman, J. L. T. Waugh and L. Pauling, *Acta Crystallogr.* **10**, 254 (1957).
3. M. Boudard, H. Klein, M. de Boissieu, M. Audier and H. Vincent, *Philos. Mag.* **A 74**, 939 (1996).
4. H. Klein, M. Audier, M. Boudard, M. de Boissieu, L. Behara and M. Duneau, *Philos. Mag.* **A 73**, 309 (1996).

5. M. Feuerbacher, C. Thomas, K. Urban, in *Quasicrystals, Structure and Physical Properties*, Ed. H.-R. Trebin (Wiley-VCH, Weinheim, 2003), p. 2.

6. G. Kreiner, H. F. Franzen, *J. Alloys & Comp.* **261**, 83 (1997).

7. S. Samson, in *Developments in the Structural Chemistry of Alloy Phases*, Ed. B. C. Giessen (Plenum, New York, 1969), p. 65.

8. R. Cerny, M. Francois, K. Yvon, D. Jaccard, E. Walker, V. Petricek, I. Cisarova, H. U. Nissen and R. Wessiken, *J. Phys.: Condens. Matter* **8**, 4485 (1996).

9. A. Smontara, A. Bilušić and H. Berger, in *Thermal Conductivity* **24**/*Thermal Expansion* **12**, Eds. P. Gall and D. A. Apostolescu (Technomic, Lancaster-Basel, 1998), p. 182.

10. A. Bilušić, A. Smontara, D. Stanić and D. Pajić, *Strojarstvo* (2008), in press.

11. R. Bergman, in *Thermal conduction in solids* (Oxford University Press, New York, 1978), p. 3.

12. H. Bougrine and M. Ausloos, *Rev. Sci. Instrum.* **66**, 199(1995).

13. E. J. W. Verwey, *Nature* (London) **144**, 327 (1939).

14. D. J. Huang, C. F. Chang, H. T. Jeng, G. Y. Guo, H. J. Lin, W. B. Wu , H. C. Ku, A. Fujimori, Y. Takahashi and C. T. Chen, *Phys. Rev. Lett.* **93** 077204 (2004).

15. G. A. Slack, *Phys. Rev.* **126**, 427 (1962).

16. N. W. Ashcroft and N. D. Mermin, *Solid State Physics* (Saunders College Publishing, 1976), p. 657.

17. R. Bergman, in Ref. 11, p. 152.

18. A. Smontara, I. Smiljanić, A. Bilušić, B. Grushko, S. Balanetskyy, Z. Jagličić, S. Vrtnik and J. Dolinšek, *J. Alloys & Comp.* **450**, 92 (2008).

19. I. Smiljanić, A. Smontara, A. Bilušić, J. Lukatela, D. Stanić, N. Barišić, J. Dolinšek, M. Feuerbacher and B. Grushko, *Philosophical Magazine.* (2008), in press.

20. A. Bilušić, A. Smontara, J. Dolinšek, P. McGuiness and H. R. Ott, *J. Alloys & Comp.* **432**, 1 (2007).

21. J. Dolinšek, T. Apih, P. Jeglič, I. Smiljanić, Ž. Bihar, A. Bilušić, A. Smontara, Z. Jagličić and M. Feuerbacher, *Intermetallics.* **15**, 1367 (2007).

22. Ž. Bihar, A. Smontara, A. Bilušić, J. Lukatela and J. Dolinšek, *Strojarstvo.* **48**, 13 (2006).

23. A. Smontara, I. Smiljanić, A. Bilušić, Z. Jagličić, M. Klanjšek, J. Dolinšek, S. Roitsch and M. Feuerbacher, *J. Alloys & Comp.* **430**, 29 (2007).

24. J. Dolinšek, Z. Jagličić and A. Smontara, *Philos. Mag. Lett.* **86**, 671 (2006).

25. J. Dolinšek, P. Jeglič, P. McGuiness, Z. Jagličić, A. Bilušić, Ž. Bihar, A. Smontara, C. Landauro, M. Feuerbacher, B. Grushko and K. Urban, *Phys. Rev.* **B 72**, 064208(2005).

26. Ž. Bihar, A. Bilušić, J. Lukatela, A. Smontara, P. P. Jeglič, P.J. McGuiness, J. Dolinšek, Z. Jagličić, J. Janovec, V. Demange, J. M. Dubois, *J. Alloys & Comp.* **407**, 65 (2006).

27. Ž. Bihar, in Istraživanje transportnih i magnetskih svojstava kvazikirstalin spojeva iz obitelji ξ'-Al-Pd-Mn (PhD Thesis, Sveučilište u Zagrebu 2005), p. 38.

28. S. Balanetskyy, B. Grushko, T. Ya., Velikanova, K. Urban, *J. Alloys & Comp.* **376**, 158 (2007).

29. J. Dolinšek, S. Vrtnik, M. Klanjšek, Z. Jagličić, A. Smontara, I. Smiljanić, A. Bilušić, Y. Yokoyama, A. Inoue and C. V. Landauro, *Physical Review* **B 76**, 054201 (2007).

30. C. Wälti, E. Felder, M. A. Chernikov, H. R. Ott, M. de Boissieu, C. Janot, *Phys. Rev.* **B 57**, 10504 (1998).

31. Y. Amazit, M. de Boissieu, A. Zarembowitch, *Europhys. Lett.* **20**, 703 (1992).

32. A. Bilušić, Ž. Budrović, A. Smontara, J. Dolinšek, P. C. Canfield and I. R. Fisher, *J. Alloys & Comp.* **342**, 413 (2002).

33. I. N. W. Ashcroft and N. D. Mermin, *Solid State Physics* (Saunders College Publishing, 1976), p. 505.

34. S. Legault, B. Ellman, J. Ström-Olsen, L. Taillefer, T. Lograsso and D. Delaney, *Quasicrystals, Proceedings of the 6th International Conference*, Tokyo, Japan 1997 (World Scientific, Singapore, 1998), p. 475.

35. K. Giannò, A. V. Sologubenko, M. A. Chernikov, H. R. Ott, P. C. Canfield and I. R. Fisher, *Phys. Rev.* **B 62**, 292 (2000).

36. M. Matsukawa, M. Yoshizawa, K. Noto, Y. Yokoyama, A. Inoue, *Physica* **B 263-264**, 146 (1999).

37. M. Feuerbacher, M. Heggen and K. Urban, *Mater. Sci. Eng. A* **375**, 84 (2004).

38. S. Balanetskyy, B. Grushko and T. Ya. Velikanova, *Z. Kristall.* **219**, 548 (2004).

39. R. Mévrel, J.-C. Laizet, A. Azzopardi, B. Leclercq, M. Poulain, O. Lavigne and V. Demange, *J. Eur. Cer. Soc.* **24**, 3081 (2004).

40. T. Takeuchi, N. Nagasako, R. Asahi and U. Mizutani, *Phys. Rev.* **B 74**, 054206 (2006).

CHAPTER 4

THERMOELECTRIC MATERIALS

Silke Paschen

Institute of Solid State Physics, Vienna University of Technology
Wiedner Haupstrasse, 8-10, 1040 Vienna, Austria
E-mail: paschen@ifp.tuwien.ac.at

Thermoelectric materials can be used for refrigeration and power generation. To date efficient materials for applications in the vicinity of room temperature exist and thermoelectric modules fabricated from these are commercially available. However, both at low and high temperatures, efficiencies are still too low and much effort is put worldwide in optimizing known and finding new materials. After a brief introduction to thermoelectric phenomena and a presentation of the state-of-the-art in traditional Bi-Te based thermoelectrics, focus will be on material classes considered more recently for thermoelectric applications. Special emphasis will be on "Cage Compounds" as one important representative of the class of "Complex Metallic Alloys".

1. Introduction

Thermoelectric (TE) materials convert heat into electric current and vice versa and can, as such, be exploited for power generation or refrigeration. Potential areas of application for the former are all kind of industrial processes, automotive exhaust, domestic heating, and many others. The use of TE power generators would not only increase the overall efficiency but also actively counteract global warming. Thus, a breakthrough in the efficiency of TE materials is a key contribution to an environment friendly and sustainable energy system for the coming generations.

An equally important issue is "green" cooling. When used in solid state Peltier coolers thermoelectric materials cool without any harmful liquid refrigerants, without movable parts, and completely silently.

Promising application areas are the replacement of compression based refrigerators, cooling of CPUs and other normal or even superconducting electronic components, of infrared detectors, and so on. With the miniaturization of electronic de-vices the dissipated power density increases to an extent which will shortly require cooling efficiencies that exceed those of presently available technologies.

Currently broad applications of thermoelectric materials are inhibited by their limited efficiency. The optimization of materials for higher efficiency is challenging since the three physical quantities determining the efficiency (thermopower, thermal conductivity, electrical conductivity) are generally interdependent in a way that optimizing one will adversely affect another. Thus, the strategy governing the search for better materials has long been to find the best compromise. In the mid-1990's the field was given new impetus by the advent of nanotechnology which could, in some cases, decouple the physical quantities from each other. Today thermoelectric materials build a highly active research field both in academia and industry.

A number of thermoelectric materials are well established and are already today used in commercial applications. At temperatures around room temperature Bi_2Te_3-based materials are universally used in thermoelectric refrigeration. Between about 100 and 600°C lead tellurides are the materials of choice and, at even higher temperatures, SiGe is employed. The thermoelectric figure-of-merit of these and other less well established materials is shown in Fig. 1.

This chapter deals with a relatively new class of materials with promising thermoelectric properties at high temperatures, cage compounds. Also, attempts to improve their properties at low temperatures by the introduction of effects of strong electronic correlations will be discussed shortly. Cage compounds are considered to be members of a larger class of materials called "complex metallic alloys (CMAs)", which are the topic of the European Network of Excellence CMA. Thermoelectric properties of non-caged CMAs are covered in Chapter XX. Other promising materials classes such as oxides or low-dimensional systems are beyond the scope of these lecture notes.

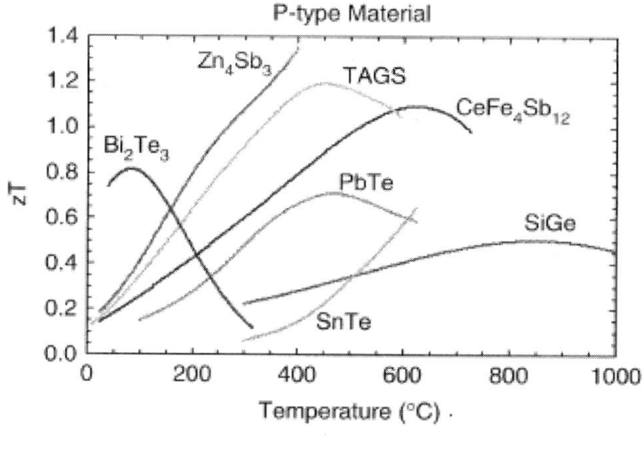

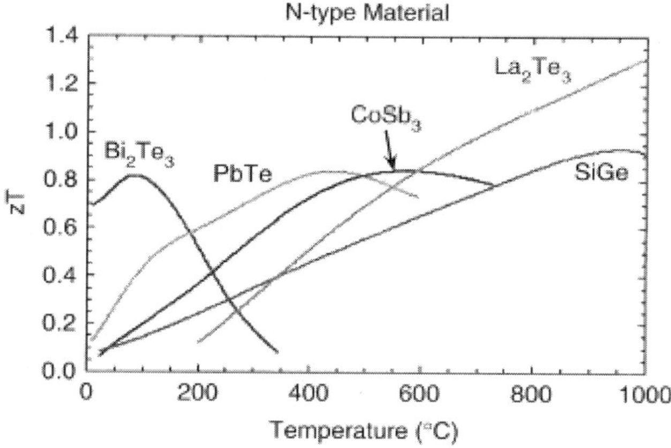

Fig. 1. Dimensionless thermoelectric figure-of-merit of established and more recently investigated p- and n-type materials vs temperature[1] (upper panel and lower panel, respectively).

2. Cage Compounds

2.1. *Definitions*

What are cage compounds? This notion is not yet well established in the literature and various definitions are conceivable which shall be explained in more detail below:

2.1.1. *Guest/host atoms*

Guests are situated in cages made up by the host species. An example is shown in Fig. 2.

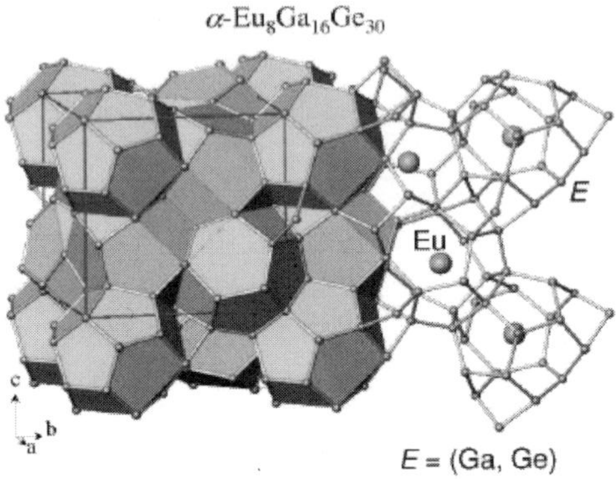

Fig. 2. Cage compound defined as a guest – host system[2] (Eu are guest atoms).

2.1.2. *Coordination number (c.n.)*

In a cage compound the guest atom has a large c.n. with the host atoms. In the clathrate β-Eu$_8$Ga$_{16}$Ge$_{30}$, for instance, there are two guest sites. Guest 1 in the smaller cage has c.n. = 20 while guest 2 in the larger cage has c.n. = 24 (cf. Fig. 3).

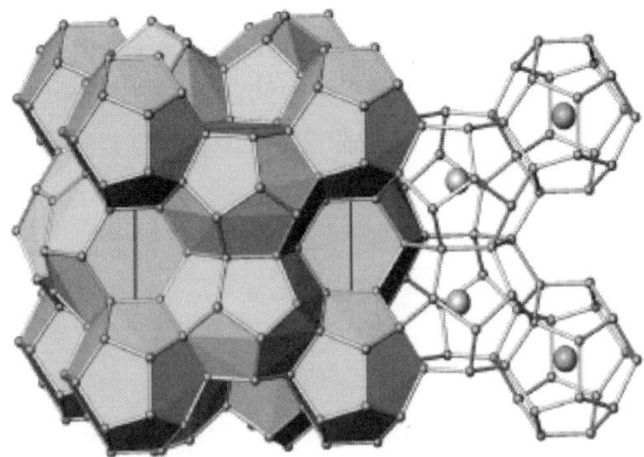

Fig. 3. Large coordinations numbers in the cage compound β-Eu$_8$Ga$_{16}$Ge$_{30}$.

2.1.3. *Bond length/strength*

Bonds between guest and host species are longer (bond strength weaker) (cf. Fig. 3) than in related non-caged compounds (cf. Fig .4). While in β-$Eu_8Ga_{16}Ge_{30}$ the shortest Eu–Ga/Ge distance is 3.41 Å or 3.59 Å for the smaller and larger cage, respectively, the shortest Eu–Ge distance in $EuGe_2$ is 3.11 Å (cf. Fig.4).

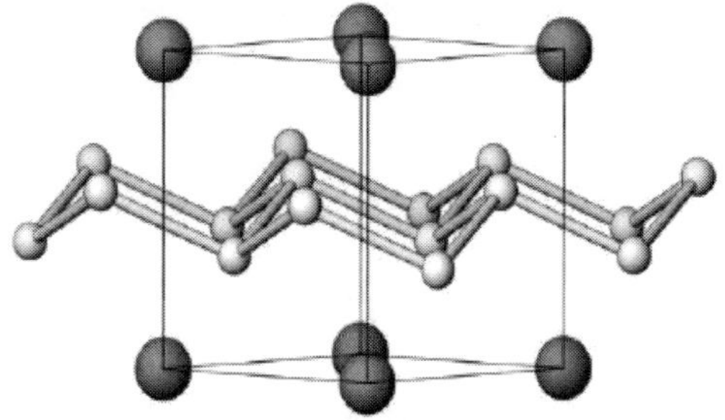

Fig. 4. Crystal Structure of $EuGe_2$.[3]

2.1.4. *Empty host*

The host can, at least in some cases, exist without the guest. This is very typical for the skutterudites which are well known both for their empty (Fig. 5) and filled (Fig. 6) form.

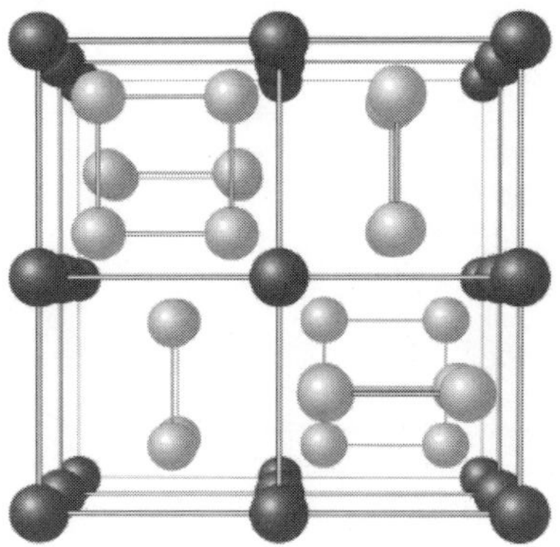

Fig. 5. Empty skutterudite, e.g. $CoAs_3$.

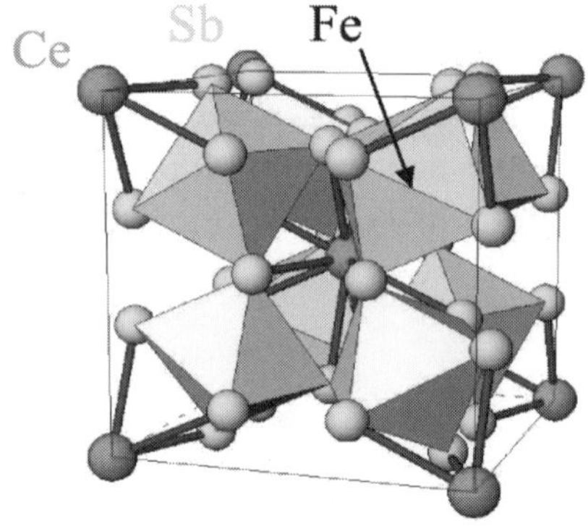

Fig. 6. Filled skutterudite, e.g., CeFe4Sb12, where Ce is the guest.

In clathrates the existence of empty variants appears to be much more subtle. While in so-called type-I clathrates no partial filling of the cages has ever been observed, the type-II structure is more disposed to exist in empty form (Fig.7).

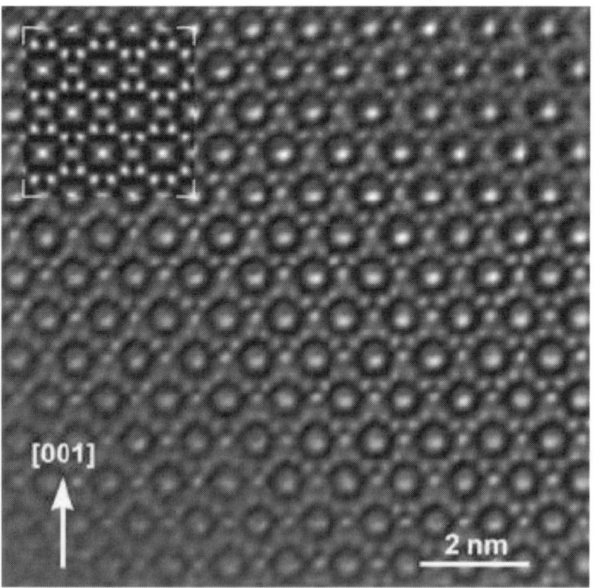

Fig. 7. TEM image of the empty type-II clathrate $U_{24}Ge_{136}$.[4] (U denotes a void).

2.2. *Examples*

Cage structures occur in various classes of materials. The below list is certainly not exhaustive.

2.2.1. *Filled skutterudites*

The structure of filled skutterudites results from the one of binary (or empty) skutterudites (cf. Fig. 5). These latter consist of eight formula units per unit cell and have a cubic crystal structure of space group $Im\bar{3}$. The large voids in 2(a) position can accommodate interstitially various kinds of atoms, which may be represented by the notation $U_{24}T_8X_{24}$, where U = void. The transition metal atoms T occupy the 8(c) site, the atoms X the 24(g) site. The interstitial ions are enclosed in an irregular dodecahedral cage of X atoms.

Skutterudites of many different compositions exist. The prototypical binary skutterudite is CoAs3 (Fig. 5). Other binary skutterudites can be created by replacing Co and As by the isoelectronic elements Rh and Ir (Fig. 8) or As and Sb (Fig. 9), respectively, resulting in the possible compositions given in the first column of the upper table of Table 1). Many other compositions, in particular ternary and quatenary compounds, can be designed by respecting the overall charge balance, e.g., if one atom per formula unit is replaced by another one with one extra valence electron then another atom has to be replaced by an atom with one less valence electron (Fig. 1).

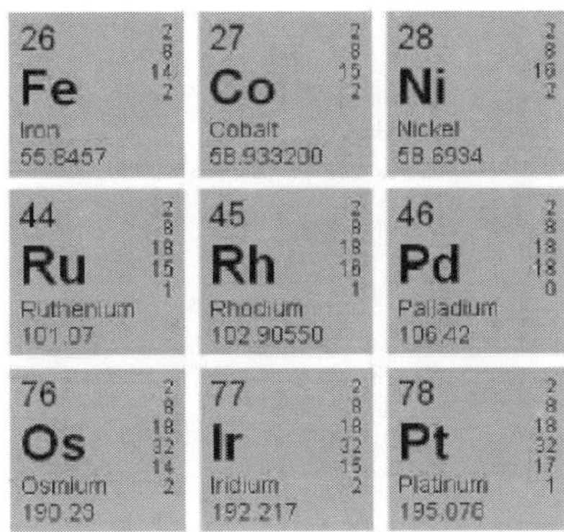

Fig. 8. Neighbours of Co in the periodic table.

                                                        Silke Paschen

Table 1. Possible compositions of skutterudites
(by courtesy of A. Grytsiv, unpublished).

unfilled skutterudites						
binary			ternary			
T_4X_{12}	T'_4X_{12}	T''_4X_{12}	$T_2X'_6X''_6$	$T'_2T''_2X_{12}$	$T'_4X_8X''_4$	$T''_4X_8X'_4$
Co_4P_{12}	"Fe_4Sb_{12}"	Ni_4P_{12}	$Co_4Ge_6Te_6$	$Fe_2Ni_2Sb_{12}$	$Fe_4Sb_8Se_4$	$Ni_4P_8Ge_4$
Co_4As_{12}		Pd_4P_{12}	$Co_4Sn_6Se_6$	$Fe_2Ni_2As_{12}$	$Fe_4Sb_8Te_4$	$Ni_4Bi_8Ge_4$
Co_4Sb_{12}			$Co_4Sn_6Te_6$	$Fe_2Pd_2Sb_{12}$	$Ru_4Sb_8Se_4$	$Pt_4Sb_{7.2}Sn_{4.8}$
Rh_4P_{12}			$Co_4Ge_6S_6$	$Fe_2Pt_2Sb_{12}$	$Ru_4Sb_8Te_4$	$Ni_4Sb_8Sn_4$
Rh_4As_{12}			$Co_4Ge_6Se_6$	$Ru_2Ni_2Sb_{12}$	$Os_4Sb_8Te_4$	$Ni_4As_8Ge_4$
Rh_4Sb_{12}			$Rh_4Ge_6S_6$	$Ru_2Pd_2Sb_{12}$		
Ir_4P_{12}			$Ir_4Ge_6S_6$	$Ru_2Pt_2Sb_{12}$		
Ir_4As_{12}			$Ir_4Ge_6Se_6$			
Ir_4Sb_{12}			$Ir_4Sn_6S_6$			
			$Ir_4Sn_6Se_6$			
			$Ir_4Sn_6Te_6$			

filled skutterudites		
ternary	Quaternary	
EpT_4X_{12}	$EpT'_4(X_{1-x}X'_x)_{12}$	$Ep(T_{1-x}T'_x)_4X_{12}$
T = Fe, Ru, Os **X = P, As, Sb** Ep=Ca, Sr, Ba, La, Ce, Pr, Nd K, Na, Sm, Eu, Gd, Tb, Yb, Th, U	**T' = Co, Ir** **X = Sb; X' =Ge, Sn** Ep = La, Nd, Sm, Tl $LaIr_4Sb_9Ge_3$, $NdIr_4Sb_9Ge_3$,	**T = Fe; T' = Co** **X = Sb** Ep = Tl $TlFeCo_3Sb_{12}$
$LaFe_4P_{12}$, UFe_4P_{12}, $ThOs_4As_{12}$, $YbFe_4Sb_{12}$, ...	$SmIr_4Sb_9Ge_3$, $TlCo_4Sb_{11}Sn$, $La_{0.9}Co_4Sb_{10.3}Sn_{2.44}$, ...	

metastable partially filled skutterudites:

$Ep_{1-y}Fe_4Sb_{12}$; Ep = Na, Y, Hf, Sn, Lu $Ep_{1-y}Co_4Sb_{12}$; Ep = Sn, Pb

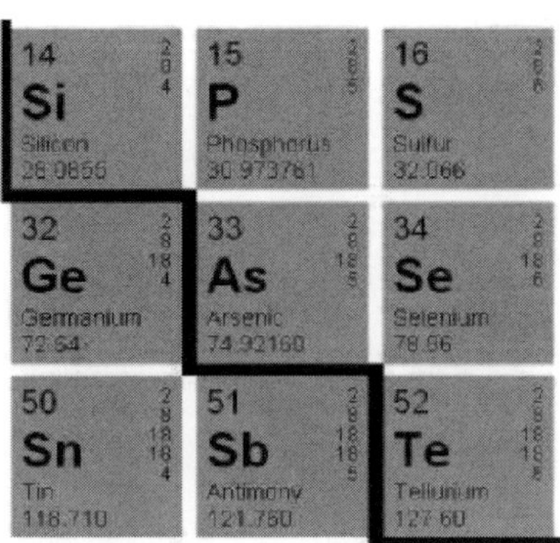

Fig. 9. Neighbours of As in the periodic table.

2.2.2. *Intermetallic clathrates*

Intermetallic clathrates are periodic solids in which tetrahedrally bonded atoms form a space-filling framework of cages that enclose guests. Clathrates have been classified with respect to the type of cage-forming polyhedra (Fig. 2).

The bonding situation may, in a first approximation, be discussed with the help of the Zintl concept: The more electropositive guest atoms (cations) donate electrons to the more electronegative cage atoms (anions) such that the latter complete their valence requirements (octet rule) when forming a covalently (sp^3-like) bonded space-filling framework (Fig. 10). E.g., for the clathrate of type-I $Ba_8Ga_{16}Ge_{30}$ this may be written as

$$(Ba^{2+})_8(Ga^{1-})_{16}(Ge^0)_{30} \tag{1}$$

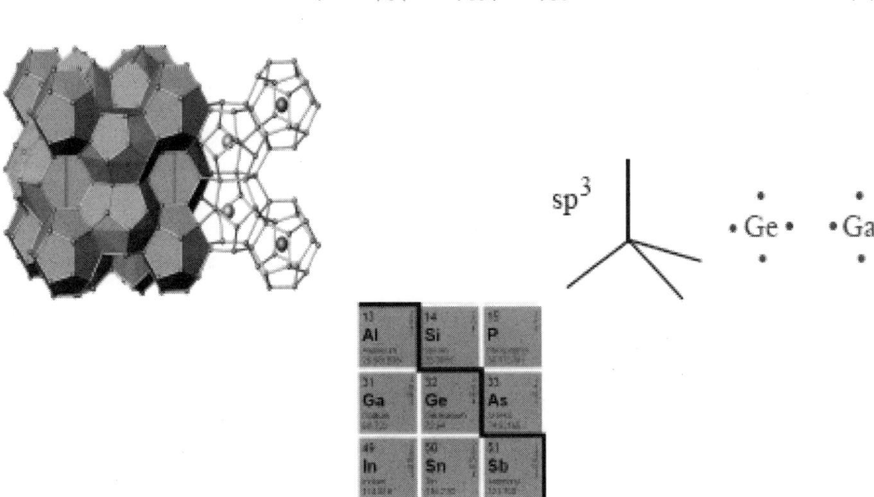

Fig. 10. Bonding situation in intermetallic type-I clathrate.

2.2.3. *Clathrate-like compounds*

Clathrate-like compounds do not fulfill the strict structural requirements of clathrates but have nevertheless a cage structure with atoms situated inside the cages (guests). Examples are, e.g., $Ce_2Ni_{36}P_{15}$ (Fig. 11) or $Ce_3Pd_{20}Si_6$ (Fig. 12) (Table 2).

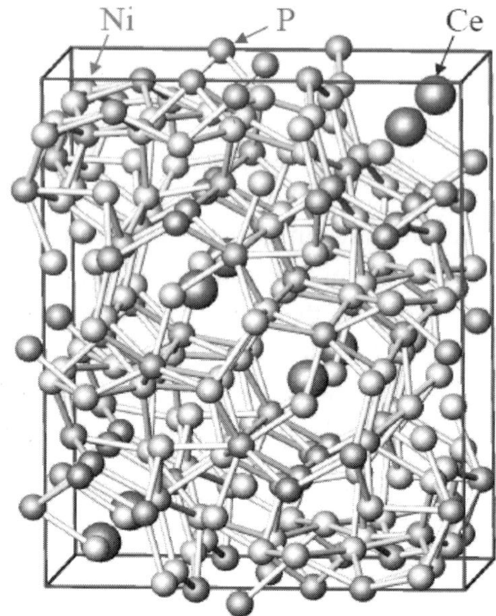

Fig. 11. Orthorhombic *Pbam* crystal structure of $Ce_2Ni_{36}P_{15}$. The cages consist of 16 Ni and 6 P atoms.[6]

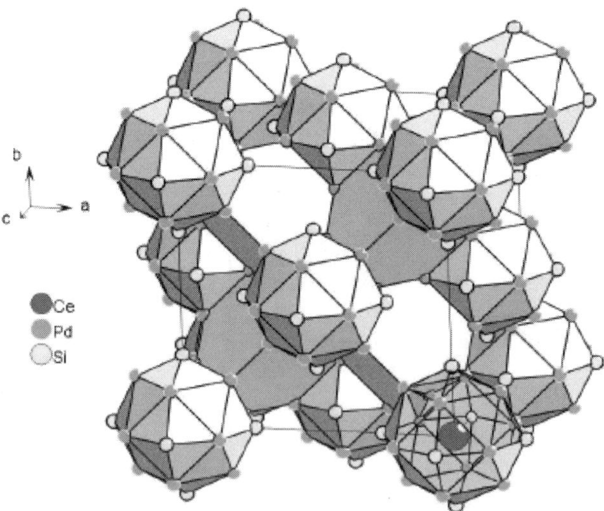

Fig. 12. $Ce_3Pd_{20}Si_6$. Cubic $Fm\overline{3}m$. Cage 1: 16 Pd, cage 2: 12 Pd, 6 Si.[7]

Table 2. Structure types of intermetallic and hydrate clatrathes:
Y=guest atoms, T=host atoms.[5]

Type	Ideal unit cell formula	Polyhedra	Space group	Intermetallic Clathrate
I	$6X^*2Y^*46T^a$	$[5^{12}6^2]_6[5^{12}]_2$	Pm-3n	K_8Ge_{46-x}, $Ba_8Al_{16}Ge_{30}$, $Eu_2Ba_6Cu_4Si_{42}$
II	$8X^*16Y^*136T$	$[5^{12}6^4]_8[5^{12}]_{16}$	Fd-3m	Na_xSi_{136}, $Cs_8Na_{16}Ge_{136}$
III	$20X^*10Y^*172T$	$[5^{12}6^2]_{16}[5^{12}6^3]_4[5^{12}]_{10}$	$P4_2/mnm$	$Cs_{30}Na_{(1.33x-10)}Sn_{(172-x)}$, $x = 9.6$
IV	$8X^*6Y^*80T$	$[5^{12}6^2]_4[5^{12}6^3]_4[5^{12}]_6$	P6/mmm	$\sim K_7Ge_{40-x}$, $x \sim 2^c$
V	$4X^*8Y^*68T$	$[5^{12}6^4]_4[5^{12}]_8$	$P6_3/mmc$	-
VI	$16X^*156T$	$[4^35^36^27^3]_{16}[4^35^4]_{12}$	I-43d	-
VII	$2X^*12T$	$[4^66^8]_2$	Im-3m	-
VIII	$8Y^*46T$	$[3^34^35^9]_8^d$	Im-3m	$Ba_8Ga_{16}Sn_{30}$, $Eu_8Ga_{16}Ge_{30}$
IX	$16Y^*8X^*100T$	$[5^{12}]_8 + [4^{10}]_4+...^e$	$P4_132$	$Ba_6Ge_{21}In_4$, Ba_6Ge_{25}

2.2.4. *Oxides*

Also in oxides cage structures can occur. An example is KOs_2O_6 which crystallizes in the β-pyrochlore structure and where K is the guest atom in cages built of Os and O (Fig. 13).

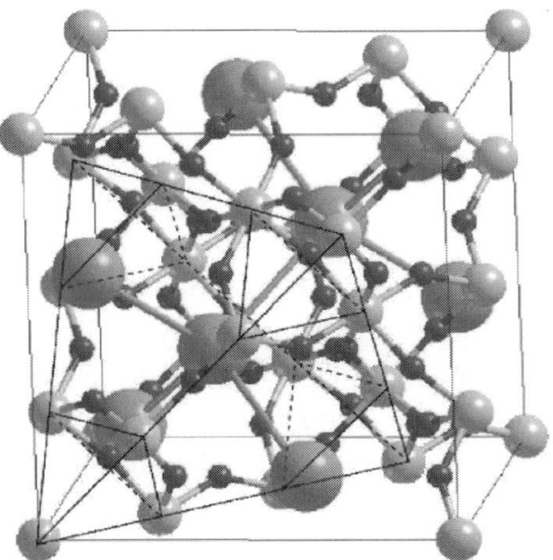

Fig. 13. β-pyrochlore structure; K, Os, O. K in cages.[8]

In $Na_2V_3O_7$, instead of approximately spherical cages there exist nanotubes formed by VO5 pyramides. Only part of the Na atoms are captured within these tubes (Fig. 14).

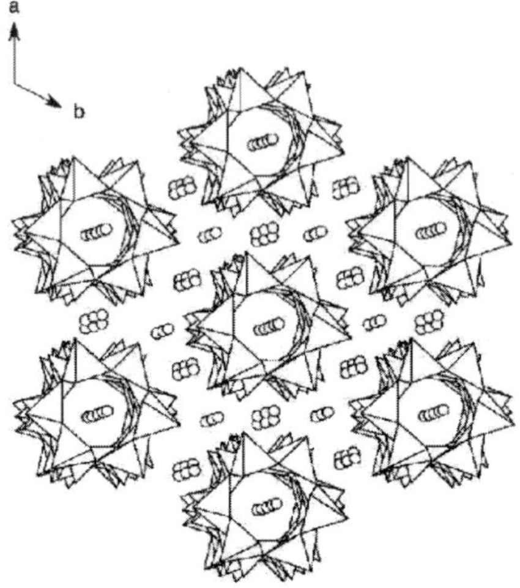

Fig. 14. $Na_2V_3O_7$ nanotubes. Square pyramides: VO_5, circles: Na. Cage (or ring) diameter ~ 5 Å.[9]

2.3. *Characteristic properties of cage compounds*

Cage compounds may as well be identified by the occurrence of the characteristic properties

2.3.1. *Rattling/tunnelling*

Rattling refers to one mass (the guest) moving incoherrently with respect to a framework (host) just as peas in a baby rattle. Tunneling is a purely quantum mechanical phenomenon where a particle can traverse an energy barrier which, classically, can only be overcome by thermal activation across the barrier. In what follows the evidence for rattling and/or tunneling from different experiments is presented.

First, neutron scattering experiments on a clathrate are presented. The neutron is a subatomic particle with no net electric charge but a magnetic moment. Neutron radiation is commonly employed in neutron scattering facilities, where the radiation is used in a similar way one uses X-rays for the analysis of condensed matter. Since they are neutral they can deeply penetrate into matter and since they have a spin they are sensitive to magnetism.

In Fig. 15 atomic displacement parameters (Ueq) and nuclear densities, both determined by neutron scattering experiments, are shown for the guest atoms X = (Ba), Sr, and Eu in $X_8Ga_{16}Ge_{30}$. Eu has the largest Ueq. The (linear) extrapolation to zero temperature suggests that at the absolute zero in temperature Eu still has a large Ueq. Since there are no temperature induced oscillations at 0 K this indicates that so-called split sites exist in $Eu_8Ga_{16}Ge_{30}$, as indeed concluded from the nuclear density maps.

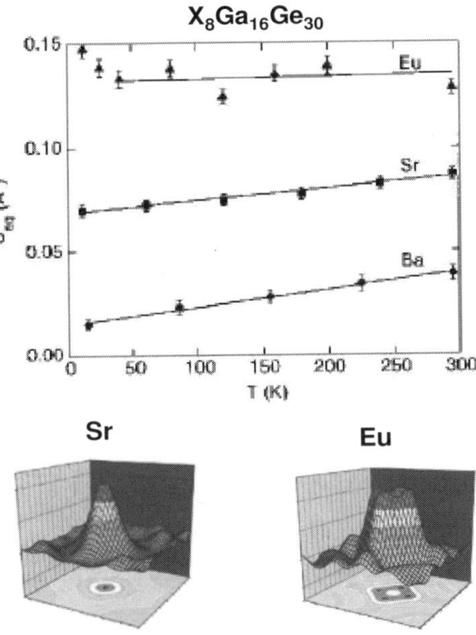

Fig. 15. Atomic displacement parameters and nuclear densities.[10]

Very direct evidence for rattling comes from Raman scattering experiments. In Raman spectroscopy monochromatic light, usually from a laser, is inelastically scattered from a solid, thereby exciting certain phonon modes or other degrees of freedom in the system. Figure 16 displays Raman spectra of various filled skutterudite compounds. Arrows show the rattling modes. With increasing cage size (from top to bottom), the rattling frequency of second-order phonons decreases. The peak intensity decreasing with lowering the temperature (not shown) indicates that (thermally activated) rattling is dominant and not (temperature independent) tunneling.

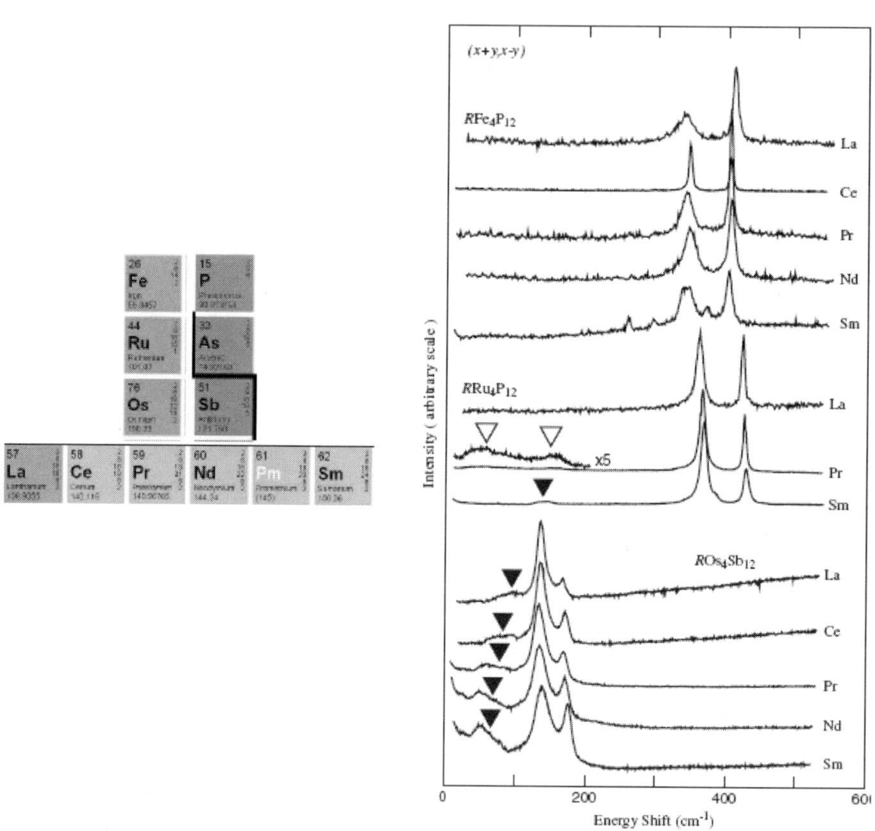

Fig. 16. Raman scattering spectra of filled skutterudites[11] and sections of periodic table.

The results of resonant ultrasound spectroscopy (Fig. 17), on the other hand, need to be compared to theoretical curves in order to be conclusive. Theoretical curves (full lines) agree well with the experiment (symbols) if a four-well potential (top left) is assumed for the guest atoms in the clathrate cages.

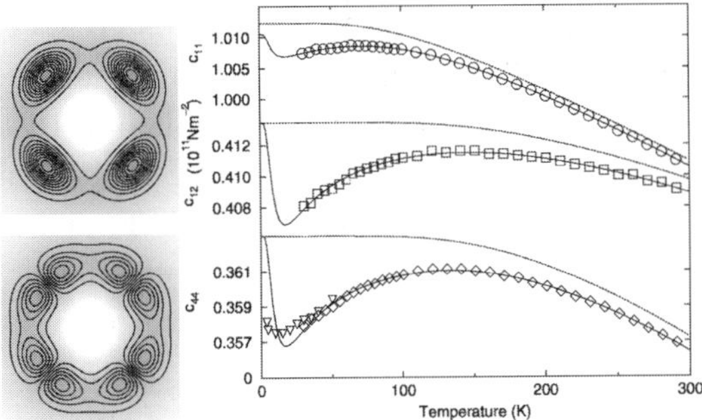

Fig. 17. Four-well potential and elastic constants of $Eu_8Ga_{16}Ge_{30}$.[12]

Direct evidence for tunnelling of guest atoms between split sites is provided by Mössbauer spectroscopy on $Eu_8Ga_{16}Ge_{30}$ (Fig. 18).

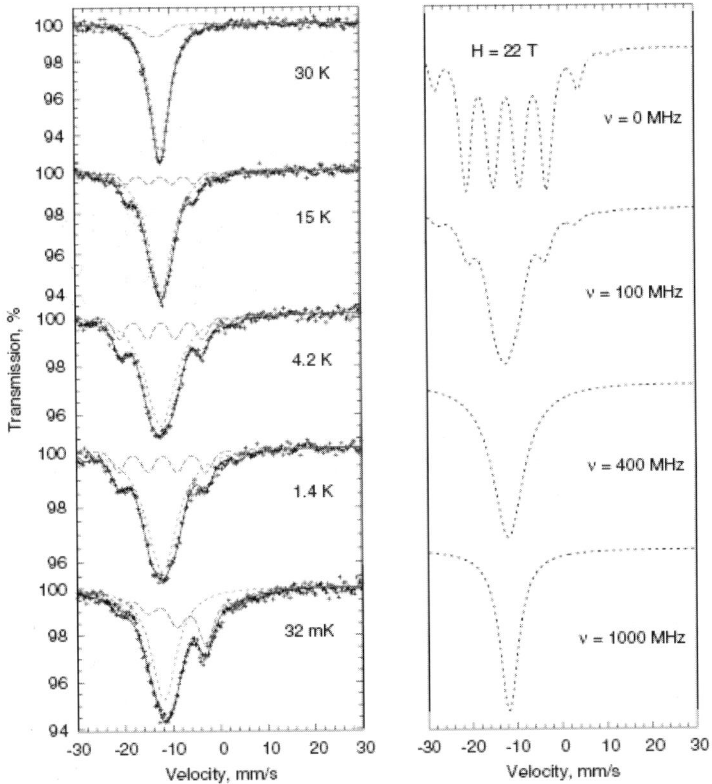

Fig. 18. Mössbauer spectroscopy on $Eu_8Ga_{16}Ge_{30}$.[13]

The panels on the right hand side are calculated spectra for divalent Eu in a hyperfine field of 22 T, fluctuating at different frequencies ν. The panels on the left hand side show the experimental results at different temperatures together with the fits representing spectra of two Eu atoms in small cages in a static hyperfine field of 20 T, and six Eu atoms in large cages tunnelling at 450 MHz between 4 split sites.

2.3.2. *Phonon glass-electron crystal*

A phonon glass – electron crystal (PGEC) (Fig. 19) is a material which conducts heat like a glass, i.e., very poorly, but electricity like a (single) crystal. It is clear from the expression of the dimensionless thermoelectric figure-of-merit:

$$ZT = (S2 \cdot \sigma/\kappa) \cdot T \tag{2}$$

that such a material would be an efficient thermoelectric since the ratio σ/κ is maximized.

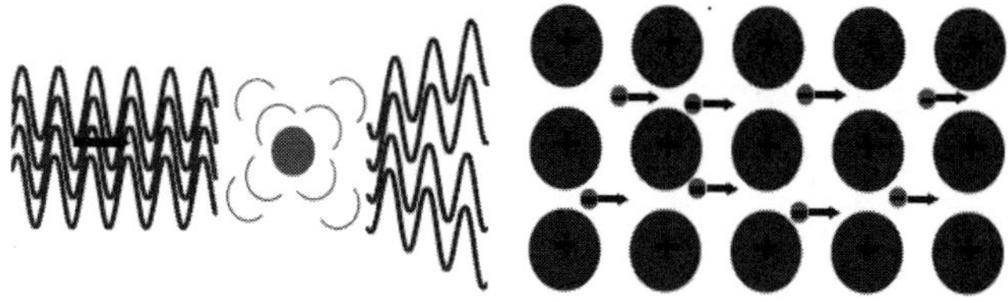

Fig. 19. Schematic representation of a "phonon glass" (! κ small) and an "electron crystal" (! σ large).

Since in most materials the (electronic) thermal conductivity κ is related to the electrical resistivity σ via the Wiedemann-Franz law ($\kappa = L$ T $\sigma(T)$ with L the Lorentz number) large σ and small κ seemingly contradict each other. This is, however, not the case in materials where the thermal conductivity is dominated by its lattice contribution κL, as in the clathrate $Ba_8Ga_{16}Ge_{30}$ (Fig. 20). The temperature dependence of κL of $Ba_8Ga_{16}Ge_{30}$ ressembles the one of vitreous SiO2, – simple window glass, which is known for its very low thermal conductivity. The fit (full line) accounts for phonon-electron scattering at low temperatures and resonant scattering off the rattling guest atoms at higher temperatures.

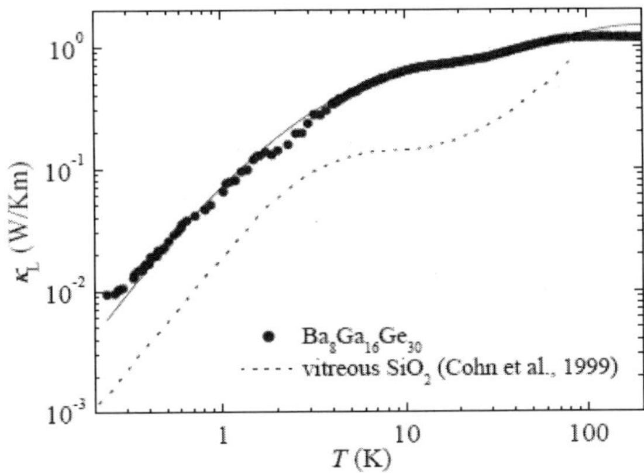

Fig. 20. Lattice thermal conductivity of the clatrathe $Ba_8Ga_{16}Ge_{30}$.[14]

The drastic lowering of the thermal conductivity by rattling guest atoms is demonstrated by Fig. 21. Upon increasing the Nd filling x in the skutterudite $Nd_xFe_4Sb_{12}$ the peak in the thermal conductivity at somewhat below 50 K is successively weakened and suppressed altogether at x ≈1.

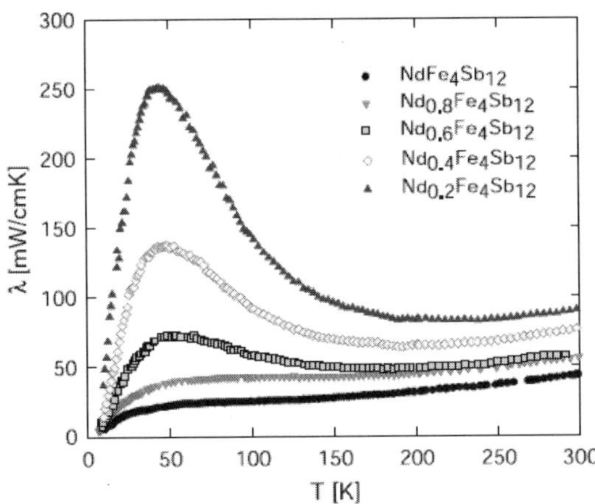

Fig. 21. Thermal conductivity of the skutterudite NdyFe4Sb12 (by courtesy of E. Bauer, unpublished).

Very instructive is also the comparison of thermal (Fig. 22) and electrical (Fig. 23) conductivities of the clathrates α- and β-$Eu_8Ga_{16}Ge_{30}$. The α-phase is a type VIII clathrate, the β-phase is of type I. Both are

 Silke Paschen

n-type conductors. The effective masses as deduced from electrical transport are m*(α) $\approx$ 1.5m$_0$ and m*(β) $\approx$ 3.2m$_0$. κL(T) appears to be dominated by phonon-electron scattering below 10 K, and by resonant scattering off the rattling guest atoms above 10 K. κL(T) being, at low temperatures, much smaller in β- then in α- Eu$_8$Ga$_{16}$Ge$_{30}$ was argued to be due to the larger effective mass of the electrons, leading to enhanced phonon-electron scattering rates.

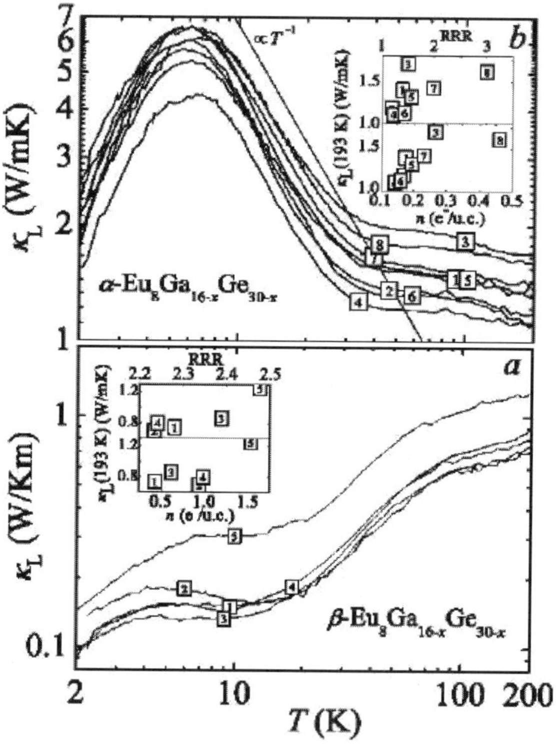

Fig. 22. Thermal conductivity of the clatrathes a- and b- Eu$_8$Ga$_{16}$Ge$_{30}$.[15]

Combining electrical resistivity and Hall effect data the electronic mean free path has been deduced for all α- and β-Eu$_8$Ga$_{16}$Ge$_{30}$ samples of Fig. 23 at 2 K: α4 : 22 Å, α8 : 83 Å, β1 : 29 Å, β5 : 37 Å. These are very short compared to typical values for clean crystalline met als/semiconductors (which can be of the order of μm). There are several possible reasons for these short mean free paths and corresponding low charge carrier mobilities: Ga/Ge disorder, deviations from the ideal Ga:Ge stoichiometry, rattling of the guest atoms. Further investigations are needed to determine the dominating effect. I any case, the low mobilities of the clathrates α- and β-Eu$_8$Ga$_{16}$Ge$_{30}$ show that the "electron crystal" has to be used with some caution for these systems.

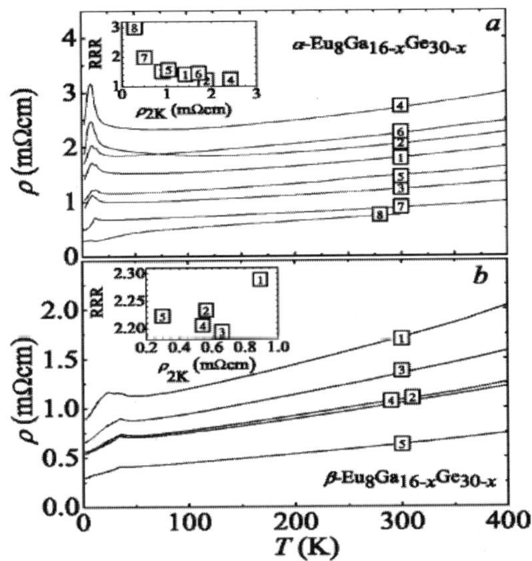

Fig. 23. Electrical resistivity of the clatrathes a- and b- Eu$_8$Ga$_{16}$Ge$_{30}$.[15]

2.4. *Tuning for optimized performance*

Many efforts worldwide aim at improving the thermoelectric properties of known materials. The attempts to do so can be subdivided as follows:

2.4.1. *Stoichiometry*

A precise control of the stoichiometry is important for many thermoelectric materials. An example is Eu$_8$Ga$_{16-x}$Ge$_{30+x}$ where x can be deliberately changed by annealing at different temperatures. The variation of x translates into a variation of the charge carrier concentration n and the thermoelectric properties, as shown in Fig. 24.

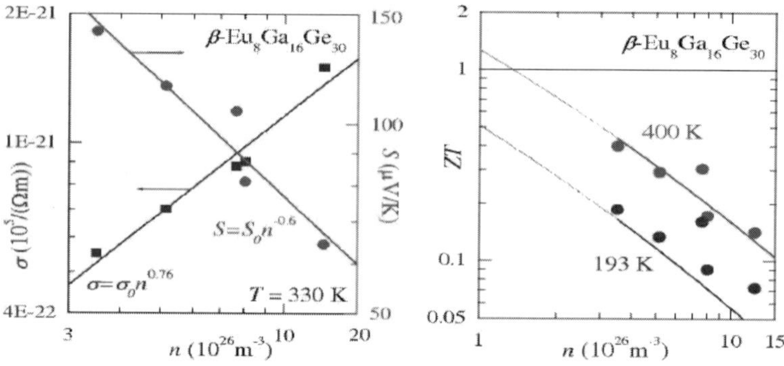

Fig. 24. Resistivity and thermopower (left) and thermoelectric figure-of-merit (right) after annealing $Eu_8Ga_{16-x}Ge_{30+x}$ at different temperatures: n ≈ x.[15,16]

2.4.2. *Doping*

In many cases doping is used to maximize ZT. Figure 25 shows such a study on the filled skutterudite $Nd_xCo_{4-y}Ni_ySb_{12}$.

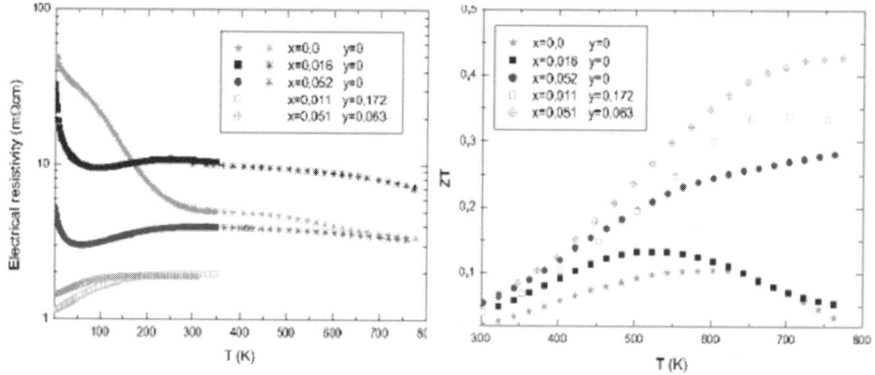

Fig. 25. Electrical conductivity (left) and thermoelectric figure-of-merit (right) of the filled skutterudite $Nd_xCo_{4-y}Ni_ySb_{12}$.[17]

2.4.3. *Substitution*

Also substitutions such as the partial replacement of Sr by Yb in the double-filled skutterudite $(Sr,Yb)_yCo_4Sb_{12}$ (Fig. 26) can lead to sizable ZT enhancements.

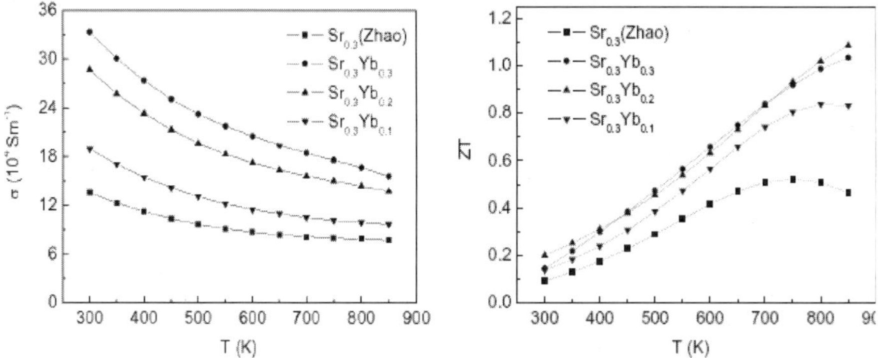

Fig. 26. Therrmoelectricfigure of merit.[17]

2.4.4. *Micro/Nanostructuring*

A technique that becomes increasingly popular is the micro- and/or nanostructuring of materials. The concept behind is that electrons are not as strongly scattered by grain boundaries as phonons and that thus the (lattice) thermal conductivity is more strongly suppressed than the electrical conductivity. Figure 27 shows, with the example of the unfilled skutterudite $CoSb_3$ processed by spark plasma sintering, that most efficient in this respect are grains of different length scales.

Fig. 27. Micro- and nanograins (left) and thermo-electric figure of merit (right) of unfilled skutteruditeCoSb$_3$ processed by spark plasma sintering.[18]

3. Strongly Correlated Cage Compounds

3.1. *The concept of strongly correlated cage compounds*

The idea to exploit the emerging class of materials, strongly correlated cage compounds, for thermoelectric applications (Fig. 28) and the underlying physical concepts (Fig. 29) shall be discussed in this section.

The motivation for using cage compounds is the phonon glass – electron crystal concept explained above and symbolized by the "vibrating" guest atoms in the clathrate structure in Fig. 28.

In addition, a new concept for increasing the thermopower – strong electronic correlations – is introduced. The symbol in Fig. 28 represents a localized magnetic moment (flash) that interacts with an ensemble of conduction electrons.

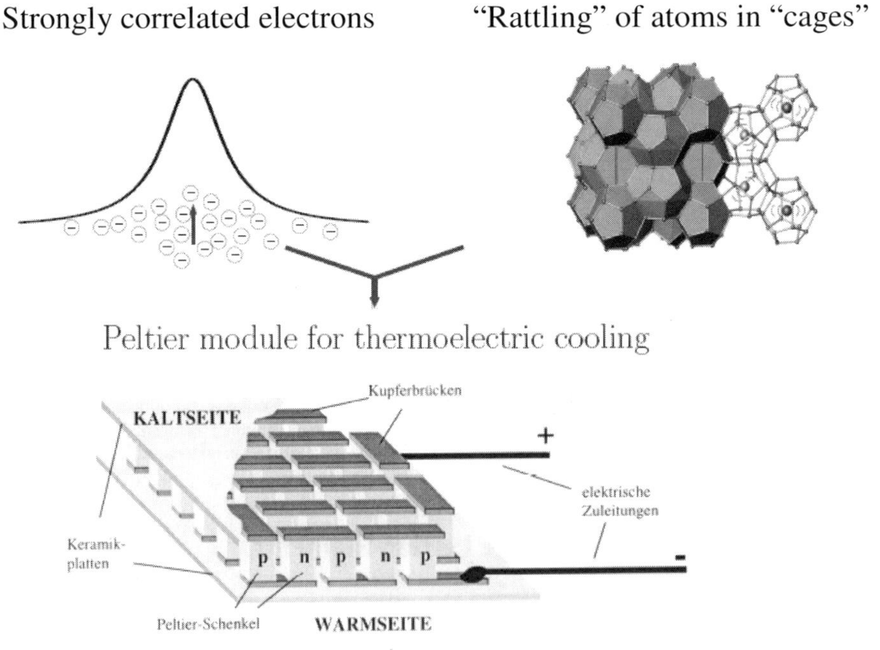

Strongly correlated electrons   "Rattling" of atoms in "cages"

Peltier module for thermoelectric cooling

Fig. 28. Exploitation of strongly correlated cage compounds for thermoelactric cooling.

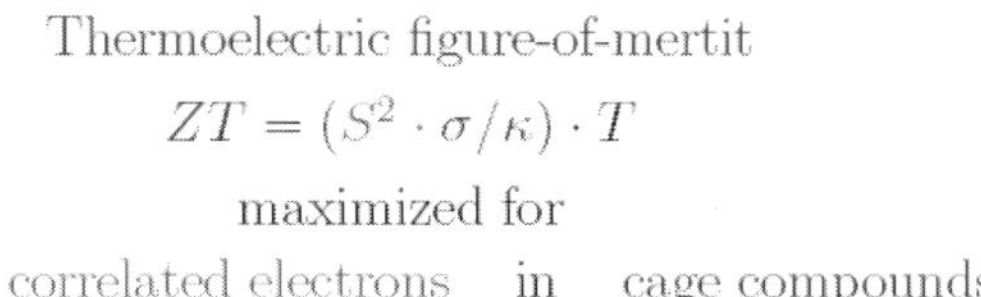

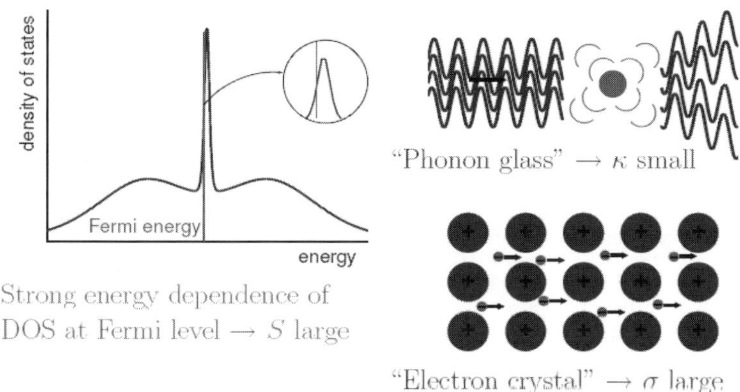

Fig. 29. The concept of large thermoelectric figure-of-merit in strongly correlated cage compounds.

3.2. *Brief introduction to strongly correlated electron systems*

Strongly correlated electron systems (SCES) are characterized by electrons strongly interacting with each other. As a consequence their effective mass is strongly enhanced (renormalized). In so-called heavy fermion systems, it is two different types of electrons that interact with each other, localized ones originating from uncompletely filled localized f (or d) shells, and itinerant ones originating from much broader bands of s, p (or d) character. At low temperatures the Kondo interaction leads to the formation of a "composit quasiparticle" consisting of the local magnetic moment collectively screened by a conduction electron cloud (Fig. 30). These quasiparticles may be described by Landau's Fermi liquid theory. However, for metallic heavy fermion systems, the prefactors in the temperature dependencies of the specific heat C, the magnetic susceptibility χ, and the electrical resistivity ρ, $\gamma0$, $\chi0$, and A, respectively (see Fig. 30), are strongly remormalized compared to the typical values in simple metals (where $m^* \approx m_e$).

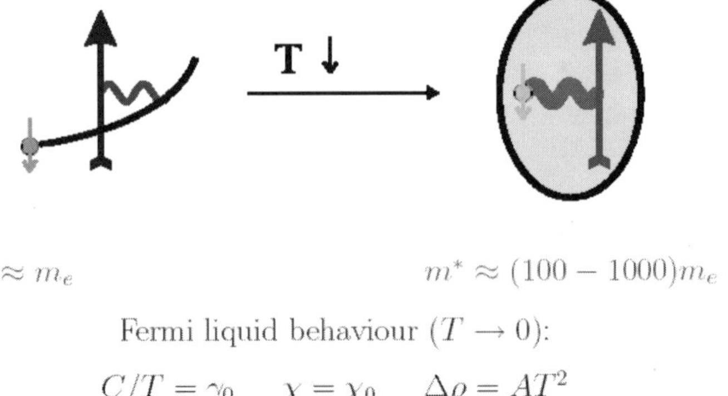

Fig. 30. Heavy quasiparticles in strongly correlated electron systems. (Graphics from Ref. 19).

As for simple (uncorrelated) materials, strongly correlated metals generally face the problem of too high (electronic) thermal conductivities. But also in SCES there exist materials with semimetallic or semiconducting properties (so-called Kondo insulators/semiconductors/ semimetals) which are better candidates for thermoelectric applications. The origin of the (pseudo) energy gap formation in these systems is the correlation effects itself. An example of a semimetallic SCES is the Kondo semimetal CeNiSn. Its pseudo energy gap ΔE is approximately $k_B \cdot$ 14 K, a very small value as compared to typical energy gaps in semiconductors (Si, Ge: $\Delta E \approx k_B \cdot$ 10 000 K).

Figure 31 shows its crystal structure, which is clearly not cage like, and the thermopower at low temperatures. The thermopower has a large peak below the gap temperature. The value at 4 K is more than an order of magnitude larger than in a typical thermoelectric clathrate. On the other hand due to the absence of a cage structure, the thermal conductivity of CeNiSn is about an order of magnitude higher.

An example of a strongly correlated semiconductor is FeSi. Its energy gap ΔE is approximately $k_B \cdot$500 K. Angle resolved photoemission spectroscopy (ARPES) experiments (Fig. 32) have revealed extremely sharp features in the electronic density of states near the Fermi level, along certain directions in reciprocal space.

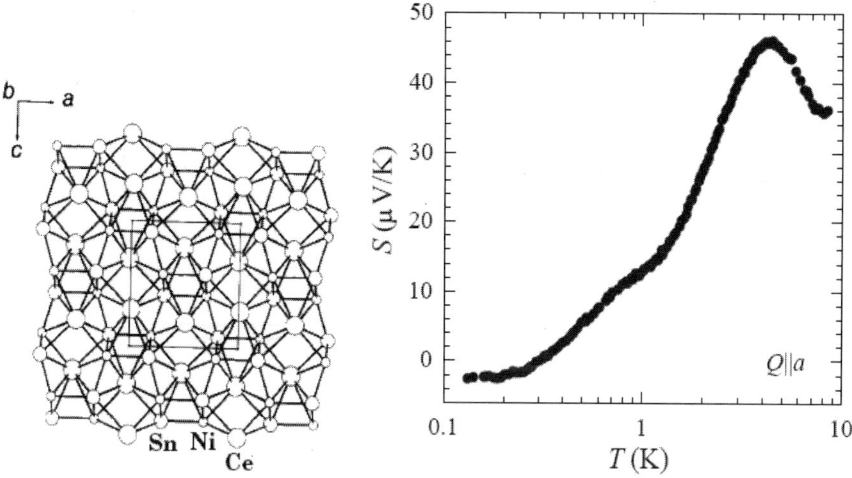

Fig. 31. Crystal structure and temperature dependence of thermopower of the Kondo semimetal CeNiSn.[20]

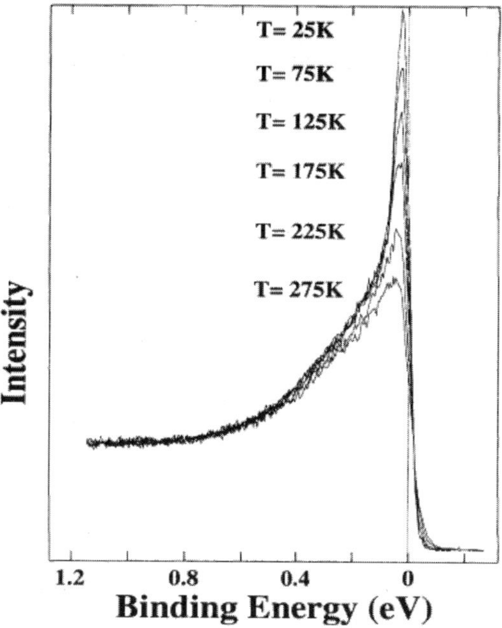

Fig. 32. Electronic density of states of FeSi determined in ARPES experiments.[21]

As expected, the thermopower of FeSi is extremely large (Fig. 33). However, since σ/κ is small, ZT reaches only a modest value of less than 0.01 at 50 K. Thus, to bring SCES to practical use, σ/κ has to be enhanced. Since in both CeNiSn and FeSi κ is dominated by the phonon contribution, it is this contribution which has to be suppressed. One possible route might be to combine, in a single substance, the advantages of cage compounds (large σ/κ) and of SCES (large S). First attempts to obtain such "strongly correlated cage compounds" are described below.

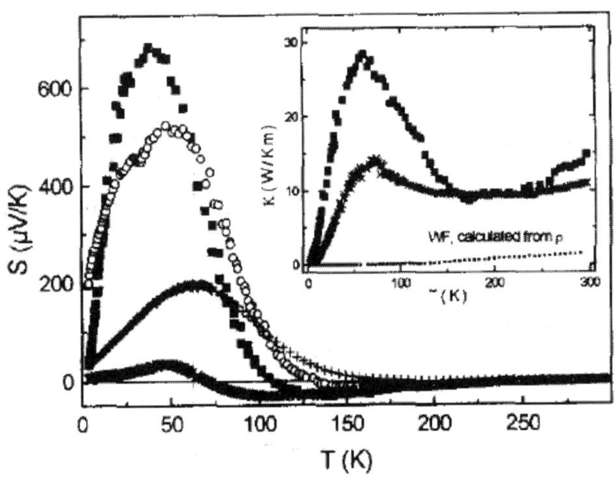

Fig. 33. Thermopower vs temperature for FeSi (full squares) and doped variants (other symbols).[22]

3.3. *Attempts to obtain strongly correlated cage compounds*

A possible route to strongly correlated cage compounds is to use rare earth elements tending to hybridization with conduction electron bands as guest atoms in cage compounds. Few such attempts are presented below.

3.3.1. *Substituted $Eu_8Ga_{16}Ge_{30}$*

Even though there are examples of Eu based compounds showing effects of strong correlations this is the exception rather than the rule. In

$Eu_8Ga_{16}Ge_{30}$ no signs of strong correlations are observed. Eu is divalent in the entire temperature range and the system orders ferromagnetically at low temperatures. Worldwide efforts to (partially) replace Eu by other rare earth elements have failed (e.g.,[23,24]). Also the seemingly successful incorporation of Ce into a Si-based type-I clathrate, $Ba_6Ce_2Au_4Si_{42}$[25], was subsequently shown to have failed.[26]

3.3.2. *The clathrate-like compound* $Ce_2Ni_{36}P_{15}$

$Ce_2Ni_{36}P_{15}$ has a complex crystal structure (Fig. 34), with Ce encapsulated as guest atom in cages of Ni and P. Electrical resistivity and specific heat show clear signs of correlation effects, with non-Fermi liquid temperature dependencies (ρ/T, $\gamma/-\ln(T)$).[27] However, due to a very small Kondo temperature and the metallic character of the system the thermoelectric performance is rather poor.

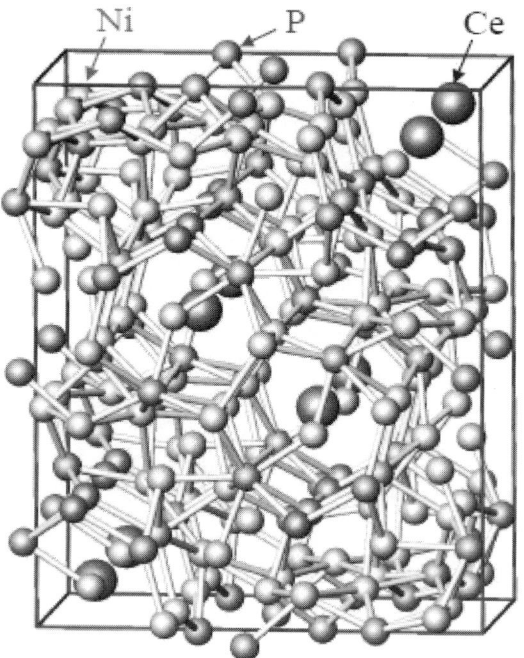

Fig. 34. Crystal structure of $Ce_2Ni_{36}P_{15}$.[27]

                    Silke Paschen

3.3.3. *The cage compound CeRu$_4$Sn$_6$*

In CeRu$_4$Sn$_6$ strong hybridization between the guest atom Ce and the host atoms leads to the opening of an energy gap of approximately 40 K. The peak in the thermopower at temperatures somewhat below, as measured on polycrystalline samples[28], is surprisingly low. One may speculate whether this is an extinction effect of thermopowers of different sign along different crystallograhpic directions. Measurements on single crystals are needed to clarify the situation. The thermal conductivity is not as strongly glass like as in some clathrates. This might be due to the smaller size of the cages.

3.3.4. *The cage compound Ce$_3$Pd$_{20}$Si$_6$*

Finally, the cage compound Ce3Pd20Si6 shall be discussed in some more detail. In the cubic crystal structure of space group Fm$\bar{3}$m, Ce is situated in two types of cages. The polyhedron 1 consists of 6 Si and 12 Pd atoms, polyhedron 2 of 16 Pd atoms (Fig. 35).

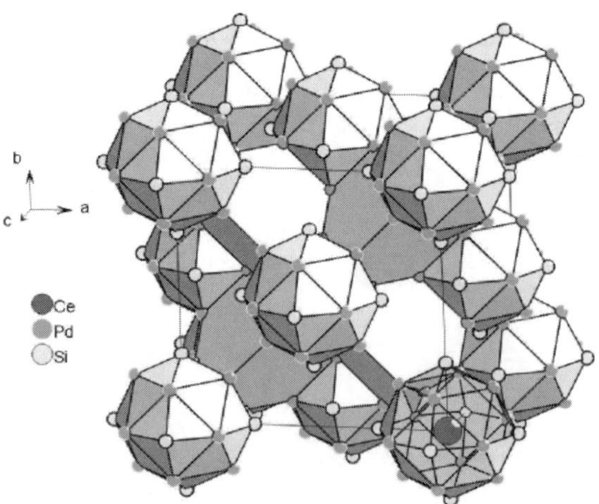

Fig. 35. Cubic crystal structure of Ce$_3$Pd$_{20}$Si6.[7]

The magnetic susceptibility is Curie-Weiss-like at high temperatures but tends to saturate to a high value at low temperature where, finally, a very weak signature of (presumably) magnetic ordering is observed

(Fig. 36), behaviour typically observed also in heavy fermion compounds. Also the electrical resistivity (not shown) with a T2 temperature dependence with a large prefactor at very low temperatures is typical for heavy fermion compounds.[29]

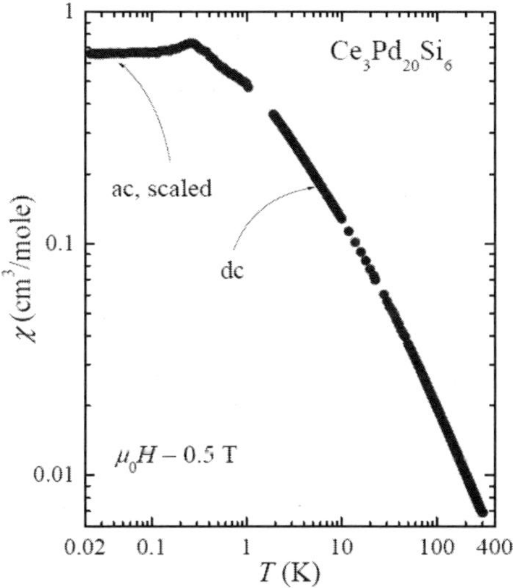

Fig. 36. Magnetic susceptibility vs temperature.[29] Similar to earlier findings.[30]

Thermopower does indeed show a relatively large value of about 45μV/K at somewhat above 20 K (Fig. 37).

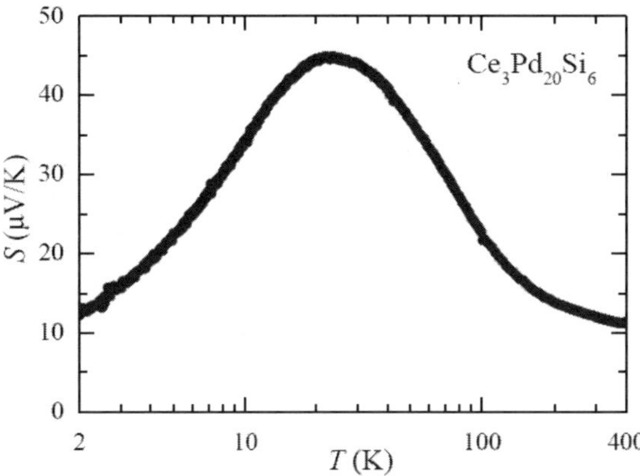

Fig. 37. Thermopower vs temperature of Ce$_3$Pd$_{20}$Si$_6$.[31]

However, similar to the case of $CeRu_4Sn_6$, also here the thermal conductivity (Fig. 38) does not show strongly pronounced glass-like features. At 20 K it has already reached a value of 3 W/Km (one third of which is due to electrons) which is distinctly higher than typically found in clathrates. Thus, also here the "rattling" of the guest atoms apprears less pronounced than in clathrates. This results in a relatively modest but nevertheless promising thermoelectric figure of merit, with a peak value of approximately 0.03 at 25 K (Fig. 39).

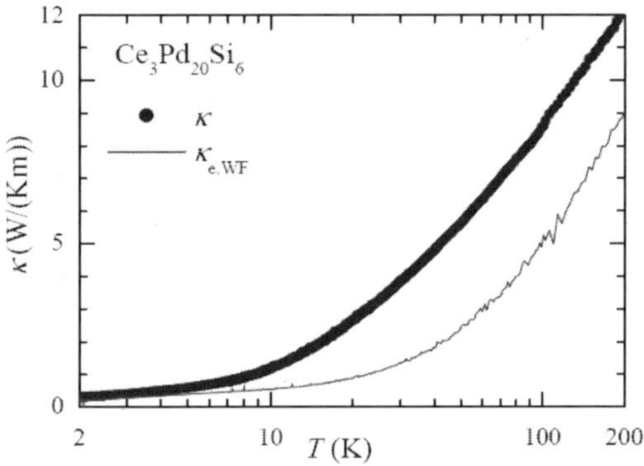

Fig. 38. Thermal conductivity vs temperature of $Ce_3Pd_{20}Si_6$. The electronic contribution as estimated from the Wiedemann-Franz law, $\kappa e,_{WF}$ is shown as solid line.[31]

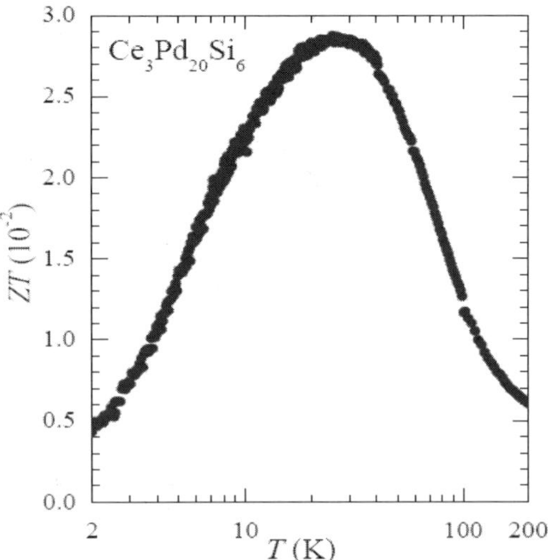

Fig. 39. Dimensionless figure of merit vs temperature of $Ce_3Pd_{20}Si_6$.[31]

Finally, Fig. 40 compares the figure of merit of $Ce_3Pd_{20}Si_6$ with other compounds (see legend). While ZT of $Ce_3Pd_{20}Si_6$ is highest below 40 K ZT clearly needs further improvement in the temperature range well below room temperature and the concept of strongly correlated cage compounds needs further exploration.

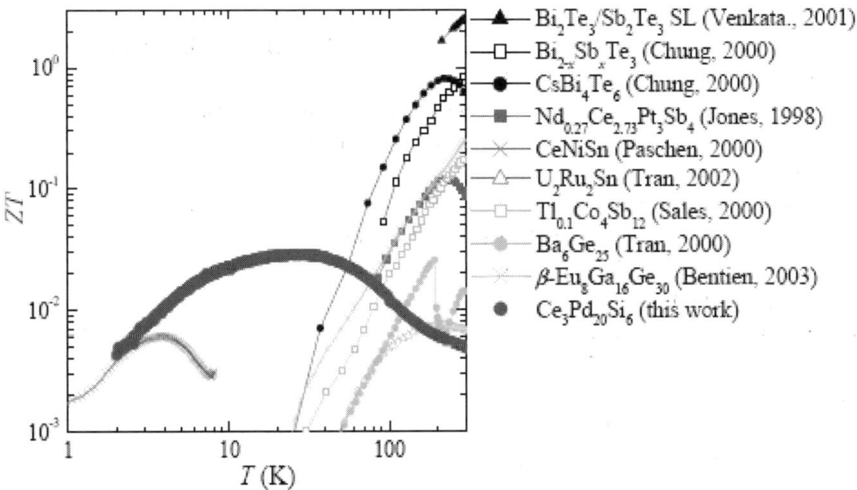

Fig. 40. Figure of merit: a comparison. Top three symbols (black): Bi-Te based materials; next three symbols (dark grey): Kondo insulators; next three (light grey): Cage compounds; lowermost data in legend (red): $Ce_3Pd_{20}S_{i6}$.

References

1. G. J. Snyder and E. S. Toberer, *Nature Materials* **7**, 105 (2008).
2. S. Paschen, W. Carrillo-Cabrera, A. Bentien, V. H. Tran, M. Baenitz, Y. Grin, and F. Steglich, *Phys. Rev.* **B 64**, 214404 (2001).
3. S. Bobev et al., *J. Solid State Chem.* **177**, 3545 (2004).
4. A. M. Guloy, R. Ramlau, Z. Tang, W. Schnelle, M. Baitinger and Y. Grin, *Nature* **443**, 320 (2006).
5. P. Rogl, in *Formation and crystal chemistry of clathrates*, in CRC Handbook of Thermoelectrics ,Ed. D. M. Rowe (CRC Press, Boca Raton, 2005), Chap. 32.
6. S. Budnyk, Ph.D. thesis, National University Lwiw, Ukraine; TU Dresden, Germany, 2003.
7. A. V. Gribanov, Y. D. Seropegin, and O. J. Bodak, *J. Alloys & Comp.* **204**, L9 (1994).

8. J. Kunes, T. Jeong and W. E. Pickett, *Phys. Rev.* **B 70**, 174510 (2004).

9. J. Choi, J. L. Musfeldt, Y. J. Wang, H.-J. Koo, M.-H. Whanghoo, J. Galy and P. Millet, *Chem. Mater.* **14**, 924 (2002).

10. B. C. Sales, B. C. Chakoumakos, R. Jin, J. R. Thompson, and D. Mandrus, *Phys. Rev.* **B 63** 245113 (2001).

11. N. Ogita, T. Kondo, T. Hasegawa, Y. Takasu, M. Udagawa, N. Takeda, K. Ishikawa, H. Sugawara, D. Kikuchi, H. Sato, C. Sekine and I. Shirotani, *Physica* **B 383**, 128 (2006).

12. I. Zerec, V. Keppens, M. A. McGuire, D. Mandrus, B. C. Sales and P. Thalmeir, *Phys. Rev. Lett.* **92**, 185502 (2004).

13. R. P. Hermann, V. Keepens, P. Bonville, G. S. Nolas, F. Grandjean, G. J. Long, H. M. Christen, B. C. Chakoumakos, B. C. Sales and D. Mandrus, *Phys. Rev. Lett.* **97**, 017401 (2006).

14. S. Paschen, V. Pacheco, A. Bentien, A. Sanchez, W. Carrillo-Cabrera, M. Baenitz, B. B. Iversen, Yu. Grin, and F. Steglich, *Physica* **B 328**, 39 (2003).

15. A. Bentien, V. Pacheco, S. Paschen, Yu. Grin, and F. Steglich, *Phy. Rev.* **B 71**, 165206 (2005).

16. V. Pacheco, A. Bentien, W. Carrillo-Cabrera, S. Paschen, F. Steglich, and Y. Grin, *Phy. Rev.* **B 71**, 165205 (2005).

17. V. Da Ros, B. Lenoir, A. Dauscher, C. Candolfi, C. Bellouard, C. Steve, E. Müller and J. Hejtmanekand, in *Proc. 22nd Int. Conf. on Thermoelectrics* (168, Vienna, Austria, ISBN 1-4244-0810-5, 2006).

18. J. X. Zhang, Q. M. Lu, X. Zhang and Q. Wei, in *Proc. 22nd Int. Conf. on Thermoelectrics* (168, Vienna, Austria, ISBN 1-4244-0810-5, 2006).

19. A. Schofield, *Contemporary Physics* **40**, 95 (1999).

20. S. Paschen, B. Wand, G. Sparn, F. Steglich, Y. Echizen, and T. Takabatake, *Phys. Rev.* **B 62**, 14912 (2000).

21. C. H. Park, Z. X. Shen, A. G. Loeser, D. S. Dessau, D. G. Mandrus, A. Migliori, J. Sarrau and Z. Fisk, *Phys. Rev.* **B 52**, 16981 (1995).

22. B. Buschinger, C. Geibel, F. Steglich, D. Mandrus, D. Young, J. L. Sarrao, and Z. Fisk, *Physica* **B 230-232**, 784 (1997).

23. S. Paschen, S. Budnyk, U. Kohler, Y. Prots, K. Hiebl, F. Steglich, and Y. Grin, *Physica* **B 383**, 89 (2006).

24. S. Paschen, C. Gspan, W. Grogger, M. Dienstleder, S. Laumann, P. Pongratz, H. Sassik, J. Wernisch, and A. Prokofiev, *J. Cryst. Growth* **310**, 1853 (2008).

25. T. Kawaguchi, K. Tanigaki, and M. Yasukawa, *Phys. Rev. Lett.* **85**, 3189 (2000).

26. V. Pacheco, W. Carrillo-Cabrera, V. H. Tran, S. Paschen, and Yu. Grin, *Phys. Rev. Lett.* **87**, 9601 (2001).

27. A. Bentien, Ph.D. thesis, TU Dresden, Germany, Shaker Verlag, Aachen, ISSN 0945-0963, 2005.

28. A. M. Strydom, Z. Guo, S. Paschen, R. Viennois, and F. Steglich, *Physica* **B 359–361**, 293 (2005).

29. S. Paschen, M. M̈uller, J. Custers, M. Kriegisch, A. Prokofiev, G. Hilscher, W. Steiner, A. Pikul, F. Steglich, and A. M. Strydom, *J. Magn. Magn. Mater.* **316**, 90 (2007).

30. J. Kitagawa, N. Takeda, and M. Ishikawa, Phys. Rev. **B 53**, 5101 (1996).

31. S. Paschen, A. Bentien, S. Budnyk, A. M. Strydom, Y. Grin, and F. Steglich, *in Proc. 22nd Int. Conf. on Thermoelectrics* (168, Vienna, Austria, ISBN 1-4244-0810-5, 2006).

CHAPTER 5

MAGNETISM OF COMPLEX METALLIC ALLOYS: CRYSTALLINE ELECTRIC FIELD EFFECTS

E. Bauer[1] and M. Rotter[2]

[1]*Institut für Festkörperphysik, Technische Universität Wien*
Wiednerhaupstrasse, 8-10, 1040, Vienna, Austria
[2]*Institut für Physikalische Chemie, Universität Wien*
Währingerstrasse, 42, 1090 Vienna, Austria
E-mail: bauer@ifp.tuwien.ac.at

One of the most important ingredients in the theory of magnetism are *crystalline electric field* (CEF) effects. Besides a deformation of the charge density of magnetic ions, CEF effects are responsible for a partial lifting of the ground state degeneracy associated with the total angular momentum of these ions. CEF effects are single ion effects, which, however, can cause dramatic modifications of physical properties when compared to those of free ions. We provide an introduction to the CEF concept for students at the undergraduate level, who have basic knowledge in quantum mechanics and solid state physics. Different point group symmetries are considered and it is shown, how CEF effects can be computed quantitatively. As a special example we use the *point-charge model* and discuss the CEF influence on the magnetic anisotropy, on inelastic neutron scattering spectra, on the magnetic contribution to the specific heat and magnetic entropy, thermal expansion, magnetostriction and transport properties, in particular on the electrical resistivity. Rare Earth elements are considered as specific examples of magnetic ions.

1. Some Aspects of the CEF Theory

In the following sections a brief overview on basic aspects of CEF theory is given and applications are mostly made in the context of Rare Earth ions which, in general, possess well defined localized magnetic moments as it is the case e.g., in filled skutterudites and in many other intermetallic compounds.

1.1. *Magnetic properties of free ions*

The Hamiltonian to describe magnetic atoms is given by

$$H_{tot} = H_0 + H_1 + H_{so} + H_{CEF} + H_{Ze} \tag{1}$$

The various contributions are described in the following. The most important part is given by the sum of the kinetic energy of each electron i and an effective radial potential $V(r_i)$.

$$H_0 = \sum_{i=1}^{N} \left(-\frac{\hbar^2}{2m} \Delta_i + V(r_i) \right) \tag{2}$$

Four quantum numbers n, l, m_l, m_s are satisfactory to describe the state of an electron in the atom ($n \dots$ principal quantum number, $l \dots$ orbital angular momentum quantum number, m_l, m_s are the respective magnetic quantum numbers).

- $l = 0, 1, 2, 3, \dots, (n-1)$
- $m_l = -l, (-l+1), \dots, 0, \dots, (l-1), l$
- $s = 1/2$
- $m_s = \pm 1/2$

The eigenstates of the Hamiltonian H_0 are fully antisymmetric linear combinations of single electron wave functions, which are usually written down in the form of Slater determinants.

The most important perturbation to the radial symmetric Hamiltonian comes from the electrostatic electron-electron interaction and is represented by H_1.

$$H_1 = \frac{1}{8\pi\epsilon_0} \sum_{i \neq j} \frac{e^2}{r_{ij}} + \frac{1}{4\pi\epsilon_0} \sum_{i=1}^{N} \left(-\frac{Ze^2}{r_i} - V(r_i) \right) \tag{3}$$

The second term in H_1 accounts for the difference of the nuclear electrostatic potential and the effective radial potential. In order to account for this interaction within a perturbation theory it is convenient to use linear combinations of Slater determinants, which are simultaneously eigenfunctions of H_0 and of the total orbital momentum operator $\vec{L} = \sum_i \vec{l}_i$ and of the total spin $\vec{S} = \sum_i \vec{s}_i$. The degeneracy of the ground state is partly lifted by considering the action of the electron-electron interaction H_1. The orbital and spin quantum numbers l and s of the ground state multiplet follow from Hund's first and second rule.[1] In the remaining part of this chapter, lower case letters denote the quantum numbers, while upper case letters refer to the corresponding quantum mechanical operators.

- 1st *rule of Hund:* The combination of electron spins $\vec{s}_i$ that gives the lowest energy (i.e. the most stable) is that with the highest value of $(2s + 1)$.
- 2nd *rule of Hund:* when the first rule has been satisfied there are in general several possible l values having the same value of $(2s + 1)$, that with the largest l will be the most stable.

Hund's rules have full theoretical justification only in limited cases, but there are very little doubts of their validity. The argument to support them is as follows.

- Two electrons in the same orbital with opposite spins must involve large electrostatic electron-electron repulsion because of the proximity of the two electrons. Energy is lowered if dual occupations are minimised, giving as many like electron spins as possible.
- Having satisfied the first condition, if the electrons orbit in the same sense electron-electron repulsive interactions are minimised because the electrons spend more of the time further apart.

Besides the kinetic- and the Coulomb energy, the spin-orbit coupling H_{so} and the CEF interaction H_{CEF} are of particular importance.

$$H_{so} = \sum_{i=1}^{N} \zeta(r_i)\vec{l}_i\vec{s}_i \tag{4}$$

$$H_{CEF} = -|e| \sum_{i=1}^{N} V_c(\vec{r}_i) \tag{5}$$

Elements of the $3d$ and the $4f$ group possess permanent magnetic moments:

- $3d$ - series $\ldots l = 2$: Fe, Co, Cr, Mn $\ldots$;
- $4f$ - series $\ldots l = 3$: Ce, Nd, $\ldots$, Tm, Yb

1.1.1. *3d-series*

While for $4f$ systems $H_{so} \gg H_{CEF}$, $3d$ systems behave opposite: $H_{so} \ll H_{CEF}$. The reason for this is (i) the fact that the $4f$ electron density is more concentrated in the inner part of the atom than that of the $3d$

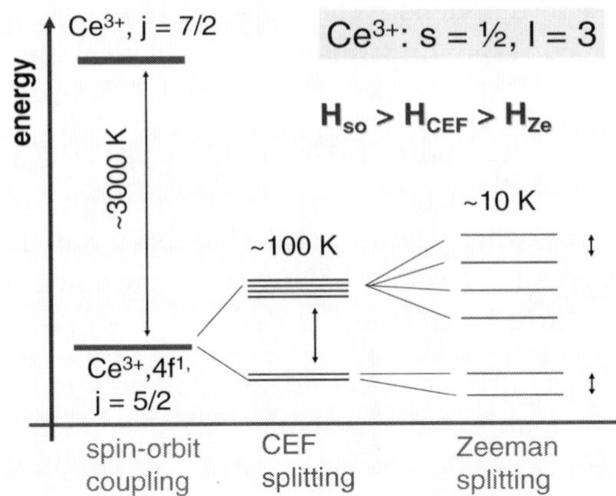

Fig. 1. General splitting scheme of a 6-fold degenerate system.

electrons and consequently the CEF experienced by the $4f$ electrons is screened by outer electrons. Furthermore (ii) the spin orbit interaction strength $\zeta(r_i)$ increases with the number of electrons and is large in the case of $4f$ electrons.

Therefore, to a good approximation magnetic ions of the $3d$ elements can be treated in the strong coupling limit, i.e. $H_{so} \ll H_{CEF}$. Usually the action of the CEF leads to a ground state with $l = 0$ (quenching of the orbital moment). This can be understood by the simple argument, that the strong CEF completely stops the orbital movement of the $3d$ electrons.

1.1.2. *4f-series*

In the following we will focus on the case of $4f$ elements, where $H_{so} \gg H_{CEF}$.[a] We now consider H_{so} as a perturbation to the ground state. It follows that m_l and m_s are no longer good quantum numbers because the spin orbit interaction does not commute with the components L_z and S_z. However, it commutes with $\vec{L}^2$ and $\vec{S}^2$ and all components of the total angular momentum $\vec{J} = \vec{L} + \vec{S}$. Therefore the multiplets of a free $4f$ ion can be characterized by the quantum numbers $|lsjm_j\rangle$. Figure 1 gives an overview on the relevant energy scales of the problem.

---

[a]Note that for most Rare Earth elements this is a good approximation. Exceptions are the Sm^{3+} and the Eu^{3+} ion with a spin orbit splitting of less than 100 meV.

The practical calculation of the total angular momentum follows then from Hund's third rule.

– *3rd rule of Hund:* The most stable configurations result from

- $j = |l - s|$ ... less than half filled shells
- $j = |l + s|$ ... more than half filled shells
- $j = s$ ... half filled shell ($l = 0$).

The corresponding magnetic moments are

$$|\vec{\mu}_L| = \mu_B |\vec{L}| = \mu_B \sqrt{l(l+1)}, \tag{6}$$

$$|\vec{\mu}_s| = g_S \mu_B |\vec{S}| = 2\mu_B \sqrt{s(s+1)}, \tag{7}$$

where μ_B is the Bohr magneton. In a classical view, the spin-orbit interaction leads to a precession of the resultant vectors $\vec{L}$ and $\vec{S}$ about their vector sum $\vec{J}$. To account for the different weight of the spin and the orbital contribution to the total angular momentum, the Landé factor g_j is introduced. In quantum mechanical terms the Landé factor follows from the application of the Wigner Eckhardt theorem to the ground state matrix elements of the total magnetic moment, which is given by:

$$\vec{\mu} = \vec{\mu}_L + \vec{\mu}_S = \mu_B(\vec{L} + 2\vec{S}) \tag{8}$$

According to the Wigner Eckhardt theorem within the ground state multiplet $|lsj, m_j = -j, \ldots, +j >$ the matrix elements of any pair of vectors (such as for instance the magnetic spin moment and the total angular momentum) are proportional to each other. Applying this relation to the total magnetic moment and the total angular momentum yields

$$|\vec{\mu}| = g_j \mu_B \sqrt{j(j+1)} \tag{9}$$

with the Landé factor

$$g_j = 1 + \frac{j(j+1) + s(s+1) - l(l+1)}{2j(j+1)} \tag{10}$$

Borderline cases: $g_j = 2$ for $l = 0$ and $g_j = 1$ for $s = 0$. This defines the magnetic moment of a single atom. Assuming for the moment, that the CEF interaction is zero, the magnetic properties of free ions can be discussed.

In the absence of external magnetic fields, all the atoms with equal magnetic moments have equal energy. If an external magnetic field $\vec{B}$ is applied, this degeneracy is lifted by the action of H_{Ze}.

$$H_{Ze} = -\vec{\mu}\vec{B} \tag{11}$$

Provided that the field is applied along the z-axis, the different eigenstates can be characterized by the magnetic quantum number m_j, the corresponding energies are

$$E_{m_j} = -\langle lsjm_j|\vec{\mu}|lsjm_j\rangle \cdot \vec{B} = -g_j\mu_B m_j B \qquad (12)$$

Then the energy depends on the thermal occupation of the $2j + 1$ sublevels, all of which differ in their m_j value.

The magnetic sub-levels are occupied according to the laws of statistical mechanics, the lowest energy having the largest population. The relative population P of a sub-level m_j is given by

$$P(m_j) = \frac{\exp(-E_{m_j}/k_B T)}{\sum \exp(-E_{m_j}/k_B T)}. \qquad (13)$$

1.2. *The CEF Hamiltonian for Rare Earth elements*

Similar to a magnetic field, an electric field is able to lift, at least partially, the $(2j + 1)$-fold degenerate ground state. Such an electric field may be created by the ions surrounding regularly a Rare Earth ion in a crystal. According to Kramers theorem,[b] there remains a certain degeneracy of the states:

- If the the electron number is odd, then the total angular momentum j is half-integer, i.e, $j = 5/2,\, 7/2,\, 9/2\ldots$, the minimum degeneracy is 2; such a state is a *doublet*. These systems are so called *Kramers ions*.
- If the the electron number is even, then the total angular momentum j is integer, i.e, $j = 4,\, 5,\, 6\ldots$, the degeneracy can be totally lifted. But in general, there is a mixture of singlet, doublet, triplet and quartet states. These systems are so called *non-Kramers ions*.

The particular scheme of a system is determined from material related properties such as charges, but also depends on both the total angular momentum and the crystal symmetry. A lower symmetry causes lower degeneracies of levels. Quartets may be found in cubic structures only.

Assume that a $4f$-ion is situated in a potential V_c, which is created from the neighbouring ions (compare Fig. 2). This potential acts in a similar way as the *Stark-effect* and causes, at least, a partial lifting of the $(2j + 1)$-fold ground state degeneracy of the free $4f$ ion. If we further assume that the $4f$ wave functions do not overlap with that of the neighbouring ions, then

---

[b]Kramers theorem follows from time inversion symmetry.[10]

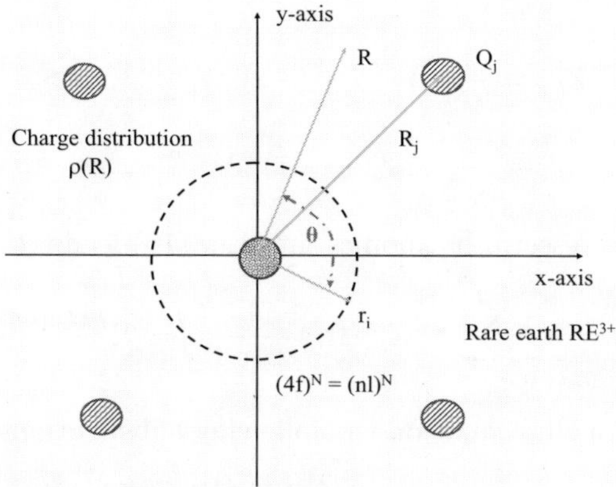

Fig. 2.   Geometric relations for the determination of the CEF.

the potential $V_c(\vec{r})$ fulfils Laplace's equation, i.e., $\nabla^2 V_c = 0$ and for the CEF contribution the following relation is valid:

$$V_c(\vec{r}) = \frac{1}{4\pi\epsilon_0} \int \frac{\rho(\vec{R})}{|\vec{r} - \vec{R}|} d^3R \tag{14}$$

with a charge density $\rho(\vec{R})$ and the series expansion

$$\frac{1}{|\vec{r} - \vec{R}|} = \frac{1}{\sqrt{(r^2 + R^2 - 2rR\cos\theta)}} = \frac{1}{R}\sum_{l=0}^{\infty}\left(\frac{r}{R}\right)^l P_l(\cos\theta) \tag{15}$$

for all $r < R$. The Legendre-polynoms $P_l(\cos\theta)$ can be expressed in terms of spherical harmonics Y_l^m ....

$$P_l(\cos\theta) = \frac{4\pi}{2l+1}\sum_{m=-l}^{l} Y_l^m(\Omega_r)Y_l^{m*}(\Omega_R). \tag{16}$$

If it is assumed that the ions $i = 1, 2, 3, \ldots$ carry point-charges q_i (*point-charge model, PCM*) we have

$$V_c(\vec{r}) = \frac{1}{4\pi\epsilon_0}\sum_i \frac{q_i}{|\vec{r} - \vec{R}_i|} \tag{17}$$

In the following we will not use the expansion Eq. (16) but instead of spherical harmonics use tesseral harmonics, which are real and defined according to Ref. 2[c] and listed explicitly in Appendix B.

---

[c]Note that there are different definitions of tesseral harmonic functions in the literature, we use here the definition given in Ref. 2.

$$Z_l^0 = Y_l^0$$

$$Z_l^m = \frac{1}{\sqrt{2}} \left[Y_l^{-m} + (-1)^m Y_l^m \right] \ldots m > 0 \tag{18}$$

$$Z_l^m = \frac{i}{\sqrt{2}} \left[Y_l^m + (-1)^m Y_l^{-m} \right] \ldots m < 0$$

Inserting the tesseral harmonics into Eq. (16) leads to

$$V_c(\vec{r}) = \sum_{l=0}^{\infty} \sum_{m=-l}^{l} r^l Z_l^m(\Omega_r) \frac{1}{2l+1} \int d^3R \frac{\rho(\vec{R}) Z_l^m(\Omega_R)}{\epsilon_0 R^{l+1}} \tag{19}$$

If on the r.h.s. of this equation the following substitutions[3,4]

$$\gamma_l^m = \frac{1}{2l+1} \int d^3R \frac{\rho(\vec{R}) Z_l^m(\Omega_R)}{\epsilon_0 R^{l+1}} \tag{20}$$

and

$$H_l^m = \sum_{i=1}^{N} r_i^l Z_l^m(\delta_i, \phi_i) \tag{21}$$

are made, the CEF Hamiltonian can be expressed as

$$H_{CEF} = -|e| \sum_{i=1}^{N} V_c(\vec{r}_i) = \sum_{l,m} \gamma_l^m H_l^m. \tag{22}$$

For a given multiplet, the matrix elements of H_{CEF} are proportional to those of *equivalent operators*, which contain J_z or $J_\pm$ (*Wigner-Eckhart-theorem*).

Example:

$$H_2^0 = \sum_i r_i^2 \sqrt{\frac{5}{4\pi}} \frac{1}{2} \left(3\cos^2 \delta_i - 1 \right) = \sum_i r_i^2 \sqrt{\frac{5}{4\pi}} \frac{1}{2} \left(\frac{3z_i^2 - r_i^2}{r_i^2} \right) \tag{23}$$

Here we have made the following assignments (spherical-polar coordinates):

$$x = r \sin \delta \cos \phi \rightarrow J_x$$

$$y = r \sin \delta \sin \phi \rightarrow J_y$$

$$z = r \cos \delta \rightarrow J_z \qquad r \rightarrow |\vec{J}|$$

as well as

$$xy \rightarrow (1/2)(J_x J_y + J_y J_x)$$

and

$$J_\pm = J_x \pm i J_y$$

For example, the matrix elements of the CEF Hamiltonian H_2^0 within the ground state multiplet $|lsjm_j\rangle$ of a Rare Earth ion can be written as

$$\langle lsjm_j|H_2^0|lsjm_j'\rangle = \langle lsjm_j|\sum_i r_i^2\sqrt{\frac{5}{4\pi}}\frac{1}{2}\left(\frac{3z_i^2 - r_i^2}{r_i^2}\right)|lsjm_j'\rangle \tag{24}$$

$$= \sqrt{\frac{5}{4\pi}}\frac{1}{2}\alpha_j\langle r_{4f}^2\rangle\langle lsjm_j|\left(3J_z^2 - j(j+1)\right)|lsjm_j'\rangle$$

with the radial matrix elements defined by

$$\langle r^l\rangle = \int |R_{4f}(r)|^2 r^{l+2}dr \tag{25}$$

and Stevens factors α_j (see Table 1). These matrix elements have been calculated for different Rare Earth atoms using the $4f$ radial wave functions $R_{4f}(r)$ as determined by the Hartree Fock method or similar, see e.g. Refs. 5 and 6.

By defining Stevens operators O_l^m according to a few rules,[7] e.g.,

$$O_2^0 - \left(3J_z^2 - j(j+1)\right) \tag{26}$$

the operators which initially were spatial dependent are now simply proportional to angular momentum operators. For each tesseral harmonic function a Stevens operator can be defined by substituting x, y and z by the components J_x, J_y and J_z, respectively and symmetrizing the resulting expression. A complete list of these Stevens operators is given in Appendix A. In addition, the coefficients of the polynoms in x,y and z in the definition of the tesseral functions (see Appendix B) are denoted by p_l^m, e.g.

$$p_2^0 = \sqrt{\frac{5}{4\pi}}\frac{1}{2} \tag{27}$$

Dropping for simplicity the $\langle lsjm_j|$ in the notation and keeping in mind, that the following equations hold only for the matrix elements of the ground state multiplet of the Rare Earth ion this yields

$$H_2^0 = \sum_i r_i^2\sqrt{\frac{5}{4\pi}}\frac{1}{2}\left(\frac{3z_i^2 - r_i^2}{r_i^2}\right) \tag{28}$$

$$\rightarrow p_2^0\alpha_j\langle r^2\rangle O_2^0(\vec{J}) \tag{29}$$

and in general

$$H_l^m = \sum_i r_i^l Z_l^m(\Omega_i) = p_l^m\Theta_l\langle r^l\rangle O_l^m(\vec{J}) \tag{30}$$

Table 1. Orbital (l), spin (s) and total angular momentum (j) quantum numbers, Landé factor g_j and Stevens factors $\alpha_j, \beta_j, \gamma_j$ for Rare Earth and actinide ions.

ion	l	s	j	g_j	$\alpha_j \times 10^2$	$\beta_j \times 10^4$	$\gamma_j \times 10^6$
Ce^{3+}	3	1/2	5/2	6/7	-5.7143	63.4921	0.0000
Pr^{3+},Ce^{2+},U^{4+}	5	1	4	4/5	-2.1010	-7.3462	60.9940
$Nd^{3+},Pr^{2+},U^{3+},Np^{4+}$	6	3/2	9/2	8/11	-0.6428	-2.9111	-37.9880
Pm^{3+},Nd^{2+}	6	2	4	3/5	0.7713	4.0755	60.7807
Sm^{3+},Pm^{2+}	5	5/2	5/2	2/7	4.1270	25.0120	0.0000
Eu^{3+},Sm^{2+}	3	3	0	$-$	0.0000	0.0000	0.0000
Gd^{3+},Eu^{2+}	0	7/2	7/2	2	0.0000	0.0000	0.0000
Tb^{3+},Gd^{2+}	3	3	6	3/2	-1.0101	1.2244	-1.1212
Dy^{3+},Tb^{2+}	5	5/2	15/2	4/3	-0.6349	-0.5920	1.0350
Ho^{3+},Dy^{2+}	6	2	8	5/4	-0.2222	-0.3330	-1.2937
Er^{3+},Ho^{2+}	6	3/2	15/2	6/5	0.2540	0.4440	2.0699
Tm^{3+},Er^{2+}	5	1	6	7/6	1.0101	1.6325	-5.6061
Yb^{3+},Tm^{2+}	3	1/2	7/2	8/7	3.1746	-17.3160	148.0001

This leads to the famous notation

$$H_{CEF} = \sum_{l,m} B_l^m O_l^m \tag{31}$$

with

$$B_l^m = -|e| p_l^m \gamma_l^m \langle r_{4f}^l \rangle \Theta_l \tag{32}$$

$$\Theta_l = \begin{cases} \alpha_j \ \dots \ l = 2 \\ \beta_j \ \dots \ l = 4 \\ \gamma_j \ \dots \ l = 6 \end{cases} \tag{33}$$

The B_l^m or γ_l^m are the CEF parameters, which can be determined from the experiment, or can be calculated from different models among them the PCM. The additional parameters and coefficients are already tabulated and given in Tables 1 and 2.

1.3. *Symmetry considerations*

1.3.1. *Disappearance of terms with $l \neq 2, 4, 6$*

A major simplification of the theory is due to the fact that for $4f$ electrons all terms with $l > 6$ disappear in Eq. (31). This can be seen by noting that the $4f$ wave functions $\langle lsjm_j|$ are a linear combination of Slater determinants of single particle wave functions of the form $\Phi_{4f}^{mm_s}(\vec{r}) =$

Table 2. List of radial matrix elements $\langle r^l \rangle$ for l = 2, 4 and 6 for Rare Earth and actinide ions[4,6,8]

ion	$\langle r^2 \rangle (\text{Å}^2)$	$\langle r^4 \rangle (\text{Å}^4)$	$\langle r^6 \rangle (\text{Å}^6)$
Ce^{3+}	0.3666	0.3108	0.5119
Pr^{3+}	0.3350	0.2614	0.4030
Nd^{3+}	0.3120	0.2282	0.3300
Pm^{3+}	0.2899	0.1991	0.2755
Sm^{3+}	0.2728	0.1772	0.2317
Eu^{3+}	0.2569	0.1584	0.1985
Gd^{3+}	0.2428	0.1427	0.1720
Tb^{3+}	0.2302	0.1295	0.1505
Dy^{3+}	0.2188	0.1180	0.1328
Ho^{3+}	0.2085	0.1081	0.1181
Er^{3+}	0.1991	0.0996	0.1058
Tm^{3+}	0.1905	0.0921	0.0953
Yb^{3+}	0.1826	0.0854	0.0863
U^{4+}	0.5718	0.5985	1.0491
U^{3+}	0.6569	0.8552	1.9882
Np^{4+}	0.5276	0.5100	0.8300
Nd^{2+}	0.3898	0.4191	0.9980
Sm^{2+}	0.3352	0.3028	0.6271
Eu^{2+}	0.3075	0.2641	0.5178
Gd^{2+}	0.2879	0.2333	0.4359
Tb^{2+}	0.2711	0.2082	0.3729
Dy^{2+}	0.2557	0.1875	0.3235
Ho^{2+}	0.2425	0.1701	0.2837
Er^{2+}	0.2307	0.1552	0.2514
Tm^{2+}	0.2198	0.1426	0.2249
Yb^{2+}	0.2100	0.1315	0.2027

$R_{4f}(r) Y_3^m(\Omega) \chi_i^{m_s}$. Therefore, the matrix elements of $r^l Z_l^m(\Omega)$ in Eq. (30) are a linear combination of integrals of the form

$$\int \Phi_{4f}^{m m_s *}(\vec{r}) r^l Y_l^{m''}(\Omega) \Phi_{4f}^{m' m'_s}(\vec{r}) r^2 dr d\Omega$$

$$= \langle r_{4f}^l \rangle \delta_{m_s m'_s} \int (-1)^m Y_3^{-m}(\Omega) Y_l^{m''}(\Omega) Y_3^{m'}(\Omega) d\Omega \qquad (34)$$

From elementary quantum mechanics it is known, that the spherical harmonics Y_l^m are eigenfunctions of the angular momentum operator. The addition of angular momentum is done according to the Clebsch Gordon expansion. From this Clebsch Gordan expansion the spherical harmonic addition relation can be derived (for a proof see for example,[9] page 1046,

Eq. (21)), which for the special case of $l = 3$ can be written as

$$Y_3^{-m}(\Omega)Y_3^{m'}(\Omega)$$

$$= \sum_{l=0}^{6} \sum_{m''=-l}^{l} \frac{7}{\sqrt{4\pi(2l+1)}} \langle 3,3;0,0|l,0\rangle \langle 3,3;-m,m'|l,m'' > Y_l^{m''}(\Omega)$$

$$(35)$$

Using this expansion the product on the right side of Eq. (34) can be written as a linear combination of products of two spherical harmonics, where l takes the values $l = 0,\ldots,6$. Moreover, keeping in mind that the Clebsch-Gordon coefficients fulfil the symmetry relation $\langle l_1, l_2; -m_1, -m_2|l,m\rangle = (-1)^{l_1+l_2+l}\langle l_1,l_2;m_1,m_2|l,m\rangle$ we note that the coefficient $\langle 3,3;0,0|l,0\rangle$ is zero for odd l. Because of the orthogonality of the spherical harmonics all terms with $l \neq 2,4,6$ disappear in Eq. (34) and consequently also in Eq. (31).

1.3.2. *Disappearance of CEF parameters due to Point Symmetry of the CEF*

If coordinate axes are properly chosen the local point symmetry at the Rare Earth site further limits the number of nonzero CEF parameters. In order to discuss this issue we refer to the definition of the coefficients γ_l^m in Eq. (20).

– *The z-axis is an p-fold axis of symmetry* keeping in mind that the tesseral harmonics are a sum of spherical harmonics and the ϕ dependence of these is given by $e^{im\phi}$ such a symmetry axis allows to split the integral in Eq. (20) into a sum of p integrals. These only differ by the factor $\exp(im2\pi/p)$. Consequently, the coefficients γ_l^m are proportional to the sum $\sum_{s=0}^{p-1} \exp(im2\pi s/p)$, which is a geometrical sum $\sum_{s=0}^{p-1} x^s = (1 - x^p)/(1-x))$. Accordingly, *the γ_l^m vanish unless m is an integer multiple of p* .

– *The y-axis is a twofold axis of symmetry* such a twofold symmetry axis is described in polar coordinates by the transformation $\Theta \to \pi - \Theta$ and $\phi \to -\pi - \phi$. Writing the integral in Eq. (20) as a sum of two integrals and applying this transformation and using the relations $Y_l^m(\Theta,-\phi) = (-1)^m Y_l^m(\Theta,\phi)$, $Y_l^m(\pi - \Theta, \phi + \pi) = (-1)^l Y_l^m(\Theta,\phi)$ leads to a factor $1 + (-1)^{l+m}$. Consequently, *all γ_l^m with $l + m$ odd vanish.*

Such symmetry considerations allow to compile Table 3, where the non-vanishing CEF parameters for all crystallographic point groups are given.

Table 3. Possible local symmetries and corresponding non zero CEF parameters. In the case when $m > 0$ and both parameters B_l^m and B_l^{-m} are nonzero, one of these B_l^m with $m > 0$ can by made zero by a rotation of the coordinate system. However, the appropriate orientation of the coordinate system in these cases is not known *a priori*. It requires the knowledge of the CEF parameters. Note, that for cubic symmetry additionally $B_4^4 = \frac{5}{2} B_4^0$, $B_6^4 = -\frac{21}{2} B_6^0$ and $B_6^2 = -B_6^6$. For the icosahedral case (not compatible with translational symmetry) $B_6^5 = -42 B_6^0$.[11]

symmetry		B_2^0	$B_2^{\pm 1}$	$B_2^{\pm 2}$	B_4^0	$B_4^{\pm 1}$	$B_4^{\pm 2}$	$B_4^{\pm 3}$	$B_4^{\pm 4}$	B_6^0	$B_6^{\pm 1}$	$B_6^{\pm 2}$	$B_6^{\pm 3}$	$B_6^{\pm 4}$	$B_6^{\pm 5}$	$B_6^{\pm 6}$
triclinic	C_i C_1	$+$	$\pm$	$\pm$	$+$	$\pm$	$\pm$	$\pm$	$\pm$	$+$	$\pm$	$\pm$	$\pm$	$\pm$	$\pm$	$\pm$
monoclinic	C_2 C_s C_{2h}	$+$		$\pm$	$+$		$\pm$		$\pm$	$+$		$\pm$		$\pm$		$\pm$
rhombic	C_{2v} D_2 D_{2h}	$+$		$+$	$+$		$+$		$+$	$+$		$+$		$+$		$+$
tetragonal	C_4 S_4 C_{4h}	$+$			$+$				$\pm$	$+$				$\pm$		
tetragonal	D_4 C_{4v} D_{2d} D_{4h}	$+$			$+$				$+$	$+$				$+$		
trigonal	C_3 S_6	$+$			$+$			$\pm$		$+$			$\pm$			$\pm$
trigonal	D_3 C_{3v} D_{3d}	$+$			$+$			$+$		$+$			$+$			$+$
hexagonal	C_6 C_{3h} C_{6h}	$+$			$+$					$+$						$\pm$
hexagonal	D_6 C_{6v} D_{3h} D_{6h}	$+$			$+$					$+$						$+$
cubic	TT_h				$+$				$+$	$+$		$+$		$+$		$+$
cubic	T_d O O_h				$+$				$+$	$+$				$+$		
icosahedral	I_h									$+$					$+$	

1.4. *Calculation of CEF splitting and 4f charge density*

In the framework of perturbation theory diagonalization of the CEF Hamiltonian leads to the energies and the eigenstates created by CEF splitting. The following eigenvalue problem has to be solved within the ground state multiplet manifold.

$$H_{CEF}|\Psi\rangle = E|\Psi\rangle \tag{36}$$

An arbitrary state of the ground state manifold can be expressed as a linear combination of the basis vectors $|lsjm_j\rangle$, i.e. $|\Psi\rangle = \sum_{m_j=-j}^{j} a_{m_j}^{\Psi}|lsjm_j\rangle$. Solving the eigenvalue problem Eq. (36) results in eigenstates $|\Gamma_i\rangle = \sum_{m_j=-j}^{j} a_{m_j}^{\Gamma_i}|lsjm_j\rangle$ and corresponding eigenvalues E_i. The notation $|\Gamma_i\rangle$ comes from group theory: it can be shown that CEF eigenstates always transform under the symmetry operations of the point group according to some irreducible representation Γ_i[10] and consequently are labelled alike.

The charge density operator of a $4f$ ion is given by elementary quantum mechanics as

$$\rho_{4f}(\vec{r}) = \sum_i -|e|\delta(\vec{r} - \vec{r}_i) \tag{37}$$

Here the index i runs from 1 to the number of $4f$ electrons. Using spherical coordinates the δ-function in Eq. (37) can be rewritten

$$\delta(\vec{r} - \vec{r}_i) = \frac{1}{r^2}\delta(r - r_i)\delta(\Omega - \Omega_i) \tag{38}$$

$$\delta(\Omega - \Omega_i) = \frac{1}{4\pi}\sum_l^{\infty}(2l + 1)P_l(cos\theta) \tag{39}$$

Substituting the Legendre Polynoms in Eq. (38) by spherical harmonics using the relation Eq. (16) we notice that the charge density operator can be written as a linear combination of products of spherical harmonics:

$$\rho_{4f}(\vec{r}) = -\frac{4\pi|e|}{r^2}\sum_i \delta(r - r_i)\sum_{l=0}^{\infty}\sum_{m=-l}^{l} Y_l^m(\Omega)Y_l^{m*}(\Omega_i) \tag{40}$$

In the calculation of the expectation value of the charge density operator for any $4f$ state the radial integrals yield the square of the radial part of the $4f$ wave function $|R_{4f}(r)|^2$:

$$\langle\rho_{4f}(\vec{r})\rangle = -4\pi|e||R_{4f}(r)|^2\sum_{l=0}^{\infty}\sum_{m=-l}^{l} Y_l^m(\Omega)\langle\sum_i Y_l^{m*}(\Omega_i)\rangle \tag{41}$$

Keeping in mind, that the spherical harmonics are linear combinations of the tesseral harmonics according to their definition in Eq. (18) the expectation value of the $4f$ charge density operator can be written as a linear combination of tesseral harmonics.

$$\langle \rho_{4f}(\vec{r}) \rangle = |R_{4f}(r)|^2 \sum_{lm} \sigma_l^m Z_l^m(\Omega) \tag{42}$$

According to Eq. (41) the coefficients σ_l^m are proportional to $\langle \sum_i Z_l^m(\Omega_i) \rangle$ and can be evaluated by applying the operator equivalent method.

In order to do this we consider the product $p_l^m \Theta_l \langle r_{4f}^l \rangle \langle O_l^m \rangle$, which according to Eq. (30) is equivalent to

$$p_l^m \Theta_l \langle r_{4f}^l \rangle \langle O_l^m \rangle = \langle \sum_i Z_l^m(\Omega_i) r_i^l \rangle \tag{43}$$

We now use the definition of Eq. (37) to proof that

$$\langle \sum_i Z_l^m(\Omega_i) r_i^l \rangle = \frac{-1}{|e|} \int \langle \rho(\vec{r}) \rangle Z_l^m(\Omega) r^l d^3r \tag{44}$$

Inserting Eq. (42) for the charge density operator on the right side of Eq. (44), and integrating keeping in mind the orthogonality of the harmonic tesseral functions leads to

$$\frac{-1}{|e|} \int \langle \rho(\vec{r}) \rangle Z_l^m(\Omega) r^l d^3r = \frac{-1}{|e|} \langle r_{4f}^l \rangle \sigma_l^m \tag{45}$$

combining Eqs. 43 and 45 we get

$$\sigma_l^m = -|e| p_l^m \Theta_l \langle O_l^m \rangle \tag{46}$$

This yields an expression for the charge density, which may be evaluated for each CEF state $|\Gamma_i\rangle$.

$$\langle \Gamma_i | \rho_{4f}(\vec{r}) | \Gamma_i \rangle = -|e| |R_{4f}(r)|^2 \sum_l^m p_l^m \Theta_l \langle \Gamma_i | O_l^m | \Gamma_i \rangle Z_l^m(\Omega) \tag{47}$$

Multiplying Eq. (47) by Boltzmann factors and summing over the different CEF states, the thermal expectation value for the charge density at any temperature can be calculated. In order to visualize the charge density, usually surfaces of constant charge density are generated and plotted.

1.5. *Ce^{3+} in cubic and hexagonal symmetries*

In the case of cubic point group symmetries T_d, O or O_h , the appropriate CEF Hamiltonian is given by

$$H_{CEF}^{cub.} = B_4^0 \left(O_4^0 + 5O_4^4 \right) + B_6^0 \left(O_6^0 - 21O_6^6 \right) \tag{48}$$

The CEF is fully determined by 2 parameters B_4^0 and B_6^0. For Ce^{3+} we have additionally $\gamma_j = 0$; i.e., in this case B_4^0 is the *only* relevant parameter.

The resulting eigenstates are a Γ_7-doublet and a Γ_8-quartet

$$\begin{aligned}
|\Gamma_7\rangle &= a|\pm 5/2\rangle - b|\mp 3/2\rangle \\
|\Gamma_8\rangle &= \begin{cases} |\pm 1/2\rangle \\ b|\pm 5/2\rangle + a|\mp 3/2\rangle \end{cases}
\end{aligned} \tag{49}$$

which can be represented in terms of J_z with $a = (1/6)^{1/2}$ and $b = (5/6)^{1/2}$. For a Ce^{3+} ion it follows immediately

$$B_4^0 = \Delta/360 \tag{50}$$

There, Δ is the energy difference between Γ_7 and Γ_8. In Fig. 3 is displayed the $4f$ charge density of a Ce^{3+} ion for cubic symmetry.

The CEF Hamiltonian for the hexagonal point groups D_6, C_{6v}, D_{3h} and D_{6h} is written as

$$H_{CEF}^{hexa} = B_2^0 O_2^0 + B_4^0 0_4^0 + B_6^0 O_6^0 + B_6^6 O_6^6 \tag{51}$$

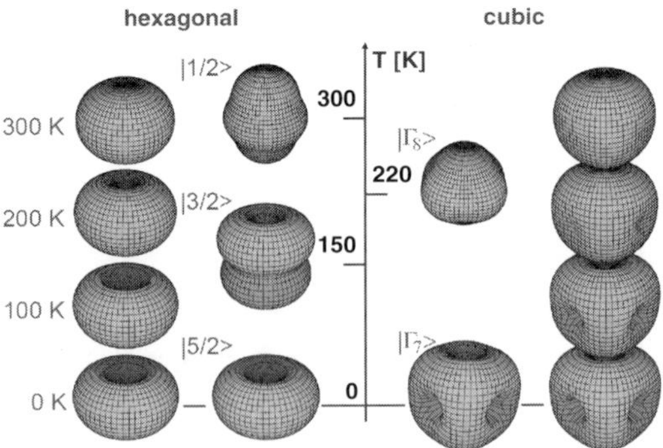

Fig. 3. Contours of the $4f$ charge density of a Ce^{3+} ion in hexagonal (left panel) and in cubic (right panel) symmetry. The middle columns represent the charge densities associated with the $|5/2\rangle$, $|3/2\rangle$, $|1/2\rangle$, as well as the $|\Gamma_7\rangle$ and $|\Gamma_8\rangle$ CEF eigenstates. Both outer columns illustrate the temperature dependent evolution of the charge density.

Again, 6-th order term vanishes in the case of cerium systems; therefore,

$$^{\text{Ce}}H_{CEF}^{hexa} = B_2^0 O_2^0 + B_4^0 0_4^0 \tag{52}$$

and the CEF in hexagonal symmetry yields "pure" eigenstates, determined from J_z. If we assume that

$$|\Gamma_7\rangle = |\pm 1/2\rangle \tag{53}$$

$$|\Gamma_8\rangle = |\pm 5/2\rangle \tag{54}$$

$$|\Gamma_9\rangle = |\pm 3/2\rangle \tag{55}$$

the energies of a CEF doublet are calculated by the application of

$$H_{CEF}|\Gamma_i\rangle = E_i|\Gamma_i\rangle \tag{56}$$

i.e.,

$$E_7 = -8B_2^0 + 120B_4^0 \tag{57}$$

$$E_8 = 10B_2^0 + 60B_4^0 \tag{58}$$

$$E_9 = -2B_2^0 - 180B_4^0. \tag{59}$$

The charge density distribution of a Ce^{3+} ion created from the CEF is also shown in Fig. 3.

1.6. *Example: an Yb ion in a hexagonal CEF*

Recall that the CEF Hamiltonian for hexagonal systems reads (see Table 3 and Eq. (51)):

$$H_{CEF}^{hexa} = B_2^0 O_2^0 + B_4^0 O_4^0 + B_6^0 O_6^0 + B_6^6 O_6^6 \tag{60}$$

where the Stevens operators are given by (see the appendix):

$$O_2^0 = [3J_z^2 - X] \quad \text{with} \quad X = j(j+1)$$

$$O_4^0 = [35J_z^4 - (30X - 25)J_z^2 + 3X^2 - 6X]$$

$$O_6^0 = [231J_z^6 - (315X - 735)J_z^4 + (105X^2 - 525X + 294)J_z^2 - 5X^3 + 40X^2 - 60X]$$

$$O_6^6 = \frac{1}{2}[J_+^6 + J_-^6].$$

In the subsequent section we outline in a step-by-step path all details to evaluate the eigenvalues and the eigenstates of the CEF levels of a typical compound based on the Rare Earth element Yb, crystallizing in a hexagonal crystal structure. Moreover, we assume that the electronic environment can be composed from a number of point-charges. Yb should be in its trivalent state, i.e, Yb^{3+}. This state can be associated with electronic configuration $4f^{13}$ revealing a total angular momentum $j = 7/2$ and an effective magnetic moment $\mu_{eff} = 4.54\ \mu_B$ (see Table 1).

Writing for simplicity m instead of m_j, the following notations and rules apply for practical calculations:

$$J_+ \equiv J_x + iJ_y \qquad J_- \equiv J_x - iJ_y$$

$$J_+|jm\rangle = [j(j+1) - m(m+1)]^{1/2}|jm+1\rangle$$

$$J_-|jm\rangle = [j(j+1) - m(m-1)]^{1/2}|jm-1\rangle$$

$$J_z|jm\rangle = m|jm\rangle$$

$$\vec{J}^2|jm\rangle = j(j+1)|jm\rangle$$

$$\langle jm|j'm'\rangle = \delta_{mm'}\delta_{jj'}$$

The individual contributions of the various O_l^m will now be evaluated by computing the respective matrix elements.

1.6.1. *The CEF Hamiltonian in matrix notation*

$\mathbf{O_2^0}$: All those states will be combined and displayed, which reveal finite matrix elements for O_2^0.

$$O_2^0 = 3J_z^2 - j(j+1)$$

$$3[\langle -7/2|J_z^2| -7/2\rangle] - 7/2(7/2+1) = 21$$

$$3[\langle -5/2|J_z^2| -5/2\rangle - 7/2(7/2+1) = 3$$

$$3[\langle -3/2|J_z^2| -3/2\rangle - 7/2(7/2+1) = -9$$

$$3[\langle -1/2|J_z^2| -1/2\rangle - 7/2(7/2+1) = -15$$

$$3[\langle 1/2|J_z^2|1/2\rangle - 7/2(7/2+1) = -15$$

$$3[\langle 3/2|J_z^2|3/2\rangle - 7/2(7/2+1) = -9$$

$$3[\langle 5/2|J_z^2|5/2\rangle - 7/2(7/2+1) = 3$$

$$3[\langle 7/2|J_z^2|7/2\rangle - 7/2(7/2+1) = -9$$

O_4^0

$$O_4^0 = [35J_z^4 - 30\vec{\boldsymbol{J}}^2 J_z^2 + 25J_z^2 - 6\vec{\boldsymbol{J}}^2 + 3\vec{\boldsymbol{J}}^4]$$

$$35[\langle -7/2|J_z^4| - 7/2\rangle] - 30\cdot(7/2)\cdot(9/2)[\langle -7/2|J_z^2| - 7/2\rangle]+$$

$$25[\langle -7/2|J_z^2| - 7/2\rangle] - 6\cdot(7/2)\cdot(9/2) + 3\cdot(7/2)^2\cdot(9/2)^2 = 420$$

$$35[\langle -5/2|J_z^4| - 5/2\rangle] - 30\vec{\boldsymbol{J}}^2[\langle -5/2|J_z^2| - 5/2\rangle]+$$

$$25[\langle -5/2|J_z^2| - 5/2\rangle] - 6\cdot(7/2)\cdot(9/2) + 3\cdot(7/2)^2\cdot(9/2)^2 = -780$$

$$\langle -3/2|\ldots| - 3/2\rangle = -180$$

$$\langle -1/2|\ldots| - 1/2\rangle - 540$$

The matrix elements associated with $+1/2$, $+3/2$, $+5/2$, $+7/2$ have the same values as the above one with $-1/2$, $-3/2$, $-5/2$, $-7/2$.

O_6^0

$$231[\langle -7/2|J_z^6| - 7/2\rangle] - 315\cdot(7/2)\cdot(9/2)[\langle -7/2|J_z^4| - 7/2\rangle]+$$

$$735[\langle -7/2|J_z^4| - 7/2\rangle] + 105\cdot(7/2)^2\cdot(9/2)^2[\langle -7/2|J_z^2| - 7/2\rangle]-$$

$$525\cdot(7/2)\cdot(9/2)[\langle -7/2|J_z^2| - 7/2\rangle] + 294[\langle -7/2|J_z^2| - 7/2\rangle]-$$

$$5\cdot(7/2)^3\cdot(9/2)^3 + 40\cdot(7/2)^2\cdot(9/2)^2 + 60\cdot(7/2)\cdot(9/2) = 1260$$

$$\langle -5/2|\ldots| - 5/2\rangle = -6300$$

$$\langle -3/2|\ldots| - 3/2\rangle = 11340$$

$$\langle -1/2|\ldots| - 1/2\rangle = -6300$$

O_6^6: The subsequent equations demonstrate how the ladder operators act on a certain state.

$$J_+^6| - 7/2\rangle \propto c_5 J_+^5| - 5/2\rangle \propto c_4 J_+^4| - 3/2\rangle \propto c_3 J_+^3| - 1/2\rangle$$

$$\propto c_2 J_+^2| + 1/2\rangle \propto c_1 J_+|3/2\rangle = c|5/2\rangle$$

$$J_+^6|-5/2\rangle \propto d_5 J_+^5|-3/2\rangle \propto d_4 J_+^4|-1/2\rangle \propto d_3 J_+^3|1/2\rangle$$

$$\propto d_2 J_+^2|+3/2\rangle \propto d_1 J_+|5/2\rangle = d|7/2\rangle$$

$$J_-^6|7/2\rangle \cdots = f|-5/2\rangle$$

$$J_-^6|5/2\rangle \cdots = g|-7/2\rangle.$$

Let us now evaluate in detail $J_+^6|j,m\rangle = J_+^6|7/2,-7/2\rangle$:

$$[j(j+1)-m(m+1)]^{1/2} = [(7/2)\cdot(9/2)-(-7/2)\cdot(-5/2)]^{1/2}$$

$$= [(63/4)-(35/4)]^{1/2} = \sqrt{7}$$

and

$$J_+^6|7/2,-7/2\rangle = \sqrt{7}J_+^5|7/2,-5/2\rangle$$

$$= \sqrt{7}[(63/4)-(15/4)]^{1/2}J_+^4|7/2,-3/2\rangle$$

$$= \sqrt{7\cdot12}[(63/4)-(3/4)]^{1/2}J_+^3|7/2,-1/2\rangle$$

$$= \sqrt{7\cdot12\cdot15}[(63/4)-(1/2)(1/2)]^{1/2}J_+^2|7/2,1/2\rangle$$

$$= \sqrt{7\cdot12\cdot15\cdot16}[(63/4)-(1/2)(3/2)]^{1/2}J_+|7/2,3/2\rangle$$

$$= \sqrt{7\cdot12\cdot15\cdot16\cdot15}[(63/4)-(3/2)(5/2)]^{1/2}|7/2,5/2\rangle$$

$$= \sqrt{7\cdot12\cdot15\cdot16\cdot15\cdot12}|7/2,5/2\rangle$$

$$= 720\sqrt{7}|7/2,5/2\rangle$$

It follows that

$$\langle5/2|J_+^6|-7/2\rangle = 720\sqrt{7}$$

$$\langle7/2|J_+^6|-5/2\rangle = 720\sqrt{7}$$

$$\langle-7/2|J_-^6|5/2\rangle = 720\sqrt{7}$$

$$\langle-5/2|J_-^6|7/2\rangle = 720\sqrt{7}$$

Adding together all terms to form 8×8 matrices reveals for the hexagonal case

$$^{7/2}H_{CEF}^{hexa}$$

$$= B_2^0 \times \begin{array}{c} \\ -7/2 \\ -5/2 \\ -3/2 \\ -1/2 \\ 1/2 \\ 3/2 \\ 5/2 \\ 7/2 \end{array} \begin{pmatrix} \scriptstyle -7/2 & \scriptstyle -5/2 & \scriptstyle -3/2 & \scriptstyle -1/2 & \scriptstyle 1/2 & \scriptstyle 3/2 & \scriptstyle 5/2 & \scriptstyle 7/2 \\ 21 & 0 & 0 & 0 & 0 & 0 & 0 & 0 \\ 0 & 3 & 0 & 0 & 0 & 0 & 0 & 0 \\ 0 & 0 & -9 & 0 & 0 & 0 & 0 & 0 \\ 0 & 0 & 0 & -15 & 0 & 0 & 0 & 0 \\ 0 & 0 & 0 & 0 & -15 & 0 & 0 & 0 \\ 0 & 0 & 0 & 0 & 0 & -9 & 0 & 0 \\ 0 & 0 & 0 & 0 & 0 & 0 & 3 & 0 \\ 0 & 0 & 0 & 0 & 0 & 0 & 0 & 21 \end{pmatrix}$$

$$+ B_4^0 \times \begin{array}{c} \\ -7/2 \\ -5/2 \\ -3/2 \\ -1/2 \\ 1/2 \\ 3/2 \\ 5/2 \\ 7/2 \end{array} \begin{pmatrix} \scriptstyle -7/2 & \scriptstyle -5/2 & \scriptstyle -3/2 & \scriptstyle -1/2 & \scriptstyle 1/2 & \scriptstyle 3/2 & \scriptstyle 5/2 & \scriptstyle 7/2 \\ 420 & 0 & 0 & 0 & 0 & 0 & 0 & 0 \\ 0 & -780 & 0 & 0 & 0 & 0 & 0 & 0 \\ 0 & 0 & -180 & 0 & 0 & 0 & 0 & 0 \\ 0 & 0 & 0 & 540 & 0 & 0 & 0 & 0 \\ 0 & 0 & 0 & 0 & 540 & 0 & 0 & 0 \\ 0 & 0 & 0 & 0 & 0 & -180 & 0 & 0 \\ 0 & 0 & 0 & 0 & 0 & 0 & -780 & 0 \\ 0 & 0 & 0 & 0 & 0 & 0 & 0 & 420 \end{pmatrix}$$

$$+ B_6^0 \times \begin{array}{c} \\ -7/2 \\ -5/2 \\ -3/2 \\ -1/2 \\ 1/2 \\ 3/2 \\ 5/2 \\ 7/2 \end{array} \begin{pmatrix} \scriptstyle -7/2 & \scriptstyle -5/2 & \scriptstyle -3/2 & \scriptstyle -1/2 & \scriptstyle 1/2 & \scriptstyle 3/2 & \scriptstyle 5/2 & \scriptstyle 7/2 \\ 1260 & 0 & 0 & 0 & 0 & 0 & 0 & 0 \\ 0 & -6300 & 0 & 0 & 0 & 0 & 0 & 0 \\ 0 & 0 & 11340 & 0 & 0 & 0 & 0 & 0 \\ 0 & 0 & 0 & -6300 & 0 & 0 & 0 & 0 \\ 0 & 0 & 0 & 0 & -6300 & 0 & 0 & 0 \\ 0 & 0 & 0 & 0 & 0 & 11340 & 0 & 0 \\ 0 & 0 & 0 & 0 & 0 & 0 & -6300 & 0 \\ 0 & 0 & 0 & 0 & 0 & 0 & 0 & 1260 \end{pmatrix}$$

$$+ B_6^6 \times \begin{array}{c} \\ -7/2 \\ -5/2 \\ -3/2 \\ -1/2 \\ 1/2 \\ 3/2 \\ 5/2 \\ 7/2 \end{array} \begin{pmatrix} \scriptstyle -7/2 & \scriptstyle -5/2 & \scriptstyle -3/2 & \scriptstyle -1/2 & \scriptstyle 1/2 & \scriptstyle 3/2 & \scriptstyle 5/2 & \scriptstyle 7/2 \\ 0 & 0 & 0 & 0 & 0 & 0 & 720\sqrt{7} & 0 \\ 0 & 0 & 0 & 0 & 0 & 0 & 0 & 720\sqrt{7} \\ 0 & 0 & 0 & 0 & 0 & 0 & 0 & 0 \\ 0 & 0 & 0 & 0 & 0 & 0 & 0 & 0 \\ 0 & 0 & 0 & 0 & 0 & 0 & 0 & 0 \\ 0 & 0 & 0 & 0 & 0 & 0 & 0 & 0 \\ 720\sqrt{7} & 0 & 0 & 0 & 0 & 0 & 0 & 0 \\ 0 & 720\sqrt{7} & 0 & 0 & 0 & 0 & 0 & 0 \end{pmatrix}$$

$$(61)$$

1.6.2. *Diagonalisation of the CEF Hamiltonian*

Combining the matrices $B_2^0 O_2^0$, $B_4^0 O_4^0$, $B_6^0 O_6^0$ and $B_6^6 O_6^6$ of Eq. (61) gives an 8×8 matrix, which can be diagonalized in order to obtain the energies of the several states in a CEF described by the CEF Hamiltonian Eq. (60). All the diagonal elements correspond to contributions from O_l^m with $m = 0$, the off-diagonal elements derive from O_l^m with $m \neq 0$. The nonzero matrix elements of H_{CEF} between the various states are then given by:

$$E_{11} = 21 \cdot B_2^0 + 420 \cdot B_4^0 + 1260 \cdot B_6^0 + 0 \cdot B_6^0$$

$$E_{22} = 3 \cdot B_2^0 - 780 \cdot B_4^0 - 6300 \cdot B_6^0 + 0 \cdot B_6^0$$

$$E_{33} = -9 \cdot B_2^0 - 180 \cdot B_4^0 + 11340 \cdot B_6^0 + 0 \cdot B_6^0$$

$$E_{44} = -15 \cdot B_2^0 + 540 \cdot B_4^0 - 6300 \cdot B_6^0 + 0 \cdot B_6^0$$

$$E_{55} = -15 \cdot B_2^0 + 540 \cdot B_4^0 - 6300 \cdot B_6^0 + 0 \cdot B_6^0$$

$$E_{66} = -9 \cdot B_2^0 - 180 \cdot B_4^0 + 11340 \cdot B_6^0 + 0 \cdot B_6^0$$

$$E_{77} = 3 \cdot B_2^0 - 780 \cdot B_4^0 - 6300 \cdot B_6^0 + 0 \cdot B_6^0$$

$$E_{88} = 21 \cdot B_2^0 + 420 \cdot B_4^0 + 1260 \cdot B_6^0 + 0 \cdot B_6^0$$

$$E_{71} = 720\sqrt{7} \cdot B_6^6$$

$$E_{82} = 720\sqrt{7} \cdot B_6^6$$

$$E_{17} = 720\sqrt{7} \cdot B_6^6$$

$$E_{28} = 720\sqrt{7} \cdot B_6^6$$

When operators of the form O_l^m with $m \neq 0$ are involved, the wave functions (or eigenvectors) may consist of linear combinations of two or more m states. The operator O_l^m mixes m states differing by m. In order to obtain the eigenstates and the eigenenergies of the above matrices, diagonalisation of the resulting 8×8 matrix has to be carried out. This, of course, has to be done by help of appropriate computer programmes. Diagonalisation rotates the coordinate axes so that the eigenvectors are no longer pure m states, but instead are linear combinations of the type $\sum a_m|m\rangle$, involving states which differ by m. The energies of the eigenstates are functions of the B_l^m.

Formally, this procedure can be outlined as follows: The application of the Hamiltonian H to the states $\Gamma = |lsjm\rangle = |m\rangle$, i.e.,

$$H|\Gamma\rangle = E|\Gamma\rangle \tag{62}$$

reveals the energy matrix

$$E_{mm'} = \langle m'|H|m\rangle \tag{63}$$

Diagonalisation of V_c is made straightforwardly employing a *unitary transformation U*, i.e.,

$$V_c^D = U^+ V_c U \tag{64}$$

yielding the eigenfunctions $|i\rangle$

$$|i\rangle = \sum_{m=-j}^{j} U_{i,m}|m\rangle \tag{65}$$

as well as the associated energy eigenvalues E_i. In a general case the eigenfunctions are mixed states of the form:

$$|i\rangle = U_{i,-j}|-j\rangle + U_{i,-j+1}|-j+1\rangle + \ldots \tag{66}$$

with $|U_{i,-j}|^2 + |U_{i,-j+1}|^2 + \cdots = 1$.

1.6.3. *Point-Charge Model A*

Let us now consider a simple example of an Yb^{3+} ion in a CEF produced by two negative point-charges $i = 1,2$ of $q_1 = q_2 = -0.2|e|$ situated at the positions $\vec{r}_i = (0,0,\pm 4\text{Å})$. In this case only the γ_l^0 with $l = 2,4,6$ are nonzero and can be calculated according to Eq. (20), which for the PCM can be rewritten as

$$\gamma_l^m = \frac{1}{2l+1} \sum_i \frac{q_i Z_l^m(\Omega_i)}{\epsilon_0 r_i^{l+1}} \tag{67}$$

Using the table given for the tesseral harmonic functions Z_l^m (see appendix) this gives

$$\gamma_2^0 = \frac{-0.2|e|}{5\epsilon_0} \sum_{i=1,2} \left(\frac{1}{4}\sqrt{\frac{5}{\pi}}[(3z_i^2 - r_i^2)/r_i^5] \right)$$

$$\gamma_4^0 = \frac{-0.2|e|}{9\epsilon_0} \sum_{i=1,2} \left(\frac{3}{16}\frac{1}{\sqrt{\pi}}[35z_i^4 - 30z_i^2 r_i^2 + 3r_i^4)/r_i^9] \right) \tag{68}$$

$$\gamma_6^0 = \frac{-0.2|e|}{13\epsilon_0} \sum_{i=1,2} \left(\frac{1}{32}\sqrt{\frac{13}{\pi}}[(231z_i^6 - 315z_i^4 r_i^2 + 105z_i^2 r_i^4 - 5r_i^6)/r_i^{13}] \right)$$

and inserting $z_{1,2} = \pm 4$ Å, $\quad r_i = 4$ Å, $\quad \epsilon_0 = 8.85419 \times 10^{-12}\frac{\text{As}}{\text{Vm}}$ and $e = 1.60217 \times 10^{-19}$ C, using Eq. (32) and looking up the Stevens factors and radial matrix elements for the Yb^{3+} ion in Tables 1 and 2, respectively, the values of the CEF parameters can be calculated.

$$B_2^0 = -|e|p_2^0\gamma_2^0\langle r_{4f}^2\rangle\alpha_j = 0.261 \text{ meV}$$

$$B_4^0 = -|e|p_4^0\gamma_4^0\langle r_{4f}^4\rangle\beta_j = -1.04 \times 10^{-4} \text{ meV} \tag{69}$$

$$B_6^0 = -|e|p_6^0\gamma_6^0\langle r_{4f}^6\rangle\gamma_j = 2.81 \times 10^{-7} \text{ meV}$$

In a similar way the CEF parameters for point-charges $+0.1|e|$ situated on the corner points of a hexagon may be calculated. Assuming the hexagon is given by $(\pm 2\sqrt{3}, 0, 0)$ Å, $(\pm\sqrt{3}, \pm 3, 0)$ Å and $(\pm\sqrt{3}, \mp 3, 0)$ Å we obtain

$$B_2^0 = 0.3013\text{meV}$$

$$B_4^0 = 1.2 \times 10^{-4} \text{ meV}$$

$$B_6^0 = 3.603 \times 10^{-7} \text{ meV} \tag{70}$$

$$B_6^6 = -8.324 \times 10^{-6} \text{ meV}$$

Taking the sum of the the CEF produced by the charges along the z-axis Eq. (69) and on the hexagon Eq. (70) we arrive at the final result of our PCM calculation, which is summarized in Table 4

Diagonalisation of Eq. (61) taking into account the CEF parameters B_l^m as given in Table 4 reveals the eigenvalues and the eigenvectors of the problem. The following eigenvalues E_i- expressed in Kelvin — with $0 \leq i \leq 8$ are derived: 138.22, 138.22, -98.6, -98.6, -59.14, -59.14, 19.54, 19.54

Table 4. Positions of point-charges and CEF parameters for an Yb^{3+} ion in a hexagonal CEF (model A).

point-charges ($|e|$)	x(Å)	y(Å)	z(Å)
-0.2	0	0	4
-0.2	0	0	-4
0.1	3.4641	0	0
0.1	-3.4641	0	0
0.1	1.73205	-3	0
0.1	-1.73205	3	0
0.1	1.73205	3	0
0.1	-1.73205	-3	0

CEF parameters B_l^m

$$B_2^0 = 0.5622 \text{ meV}$$

$$B_4^0 = 1.6087 \times 10^{-5} \text{ meV}$$

$$B_6^0 = 6.412 \times 10^{-7} \text{ meV}$$

$$B_6^6 = -8.324 \times 10^{-6} \text{ meV}$$

In general, these values are ordered such that the lowest energy is set to zero, resulting now the following sequence: 236.83, 236.83, 0., 0., 39, 46, 39.46, 118.15, 118.15.

Below, the eigenstates of these eigenvalues as derived from the diagonalisation are arranged according to the sequence of energy:

$$
\begin{aligned}
&0.00078|-5/2\rangle - 0.9999|7/2\rangle, \\
&0.9999|-7/2\rangle + 0.00078|5/2\rangle, \\
&1.|1/2\rangle, \\
&1.|-1/2\rangle, \\
&1.|3/2\rangle, \\
&1.|-3/2\rangle, \\
&-0.9999|5/2\rangle - 0.00078|7/2\rangle, \\
&0.00078|-7/2\rangle + 0.9999|5/2\rangle
\end{aligned}
\tag{71}
$$

The knowledge of the eigenvalues and the eigenstates allows the calculation of several physical properties, which will be described in the following.

Using Eq. (47) the $4f$ charge density can be calculated. Figure 4 shows the result of the Yb^{3+} ion surrounded by the point-charges producing the CEF.

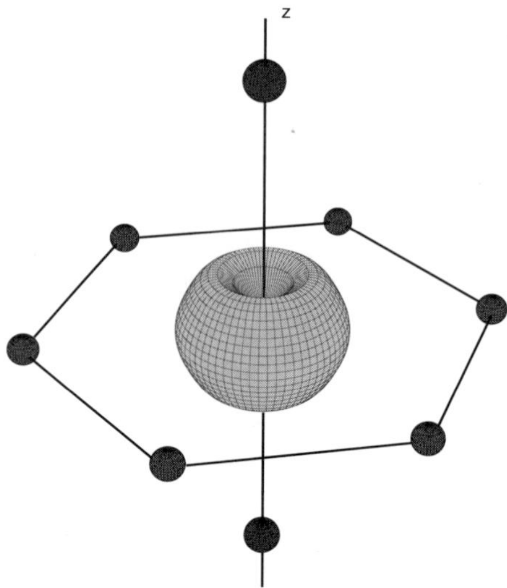

Fig. 4. **Model A** $4f$-charge density for an Yb^{3+} ion surrounded by point-charges (indicated by spheres) calculated at a temperature $T = 10$ K. The CEF parameters are taken from Table 4. The charges along the z-axis are $-0.2|e|$ (blue) and those in the hexagon are $+0.1|e|$ (red).

1.6.4. *Calculation of magnetic moments*

Magnetic moments μ can be evaluated for each individual CEF level. They are determined by the eigenstates, the Landé factor and by the operators J_z and/or $J_\pm$. Results of such calculations impressively demonstrate why only in rare cases the observed magnetisation $M = N_A\mu$ corresponds simply to a magnetic moment $\mu = g_j\mu_B j$. N_A is the Avogadro number and μ_B is the Bohr magneton. The magnetisation is derived by evaluating the appropriate matrix elements of each CEF state.

For the z-direction the magnetisation follows from $M_z = g_j\mu_B N_A\langle\Gamma_i|J_z|\Gamma_i\rangle$, while the magnetisation perpendicular to the z-axis derives from $M_x = (1/2)g_j\mu_B\langle\Gamma_i|(J_+ + J_-)|\Gamma_i\rangle$ and $M_y = -(1/2)g_j\mu_B i\langle\Gamma_i|(J_+ - J_-)|\Gamma_i\rangle$.

$\vec{J}_z$ **operator** In order to calculate the magnetic moment of simple and complicated eigenstates, the following procedures have to be kept in mind: $J_z|j, m_j\rangle = m_j|j, m_j\rangle$. The magnetic moment $\mu_z = g_j\mu_B\langle\Gamma_i|J_z|\Gamma_i\rangle$ follows then simply from $\mu_z = g_j\mu_B\langle j, m_j|J_z|j, m_j\rangle$ for a given total angular momentum j and associated magnetic quantum number m_j.

Example Considering the eigenstate $|j = 5/2, m_j = 3/2\rangle$ and $g_j = 6/7$:
$$\mu_z = g_j\mu_B\langle 5/2, 3/2|J_z|5/2, 3/2\rangle = g_j\mu_B(3/2)\langle 5/2, 3/2||5/2, 3/2\rangle = g_j(3/2)\delta_{i,j} = 6/7\mu_B \cdot 3/2 \cdot 1 = 1.285\ \mu_B$$

Ladder operators Definition: $J_\pm = J_x \pm iJ_y$
$J_\pm|j, m_j\rangle = \sqrt{j(j+1) - m_j(m_j \mp 1)}|j, m_j \pm 1\rangle$. The magnetic moment follows then via $\mu_\pm = g_j\mu_B\langle\Gamma_i|J_\pm|\Gamma_i\rangle$.

Example Considering the eigenstate $|5/2, 3/2\rangle$ and $g_j = 6/7$:
$$\mu_+ = g_j\mu_B\langle 5/2,\ 3/2|J_+|5/2,\ 3/2\rangle = g_j\mu_B\sqrt{5/2 \cdot 7/2 - 3/2 \cdot 5/2}\langle 5/2,\ 3/2||5/2,\ 5/2\rangle = g_j\mu_B\sqrt{5}\delta_{i,j} = 2.23\ \mu_B \cdot 0 = 0\ \mu_B.$$

In order to obtain a finite magnetisation, states like $\langle 5/2, 3/2||5/2, 1/2\rangle$ or $\langle 5/2, 5/2||5/2, 3/2\rangle$ or $\langle 5/2, 1/2||5/2, -1/2\rangle$ have to be combined. Very frequently, the actual wave function does not provide such eigenstates; therefore the magnetic moments derived from the $J_\pm$ operators become zero. The above example would give a magnetic moment only in the case $\mu_+ = g_j\mu_B\langle 5/2, 5/2|J_+|5/2, 3/2\rangle = g_j\mu_B \cdot \sqrt{5/2 \cdot 7/2 - 3/2 \cdot 5/2}\langle 5/2, 5/2||5/2, 5/2\rangle = g_j\sqrt{5}\delta_{i,j} = 2.23\ \mu_B \cdot 1 = 2.23\ \mu_B$.

Magnetic moments of model A As an example, we calculate the magnetic moment of the ground state of the hexagonal Yb compound, i.e., of the CEF level with lowest energy as derived by the CEF parameters given in Table 4. $\mu_z = g\mu_B(\langle 1/2|J_z|1/2\rangle = 8/7\mu_B \cdot 0.5\delta_{ij} \approx 0.571\ \mu_B$ The first eigenstate at 39.46 K with eigenfunction $|\pm 3/2\rangle$, reveals a moment $\pm 1.714\ \mu_B$. Since both states of a doublet have the same energy, they are two-fold degenerate. The application of a small external field would split the CEF states lifting therefore this degeneracy. Results of the magnetic moments associated with the various CEF states are summarised in Table 5.

1.6.5. *Effect of magnetic field on the charge density — a source of magnetostriction*

The application of a magnetic field gives also rise to a change in the $4f$ charge density, which in turn may lead to magnetostriction. In order to demonstrate this effect we show in Fig. 5 the charge density of the Yb^{3+} ion in an applied magnetic field of 40 T along the y axis.

Table 5. Magnetic moments of an Yb^{3+} ion exposed to CEF effects described by the CEF parameters in Table 4 (model A).

energy level [K]	M_z μ_B	M_+ μ_B	M_- μ_B
0	-0.42	1.71	-1.71i
0	0.42	1.71	-1.71i
39.46	-1.28	0	0
39.46	1.28	0	0
118.15	2.14	0.0017	-0.0017i
118.15	-21.14	0.0017	0.0017i
236.83	-3	-0.0017	0.0017i
236.83	3	-0.0017	-0.0017i

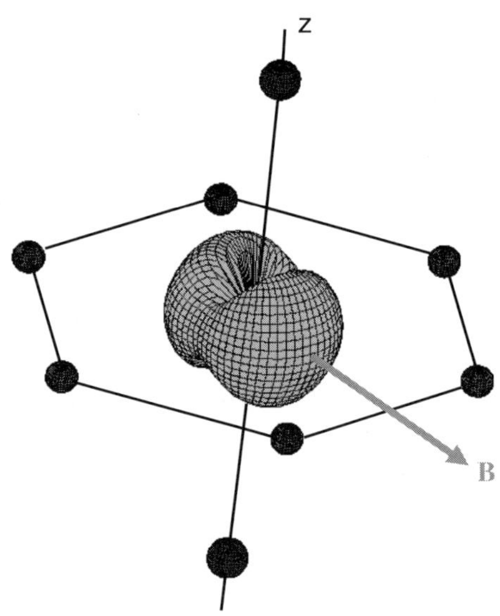

Fig. 5. Model A: $4f$-charge density for Yb^{3}+ surrounded by point-charges indicated by the spheres calculated at a temperature $T = 10$ K and an applied magnetic field of 40 T along (010).

1.6.6. *Point-Charge Model B*

In order to show, how changes of the point-charges influence the experimental results, we evaluate also another model, with different sign and distance of point-charges situated on the hexagonal axis. The point-charges and the resulting CEF parameters are summarized in Table 6.

Table 6. Positions of point-charges and CEF parameters for an Yb^{3+} ion in a hexagonal CEF (Model B).

point-charges ($|e|$)	$x(\text{Å})$	$y(\text{Å})$	$z(\text{Å})$
0.8	0	0	3
0.8	0	0	-3
0.8	3.4641	0	0
0.8	-3.4641	0	0
0.8	1.73205	-3	0
0.8	-1.73205	3	0
0.8	1.73205	3	0
0.8	-1.73205	-3	0

CEF parameters B_l^m

$$B_2^0 = -0.06364 \text{ meV}$$

$$B_4^0 = 2.713 \times 10^{-3} \text{ meV}$$

$$B_6^0 = -5.5335 \times 10^{-6} \text{ meV}$$

$$B_6^6 = -6.659 \times 10^{-5} \text{ meV}$$

Diagonalizing the Hamiltonian in a similar manner as was done in Eq. (61), the eigenstates and the eigenvalues are derived for the CEF parameters outlined in Table 6. Model B yields a rather narrow splitting with doublets at 0, 24.24, 26.86 and 55.3 K. The magnetic moment of the ground state doublet $-0.995|-5/2\rangle - 0.0306|7/2\rangle$ and $0.995|5/2\rangle + 0.0306|-7/2\rangle$ associated with J_z amounts to $\pm 2.85\ \mu_B$. Note the significant changes in both the overall splitting and the magnetic moment of the ground state triggered by changes of the point-charges. The minor separation between the first and the second excited level gives rise to the appearance of a so called quasi-quartet.

2. Physical Properties and CEF Effects

In the following section we will discuss a number of physical properties, which cannot be explained unless CEF effects are considered.

Two different scenarios have to be taken into account:

- Properties depend only on the energies and the degeneracy of the various CEF states. Examples here are the specific heat or magnetic entropy.

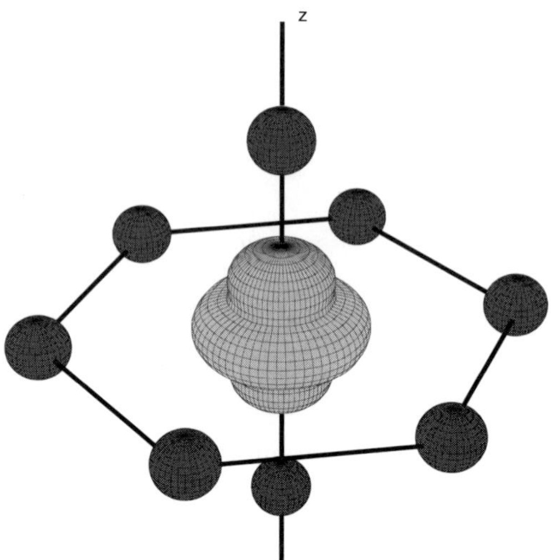

Fig. 6.   Model B: $4f$-charge density for an $Yb^{3}+$ ion surrounded by point-charges indicated by the spheres calculated at the temperature $T = 10$ K. The CEF parameters are taken from Table 6. The charges along the z-axis and in the hexagon are $0.8|e|$.

- Properties depend on the degeneracy and on the matrix elements between a certain CEF level an on the matrix elements between various CEF states, e.g., electrical resistivity, magnetic susceptibility, isothermal magnetisation, inelastic neutron scattering cross section.

2.1. *Inelastic neutron scattering*

Neutrons can be used to probe CEF levels due to their dipolar nuclear magnetic moment $\mu_n = g_n\mu_N = -1.9135\ \mu_N$, which interacts with the $4f$ electronic moment. Here, the nuclear magneton $\mu_N = e\hbar/(2m_p)$ and m_p is the proton rest mass. Note that $1\mu_B \approx 2000\mu_N$. Neutrons may induce transitions between CEF levels, thereby transferring a part of their energy to the sample. Thus measuring the energy of the neutron before and after the scattering process provides the most direct method for determining CEF excitations.

The neutron powder cross section (unpolarised neutrons) for each CEF transition $\Gamma_i \rightarrow \Gamma_{i'}$ (units are in [barn/ion/sr]) for a system of N non-interacting magnetic ions at a temperature T can be expressed as:[4]

$$\frac{d^2\sigma_{i\to k}}{d\Omega dE'} = \frac{k'}{k}S(Q,\omega) = N\frac{k'}{k}\{\frac{1}{2}g_j F(Q)\}^2 \exp(-2W(Q))\left(\frac{\hbar\gamma e^2}{mc^2}\right)^2$$

$$\times \frac{\exp(-E_i/k_B T)}{\sum_j \exp(-E_j/k_B T)} \times \frac{2}{3}\sum_{\alpha=x,y,z}|\langle\Gamma_i|J^\alpha|\Gamma_{i'}\rangle|^2\delta(E_{i'} - E_i - \hbar\omega) \quad (72)$$

Here $4\pi(\hbar\gamma e^2/mc^2)^2$ is the total magnetic cross section (=3.65 barns) and $e^2/mc^2 = 2.82$ fm is the classical electron radius; $\gamma = g_n/(2\hbar)$ is the gyromagnetic ratio. Furthermore, k'/k is the ratio of the outgoing and incoming wave vector of the neutrons, $\hbar\omega = E'-E$ the energy transfer of the neutron to the sample $F(Q)$ denotes the form factor and $\exp(-2W(Q))$ the Debye Waller factor. This differential cross-section of magnetic scattering is related to the imaginary part of the wavenumber and frequency dependent magnetic susceptibility $\chi(\vec{Q},\omega)$.

Taking into account model A with the CEF parameters given in Table 4, both the eigenvalues and the eigenstates are derived after diagonalising the appropriate Hamiltonian (see Eq. (61)). While the former define the energy separation between levels, the latter allow to calculate the matrix elements within a certain CEF multiplet and between the various levels. Carrying out these calculations according to Eq. (72) and convolute the derived data with a Gaussian resolution function gives calculated neutron inelastic neutron (INS) spectra for the hexagonal Yb compound in terms of model A as shown in Fig. 7. At low temperature only one transition can be seen, from the ground state to the first excited level. All other matrix elements in Eq. (72) are zero, hence no intensity is observed. Increasing the temperature populates excited CEF states and the neutrons can induce further transitions between excited states as can be seen in Fig. 7. Moreover, neutrons can also gain energy from the transition from thermally excited CEF levels. This leads to the peaks for negative energy transfer in Fig. 7. For the analysis of data the condition of detailed balance relating the left and the right side of the spectrum is extremely important, which can be easily seen from Eq. (72).

$$S(Q,\omega) = e^{-\hbar\omega/k_B T}S(Q,-\omega) \quad (73)$$

At this point it should be mentioned that the ground state wave function can also be determined by a diffraction experiment (neutron form factor measurement) on a single crystal using polarized neutrons and a strong magnetic field.[12,13] To understand this we note that the form factor $F(Q)$

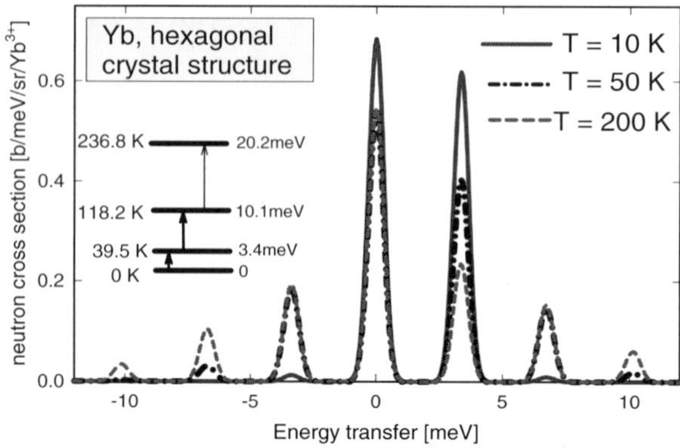

Fig. 7. Calculated neutron spectra at different temperatures for the Yb^{3+} ion in a hexagonal CEF according to model A. The inset shows the calculated CEF level scheme, arrows indicate strong transition matrix elements $\sum_{\alpha=x,y,z} \langle \Gamma_i | J_\alpha | \Gamma_k \rangle$.

is essentially the Fourier transform of the $4f$ spin density as calculated by Eq. (47).

Doing inelastic polarized neutron scattering on single crystals permits the determination of individual transition matrix elements $|\langle \Gamma_i | J^\alpha | \Gamma_{i'} \rangle|^2$, $\alpha = x, y, z$.[14]

2.2. *The Schottky contribution to the specific heat*

The so called Schottky contribution to the specific heat is originated from the thermal population of the various CEF levels and thus results from the increase of the magnetic energy of the system.

$$F = \frac{R \sum_{r=0}^{m} \Delta_r g_r \exp\left(\frac{-\Delta_r}{k_B T}\right)}{\sum_{r=0}^{m} g_r \exp\left(\frac{-\Delta_r}{k_B T}\right)} \tag{74}$$

g_r is the degeneracy of the r-th level, Δ_r the energy difference to the ground state and R the gas constant ($R = 8.314$ J/molK).

The specific heat can then be calculated from Eq. (74) employing

$$C_{V,P} = -T \frac{\partial^2 F(T)}{\partial T^2}\Big|_{V,P} = T \frac{\partial^2 (k_B T \ln Z)}{\partial T^2}\Big|_V = \frac{\partial U(T)}{\partial T}\Big|_{V,P} \tag{75}$$

where the partition function Z is defined as $F = -k_B T \ln Z$. Z represents the accessible states at a given temperature, hence, the calorimetric technique can be considered as a type of *thermal spectroscopy*. There, the heat

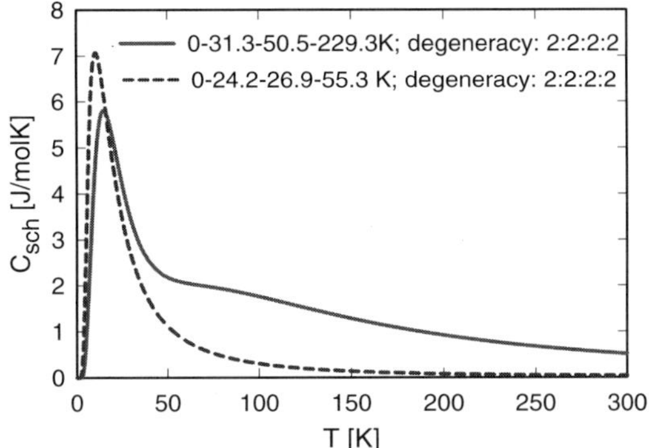

Fig. 8.   Temperature dependent specific heat corresponding to a hexagonal Yb compound calculated for the CEF parameters of model A (Table 4) and model B (Table 6). The sequence of numbers gives the respective CEF splitting and the degeneracy.

dQ is the excitation energy absorbed by the system. The initial state is characterised by the temperature T_i and the final one by $T_f = T_i + dT$. The system remains in its final state, because it is always in thermal equilibrium.[15]

For solids, the heat capacity taken at constant pressure equals about that taken at constant volume, i.e., $C_P \approx C_V$.

Figure 8 shows the CEF contribution to the specific heat corresponding to the CEF level splittings and degeneracies according to model A and B.

In particular, a 2-level system is simply described by:

$$C_{Sch} = \frac{R\left(\frac{\Delta}{k_B T}\right)^2 g \exp\left(\frac{\Delta}{k_B T}\right)}{\left(1 + g \exp\left(\frac{-\Delta}{k_B T}\right)\right)^2}. \tag{76}$$

There, g is the ratio of the degeneracies of the excited level to that of the ground state. Some general trends of the CEF contribution of 2-level systems are summarised in Table 7. If the levels are well separated, the maximum in the specific heat appears *roughly* at half of the splitting of the states.

2.3. *Magnetic entropy*

Magnetic entropy is one of the most important physical quantities and can be derived without an application of models and approximations. The

Table 7. Temperature T_{max} of the maximum of the Schottky anomaly c_{Sch} and entropy as a function of the degeneracy ratio g and CEF splitting Δ.[15]

excitation	g	c_{Sch}^{max} [J/molK]	T_{max}/Δ	entropy
Quartett $\rightarrow$ Dublett	0.5	2.00	0.448	$R\ln(3/2)$
Dublett $\rightarrow$ Dublett	1	3.64	0.417	$R\ln 2$
Dublett $\rightarrow$ Quartett	2	6.31	0.377	$R\ln 3$

magnetic entropy measures the degrees of freedom of a particular spin system. In the most simplest scenario, an electron spin $1/2$ has two degrees of freedom, i.e, spin$\uparrow$, and spin$\downarrow$. The entropy associated with the lifting of the degeneracy of these degrees of freedom derives from the well known statistical thermodynamics of Boltzmann and amounts to $S_{mag} = R\ln 2$.

From thermodynamics it is well known that the entropy S is defined as $S = k_B \ln Z$ and can be derived by integrating C_{mag}/T, i.e.,

$$S_{mag}(T) = S_0 + \int \frac{C_{mag}(T)}{T} dT. \tag{77}$$

Figure 9 shows the magnetic entropy for both, model A and model B. Due to the much smaller overall CEF splitting, $R\ln 4$ is recovered at much lower temperatures than the entropy of of model A, where $R\ln 4$ is not regained well below room temperature. If $S_{mag}(T)$ exhibits regions

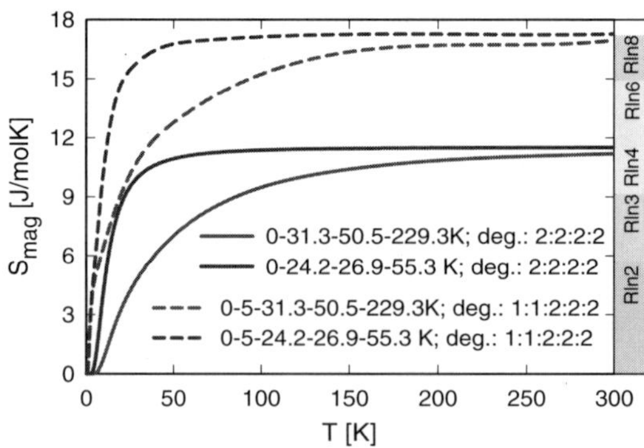

Fig. 9. Temperature dependent magnetic entropy S_{mag} corresponding to a hexagonal Yb compound calculated for the CEF parameters of model A (Table 4) and model B (Table 6), solid lines. The dashed lines represent the magnetic entropy after lifting the ground state degeneracy by Kondo interaction with $T_K = 5$ K. The sequence of numbers gives the respective CEF splitting and the degeneracy.

with plateau-like dependences, a substantial splitting is expected between levels and a steady increase of $S_{mag}(T)$ refers to relatively narrow distances between levels.

In a very general manner, the entropy well above a certain CEF level follows from $R\ln(N_o/N_g)$ where N_o is the number of states states already populated and N_g is the ground state degeneracy. Having populated e.g., 3 CEF doublets would result in $N_o = 6$ and in the case of 4 doublets, coinciding with Yb in a hexagonal CEF, $N_o = 8$. The entropy for the former is then $R\ln(6/2) = R\ln 3$, while for the latter one observes $R\ln(8/2) = R\ln 4$, in agreement to the results of Fig. 9.

It is important to note that a contribution of a ground state doublet to the specific heat and thus also to the magnetic entropy, depends on the availability of external mechanisms, lifting the ground state degeneracy. Such mechanisms may be the Kondo effect, long range magnetic order or a crystal distortion which lowers the symmetry. A doublet as ground state *does not* automatically contribute $R\ln 2$ to the measured entropy. Only mechanisms like those indicated above, will lift the degeneracy at low temperatures and thus allow to measure the entropy for this level!

Let us assume a Kondo interaction strength of $T_K = 5$ K as typical figure of Yb intermetallics. As indicated above, this phenomenon lifts the degeneracy of the ground state doublet and $R\ln 2$ is regained for $T \gg T_K$. Applying the Kondo effect to both data sets in Fig. 9 shows that at sufficiently high temperatures $R\ln 8$ is recovered.

Figure 10 shows the calculated magnetic contribution to the specific heat of a well studied Pr compound, i.e, $PrNi_5$, well in agreement with experimental data. This hexagonal compound does not show magnetic order, although the total angular momentum $j = 4$ of Pr^{3+} refers to a magnetic moment $M = g_j j = 3.2\,\mu_B$, in fact, a very large moment. The reason for the nonmagnetic ground state is CEF splitting on the non-Kramers ion Pr^{3+}, with the possibility to create non-magnetic CEF states. The expression *non-magnetic* should be considered with care: In CEF physics nonmagnetic means that the sum of the matrix elements of a certain state, particularly of the ground state becomes zero. We will come back to this scenario in the chapter concerning magnetization and susceptibility. The relevant CEF parameters B_l^m of $PrNi_5$ are[16] $B_2^0 = 5.84$ K, $B_4^0 = 4.53 \times 10^{-2}$ K, $B_6^0 = 8.86 \times 10^{-4}$ K, $B_6^6 = 3.14 \times 10^{-2}$ K.

Diagonalizing the appropriate Hamiltonian (a 9×9 matrix) reveals the following eigenvalues (in bracket is the degeneracy), numbers are rounded. $E_i = 0\ K[1],\ 24\ K[1],\ 40\ K[2],\ 49\ K[2],\ 158\ K[1],\ 333\ K[2]$.

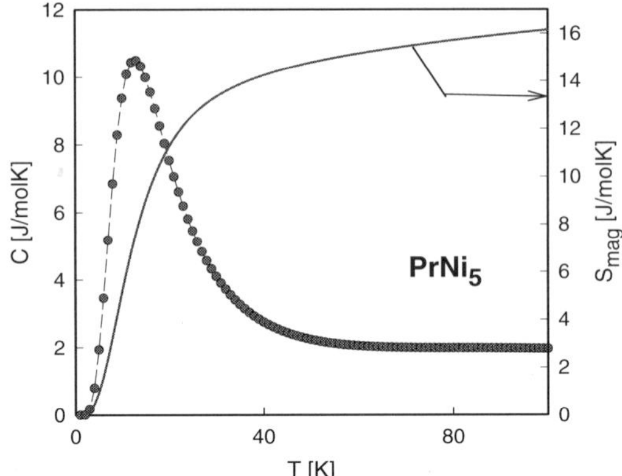

Fig. 10. Calculated temperature dependent specific heat C_{mag} (left axis) and temperature dependent magnetic entropy S_{mag} (right axis) of PrNi5. The CEF parameters are given in the text.

The eigenstates corresponding to these eigenvalues and the magnetic moments associated are (in rounded numbers): $0.71|-3\rangle + 0.71|+3\rangle$, $[0 \ \mu_B]$; $1|0\rangle$, $[0 \ \mu_B]$; $1|1\rangle$, $[-0.8 \ \mu_B]$; $1|-1\rangle$, $[0.8 \ \mu_B]$; $-0.97|-2\rangle + 0.21|+4\rangle$, $[1.38 \ \mu_B]$; $0.21|-4\rangle - 0.97|+2\rangle$, $[-1.38 \ \mu_B]$; $0.71|-3\rangle + 0.71|+3\rangle$, $[0 \ \mu_B]$; $0.97|-4\rangle + 0.21|+2\rangle$, $[2.98 \ \mu_B]$; $0.21|-2\rangle + 0.97|+2\rangle$, $[-2.98 \ \mu_B]$.

The anomaly in specific heat at low temperatures is purely of CEF origin and should not be confused with certain classes of phase transitions.

2.4. *Magnetisation and magnetic susceptibility*

To calculate magnetic properties of a single Rare Earth ion subjected to both the CEF and a magnetic field, H_i^{eff}, the CEF and Zeeman Hamiltonian Eq. (78) of the ion i has to be diagonalized together:

$$\mathcal{H} = B_l^m O_l^m(\mathbf{J}_i) - g_{Ji}\mu_B \mathbf{J}_i \mathbf{H}_i^{eff}. \tag{78}$$

It allows to calculate the expectation value of the angular momentum $\langle \mathbf{J}_i \rangle$ according to

$$\langle \mathbf{J}_i \rangle = \sum_{\Gamma} n_{\Gamma} \langle \Gamma | \mathbf{J}_i | \Gamma \rangle \tag{79}$$

with

$$n_\Gamma = \frac{\exp(-E_\Gamma/kT)}{z} \tag{80}$$

$$z = \sum_\Gamma \exp(-E_\Gamma/kT) \tag{81}$$

The operator $\vec{J}_i$ represents J_z and $J_\pm$; z is the partition sum, $|\Gamma\rangle$ the eigenstate corresponding to the eigenvalue E_Γ of the Hamiltonian Eq. (78).

2.4.1. *Isothermal Magnetisation*

We have shown in Section 1.6.4 how a magnetic moment of a certain eigenstate can be calculated. In most of the cases, however, the eigenstates may become quite complicated, consisting of a linear combination of the $|\pm 1/2\rangle$, $|\pm 1\rangle$, $|\pm 3/2\rangle \ldots$, i.e, $\alpha|\pm 1/2\rangle + \beta|\pm 1\rangle + \gamma|\pm 3/2\rangle \ldots$. Normalization of the eigenstates is ensured by requiring $(\alpha + \beta + \gamma + \ldots)^2 = 1$. α, β $\ldots$ can be both positive and negative. The matrix elements of all individual states have to be summed up to generate the total magnetic moment of the system with respect to a specific direction of the external magnetic field. Mathematically this follows from

$$\vec{\mu} = g_j\mu_B\langle\Gamma_i|\vec{J}|\Gamma_i\rangle. \tag{82}$$

This value has to be calculated at each temperature and each field. Temperature also causes that higher levels of the CEF scheme become populated and thus contribute to the mean value of $\vec{\mu}$.

Examples of isothermal magnetisation curves are shown in Fig. 11 for the hexagonal Yb compound with CEF parameters given in Table 4 for model A and in Table 6 for model B. Note that CEF splitting is responsible for the fact that the expected value $M = N_A g_j \cdot j$ is not reached unless very high external magnetic fields are applied. The values where saturation is reached are different for the different directions of the crystal.

The magnetic field not only causes a polarisation against thermal disorder, but further splits degenerate levels. A triplet, for example, is lifted into 3 singlets according to Zeeman's law.

A rather exciting phenomenon in this context is so called *level crossing*, see Fig. 12. This means that, while certain levels are shifted to higher energies as the field strength increases, others become lowered. A consequence can be that a level with a large magnetic moment is lowered, finally crossing the split levels of the ground state. Isothermal magnetisation can then exhibit sharp, field induced changes of the magnetic moment. A very

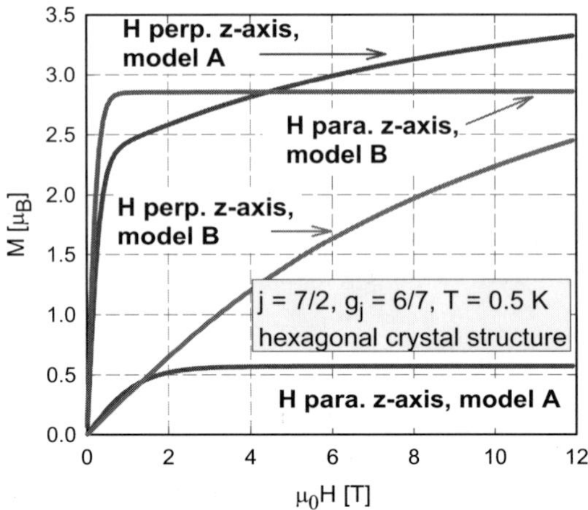

Fig. 11. Isothermal magnetization of an Yb compound ($j = 7/2$) with hexagonal crystal structure at $T = 0.5$ K for CEF parameters of model A (Table 4) and model B (Table 6).

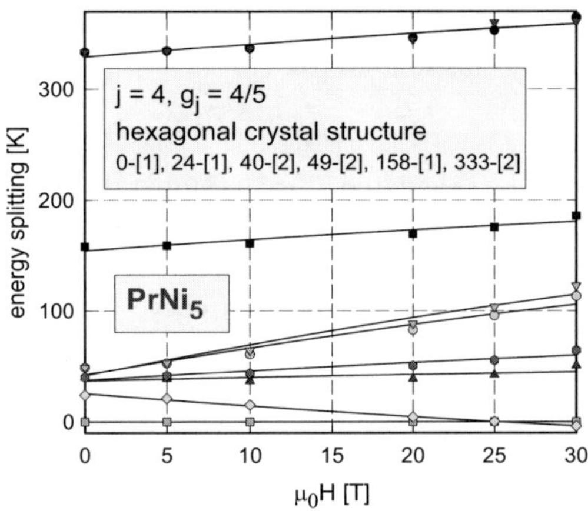

Fig. 12. Field dependence of the PrNi$_5$ eigenstates. Note the lifting of the degeneracy of states and the level crossing at about 24 T. The sequence of numbers refers to the energies of the levels in Kelvin (rounded) and the number in brackets is the degeneracy both at zero external fields.

famous example was already mentioned (PrNi$_5$): field dependent calculations of the isothermal magnetisation are shown in Fig. 13, exhibiting a metamagnetic-like transition around 23 T, well in agreement to experimental results. Slight changes of the B_l^m however, can cause dramatic changes

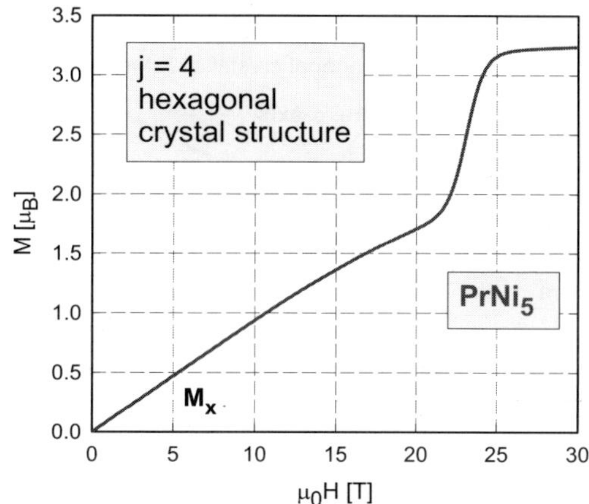

Fig. 13. Field dependent magnetisation of PrNi$_5$ with fields applied along the $\vec{x}$-direction calculated at $T = 0.5$ K. $\vec{x}$, $\vec{y}$, $\vec{z}$ correspond to the crystal directions $\vec{a}$, $\vec{b}$ and $\vec{c}$, respectively.

of this transition. In real materials, interactions between ions may occur, modifying final results, too.

2.4.2. *Temperature dependent susceptibility*

When a magnetic field is applied along the direction α, the magnetic susceptibility χ_α^{CEF} follows from the Van Vleck formula and can be expressed as,[17]

$$\chi_\alpha^{CEF} = \frac{N_A (g_j \mu_B)^2}{\sum_n \exp(-E_n/k_B T)}$$
$$\times \sum_{r,s} |\langle \Gamma_r | J_\alpha | \Gamma_s \rangle|^2 \exp\left(\frac{-E_r}{k_B T}\right) \frac{\exp((E_r - E_s)/k_B T) - 1}{E_r - E_s}. \quad (83)$$

N_A is the Avogadro number, E_r is the energy of the r-th state, g_j is the Landé factor and $\langle r | J_\alpha | s \rangle$ is the matrix element between the r- and s-state of a CEF scheme.

The magnetic susceptibility $\chi(T)$ as outlined in Eq. (83) modifies the simple Curie law by taking into account both, the matrix elements between the different states of a CEF split total angular momentum as well as the probability that a certain CEF level is populated at a distinct temperature. Results of calculations for the hexagonal $j = 7/2$ compound, model A, are shown in Fig. 14 for fields applied in the basal plane and along the c-axis,

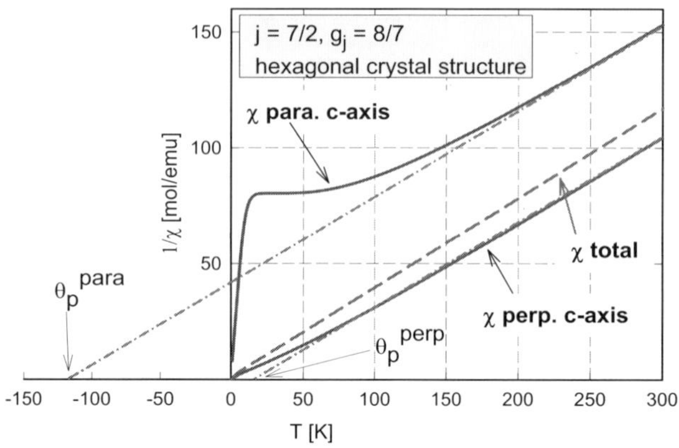

Fig. 14. Temperature dependent magnetic susceptibility χ of an Yb compound ($j = 7/2$), model A, with hexagonal crystal structure plotted as $1/\chi$ vs. T calculated for both field directions, i.e., $H//\vec{c}$ and $H \perp c$. The doublet CEF states are at 0, 31.3, 50.5, and 229.2 K. The sum curve, $\chi_{tot} = 1/3\chi_c + 2/3\chi_a$ reveals at high temperatures the effective magnetic moment of Yb^{3+}, $\mu_{eff} = 4.54\ \mu_B$.

respectively. The sum of both contributions, i.e., $\chi = 1/3\chi_{para} + 2/3\chi_{perp}$ corresponds to a measurement performed on polycrystalline materials. At very high temperatures, the slope of the averaged curve coincides with the effective moment $\mu_{eff} = 4.54\ \mu_B$ of a free, trivalent Yb ion.

At high temperatures ($k_B T \gg \Delta_{CEF}$) one can expand the susceptibility as a power series of inverse temperature, given as

$$\frac{1}{\chi} = \frac{1}{C}(T + \theta_\alpha), \tag{84}$$

where C is the Curie constant ($C = \mu_{eff}^2 N_A/(3k_B)$ and $\mu_{eff} = g_j \mu_B \sqrt{j(j+1)}$). The paramagnetic Curie temperature θ_α in the α direction is given by

$$k_B \theta_\alpha = \frac{\sum_r \langle \Gamma_r | \left[J_\alpha^2 - \frac{j(j+1)}{3} \right] H_{CEF} | \Gamma_r \rangle}{\sum_r \langle \Gamma_r | J_\alpha^2 | \Gamma_r \rangle} \tag{85}$$

The contribution can now be calculated for each term in the Hamiltonian. Among all the terms in the CEF potential, only O_2^0, O_2^0 and O_2^1 terms will contribute to θ_α. The principal values of the paramagnetic Curie temperatures due to CEF effects follow then from

$$k_B \theta_z = \frac{-1}{5} B_2^0 (2j - 1)(2j + 3)$$

$$k_B \theta^y_x = \frac{1}{10}(B^0_2 \mp B^2_2)(2j-1)(2j+3)$$

For a hexagonal or tetragonal system, the susceptibility is isotropic in the basal plane and the anisotropy between the basal plane and the z-axis allows to directly derive the value of the second order CEF parameter B^0_2:

$$\theta_x - \theta_z = \frac{3}{10}B^0_2(2j-1)(2j+3) \tag{86}$$

Eq. (86) allows reliable determination of B^0_2, once single crystal measurements of the magnetisation are available for a certain material.

Figure 15 shows the calculated temperature dependent susceptibility of PrNi$_5$. Note that the $1/\chi$ plot for both directions has finite values at $T = 0$. This refers to a nonmagnetic ground state - in contrast to the previous Yb compound - as a result of CEF effects exerted to the non-Kramers ion Pr. If the ground state is magnetic, $1/\chi$ would diverge. Moreover, $1/\chi$ exhibits strongly curved temperature dependences and thus deviates from the simple Curie-like behaviour, i.e., $1/\chi \propto T$.

2.5. *Electrical resistivity*

Scattering of electrons on ions bearing a permanent magnetic moment consists of contributions due to the standard (spherical) potential, but there is

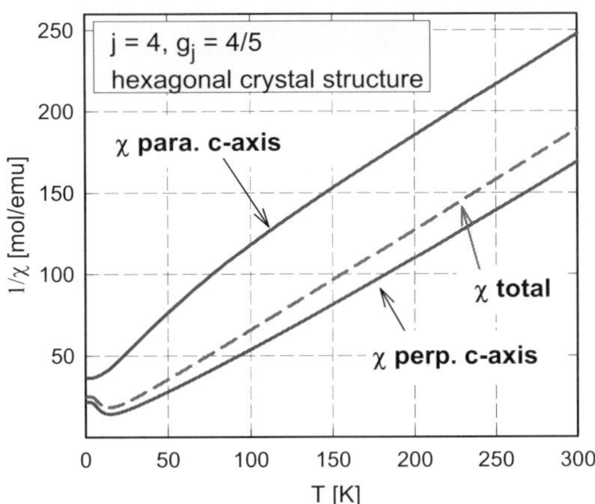

Fig. 15.   Temperature dependent magnetic susceptibility of PrNi$_5$ calculated for both field directions, i.e., $\vec{H}//\vec{c}$ and $\vec{H} \perp \vec{c}$. The sum curve, $\chi_{tot} = 1/3\chi_c + 2/3\chi_a$ reveals at high temperatures the effective magnetic moment of Pr^{3+}, $\mu_{eff} = 3.58\ \mu_B$.

additionally the interaction of the conduction electron spin with the magnetic moment of the ion.

Scattering can be elastic or inelastic; in both cases, the scattering cross section and thus the efficiency of scattering is different for spin↑ and spin↓ electrons.

If only local magnetic moments are considered, as it is almost perfectly the case for the $4f$ moments of the Rare Earth elements, perturbation type calculation in the scope of the Heisenberg model

$$H = -\mathcal{J} \cdot \vec{S} \cdot \vec{s} \tag{87}$$

can be done. To be more specific, we assume that the exchange interaction between the conduction electron at the site $\vec{r}$ with spin $\vec{s}$ and the localized magnetic ion at the site $\vec{R}$ with spin $\vec{S}$ is of importance only, if the conduction electron is in the proximity of the magnetic ion. This is expressed mathematically as a δ-function, i.e., the potential of the interaction reads

$$V(\vec{r}) = -\mathcal{J}\delta(\vec{r} - \vec{R})\vec{s}\vec{S}(1/\hbar^2).$$

where $\mathcal{J}$ is the exchange integral or exchange constant ($\mathcal{J}$ has the dimension *energy × volume*). With $\boldsymbol{J} = \boldsymbol{L} + \boldsymbol{S}$ it follows that $\boldsymbol{S} = (g-1)\boldsymbol{J}$. Within the standard notation, g_j is the Landé-factor and j the total angular momentum. Thus

$$V(\vec{r}) = -(1/\hbar^2)\mathcal{J}(g-1)\delta(\vec{r} - \vec{R})\vec{s}\vec{j}. \tag{88}$$

Note that $\vec{s}\vec{J}$ can be written as

$$\vec{s} \cdot \vec{j} = s_z J_z + \frac{1}{2}(s_+ J_- + s_- J_+)$$

A conduction electron in the state $\vec{k}$ moving in the vicinity of the magnetic ion is scattered by the potential given in Eq. (88) into a new state $\vec{k}'$. The scattering occurs without spin flip if the initial and the final state are connected by $s_z J_z$, but a spin flip occurs if they are connected by the terms $s_\pm J_\pm$.[18] The scattering probability is proportional to the absolute square of matrix elements $M_{ii'}(\vec{k} \to \vec{k}')$ connecting initial (i) and final (i') states of the system.

The electrical resistivity ρ, following the *Drude model* can be expressed as

$$\rho = \frac{3\pi^2 m}{e^2 k_F^3} \frac{1}{\tau(\epsilon)}$$

where m is the charge carrier mass, e the electron charge, k_F the Fermi wave vector and $\tau(\epsilon)$ the relaxation time. Textbooks on *Transport phenomena*, e.g., by J.M. Ziman[19] demonstrate that

$$\frac{1}{\tau_\pm} = \frac{mk_F}{\pi\hbar^3}\left[\sum_{i,i'} N_i|M_{ii'}(\vec{k}_\pm \to \vec{k}'_\pm)|^2 \frac{2}{1+\exp(-E_{ii'}/k_BT)}\right.$$
$$\left. + \sum_{j,j'} N_j|M_{jj'}(\vec{k}_\pm \to \vec{k}'_\mp)|^2 \frac{2}{1+\exp(-E_{jj'}/k_BT)}\right] \qquad (89)$$

where $\tau_\pm$ are the relaxation times for spin up $(+)$ and spin down $(-)$ electrons; $E_{ii'}$ and $E_{jj'}$ are the energies gained by the electrons in the particular scattering process. N_i and N_j are the number of scattering centres per unit volume producing the collision processes.

In the absence of electric and magnetic fields, all the m_j substates are degenerate and thus they are equally probable. $E_{ii'} = E_{jj'} = 0$ since all collisions are elastic.

The electrical resistivity for the paramagnetic temperature range $T > T_{ord}$ is then calculated from:

$$\rho_{mag}(T) = \frac{3\pi N m^*}{2\hbar e^2 \epsilon_F}|\mathcal{J}|^2\frac{1}{4}(g_j - 1)^2 \cdot j(j+1), \qquad (90)$$

indicating a temperature independent expression. For $T > T_{ord}$ the spins are independently from each other. ρ_{mag} is determined by the so called deGennes-factor $(g_j-1)^2 \cdot j(j+1)$, the coupling constant $\mathcal{J}$ and the effective mass m^* of the charge carriers.

CEF fields modelled by the potential V_c, however, cause that scattering can happen in more "channels", represented by the various multiplets which are created by the regularly arranged charges around the magnetic ions. This causes an increase of the spin-disorder resistivity as more and more levels become active (i.e., populated) as the temperature increases.

The matrix elements $M_{ii'}$ and $M_{jj'}$ have now to be taken between the different CEF states. The probability p_i of a magnetic ion being in a certain CEF state with energy E_i follows via the Boltzmann statistics from

$$p_i = \frac{N_i}{N} = \frac{\exp(-E_i/k_BT)}{\sum_j \exp(-E_j/k_BT)}.$$

The transition between different states refers to inelastic scattering, hence, the $E_{ii'}$ and $E_{jj'}$ will not all be zero.

Altogether, adding the CEF effect to Eq. (90), the electrical resistivity in presence of CEF splitting becomes modified and may be expressed as

$$\rho_{spd}(T) = \frac{3\pi N m}{\hbar e^2 \epsilon_F} \mathcal{J}^2 (g_j - 1)^2 \sum_{m_s, m_s', i, i'} \langle m_s'; \Gamma_{i'} | \vec{s}\vec{J} | m_s; \Gamma_i \rangle^2 p_i f_{ii} \qquad (91)$$

m_s and m_s' are the spins of the conduction electrons in the initial and the final states. Γ_i and $\Gamma_{i'}$ are the CEF states with energies E_i and E_i'. The matrix elements are between the simultaneous eigenstates for the local-moment-conduction-electron system; the f_{ii} are given by

$$f_{ii'} = \frac{2}{1 + \exp[-(E_i - E_i')/k_B T]}.$$

Note that similarly to the case of inelastic neutron scattering transition probabilities between CEF states are of importance. Standard selection rules are such that matrix elements have finite values only if $m_j \rightarrow m_j \pm 1$ (dipole exchange selection). If such states are not available, scattering between these states will not occur, hence does not contribute to the electrical resistivity. For temperatures high compared to the overall CEF splitting, Eq. (90) is recovered, resulting in a temperature independent expression.

Figure 16 shows the remarkable influence of CEF effects on the spin-disorder resistivity. Data from the previous hexagonal Yb compounds, model A and B are used. The thermal population in the context of transition probabilities between CEF states at different CEF levels yields a strong

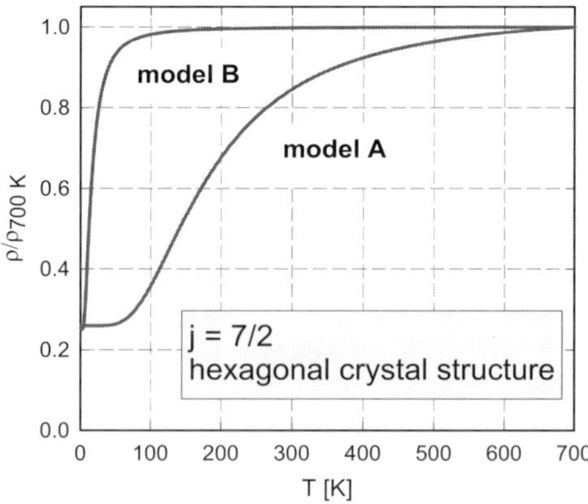

Fig. 16. Electrical resistivity of an Yb compound ($j = 7/2$) in a hexagonal crystal structure for CEF parameters of model A (Table 4) and model B (Table 6).

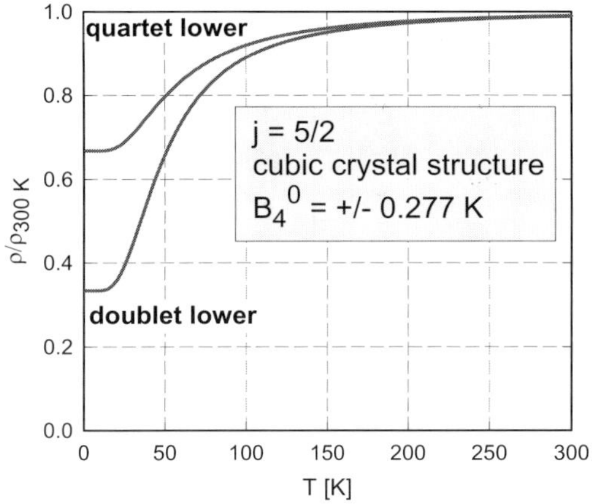

Fig. 17.   Electrical resistivity of a cerium compound ($j = 5/2$) in cubic environment. $B_4^0 = \pm\Delta/360$ [K] with $\Delta = 100$ K.

temperature dependence, which *should not* be confused with e.g., T^2 dependences derived from Fermi liquid behaviour or similar interactions!

In order to demonstrate the influence of the degeneracy of a CEF level regarding the electrical resistivity, a simple example of an $j = 5/2$ in cubic symmetry is shown in Fig. 17. A quartet as ground state, in comparison to the doublet, causes a higher scattering rate, hence the resistivity for $T \to 0$ is larger, as obvious from Fig. 17. Because the overall CEF splitting $\Delta = 100$ K, the resistivity around room temperature becomes almost constant.

2.6.   *Thermal Expansion and Magnetostriction*

As pointed out already in the context of the Yb example, the change of temperature and magnetic field will lead to a change in the charge density of the $4f$ electrons, which in turn will influence the lattice. Figure 18 shows the basic mechanism[20] for a simple example.

A CEF is produced according to Coulomb's law by two positive charges (situated above and below the rare earth). For temperatures higher than the overall CEF splitting Δ_{cf} of the $4f$ ground state multiplet the $4f$ charge density is spherical symmetric (Fig. 18(a) left). When the temperature is lower than the CEF splitting, only the low energy CEF states are thermally populated. This leads to a continuous deformation of the $4f$ charge density with decreasing temperature. The shape of the deformation resembles the

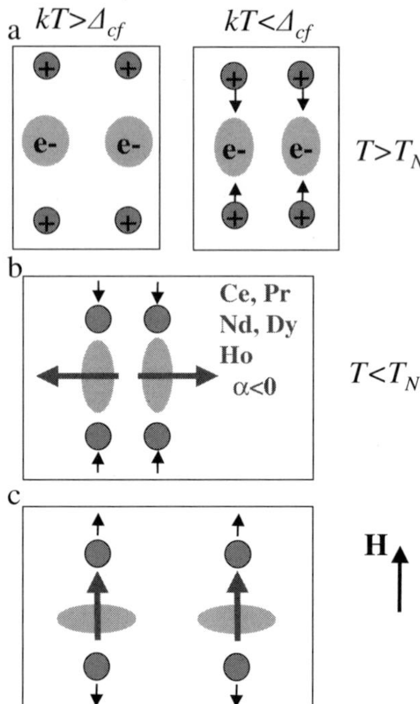

Fig. 18.   CEF mechanism (a to c) for magnetoelastic strains. For a description see text.

geometry of the CEF, in our example the two positive charges produce
an ellipsoidal (cigar) shape. The CEF striction arises from the fact, that
a thermally (or magnetic field induced) change of the $4f$ charge density
shape changes the force between the $4f$ ion and the positive charges which
produce the CEF. A magnetostrictive strain of the crystal lattice results as
indicated by the small arrows in Fig. 18(a)-right. For this magnetostrictive
effect no long range order is necessary and it is called CEF influence on the
thermal expansion.[21–23]

We now consider what happens, when the system orders antiferromagnetically. The shape of the $4f$ charge density is closely associated with
magnetic anisotropy. In CEF theory this anisotropy is governed by Stevens
factors α_j, β_j and $\gamma_j{}^2$. In our simple example of Fig. 18(a) the magnetic
easy axis will be vertical, i.e. along the revolution axis of the ellipsoid, if
the sign of the Stevens factor α_j is *positive* (such as for Sm^{3+}, Er^{3+}, Tm^{3+},
Yb^{3+}). Conversely, if the sign of the Stevens factor α_j is *negative* (such as
for Ce^{3+}, Pr^{3+}, Nd^{3+}, Tb^{3+}, Dy^{3+}, Ho^{3+}) the magnetic easy axis will be
horizontal.

In Fig. 18(b) we consider rare earth with *negative* Stevens factors α_j: Below the Néel temperature T_N the magnetic moments order antiferromagnetically as indicated by the large arrows. The deformation of the $4f$ charge density already incipient above T_N is increased by the appearance of the magnetic moment and consequently leads to a spontaneous magnetostrictive effect below T_N as indicated by the small arrows.

Due to the spin orbit coupling the orientation of the magnetic moment (easy axis) is coupled to the orientation of the $4f$ charge (Fig. 18(c)). The application of a magnetic field normal to the easy axis reduces the ellipsoidal deformation of the charge density and reverses the spontaneous magnetostrictive effect. If the magnetic field is strong enough, the charge density may even turn from an cigar shape into a pancake as shown in Fig. 18(c). In a similar way the CEF striction can be analysed for more complex configurations, where the shape of the charge density is more complicated.

Formally the CEF striction can be described by considering the dependence of the CEF parameters on the strain ϵ and the elastic energy $E_{\rm el}$:

$$\mathcal{H}_{\rm cf} = \sum_{lm,\mathbf{i}} B_l^m(\mathbf{i}, \bar{\bar{\epsilon}}) O_l^m(\mathbf{J_i}) \tag{92}$$

$$E_{\rm el} = \frac{V}{2} \sum_{\alpha\beta} c_{\alpha\beta} \epsilon^\alpha \epsilon^\beta \tag{93}$$

If a magnetic field is applied, in addition the Zeman energy has to be considered:

$$\mathcal{H}_{\rm Ze} = -\sum_{\mathbf{i}} g_j \mu_B \vec{J_{\mathbf{i}}} \vec{B}. \tag{94}$$

Starting point for the further analysis is the Taylor expansion of the CEF parameters with respect to the components of the strain tensor $\bar{\bar{\epsilon}}$, leading to the so-called magnetoelastic Hamiltonian.

$$\sum_{lm,\mathbf{i}} B_l^m(\mathbf{i}, \bar{\bar{\epsilon}}) O_l^m(\mathbf{J_i}) \approx \sum_{lm,\mathbf{i}} B_l^m(\mathbf{i}, \bar{\bar{\epsilon}}=0) O_l^m(\mathbf{J_i}) + \sum_{lm,\mathbf{i}} \epsilon^\alpha B_{l(\alpha)}^m(\mathbf{i}) O_l^m(\mathbf{J_i}) + \cdots \tag{95}$$

with the magnetoelastic constants

$$B_{l(\alpha)}^m(\mathbf{i}) = \left[\frac{\partial B_l^m(\mathbf{i}, \bar{\bar{\epsilon}})}{\partial \epsilon^\alpha} \right]_{\epsilon=0}. \tag{96}$$

Usually the analysis is limited to the first order in the strain (harmonic approximation) and second order terms (anharmonic coupling) are neglected.

230   *Ernst Bauer and Martin Rotter*

By definition the magnetic free energy is given by

$$F_m = -k_B T \ln Z \tag{97}$$

with the partition sum

$$Z = Tr\{e^{-\mathcal{H}/k_B T}\}. \tag{98}$$

Here k_B denotes the Boltzmann constant. In a first order approach the trace in Eq. (98) is calculated using the states of the unperturbed system, i.e. without taking into account the magnetoelastic interactions, by putting $\bar{\bar{\epsilon}} = 0$ in Eq. (92).

Inserting the Hamiltonian $\mathcal{H} = \mathcal{H}_{\mathrm{CEF}} + E_{\mathrm{el}} + \mathcal{H}_{\mathrm{Ze}}$ (Eq. (92)) into Eq. (97) and Eq. (98), calculating the derivative of the magnetic free energy F_m with respect to the strains ϵ^α yields the final result

$$\epsilon^\alpha_{CEF} = -\frac{1}{V} \sum_{\beta,\mathbf{i}} s^{\alpha\beta} B^m_{l(\alpha)}(\mathbf{i}) \langle O^m_l(\vec{J_i}) \rangle_{T,\vec{B}} \tag{99}$$

Eq. (99) shows that the complete temperature and field dependence of the strains can be calculated from thermal expectation values of Stevens operator equivalents $\langle O^m_l(\vec{J_i}) \rangle_{T,\vec{B}}$. The magnetoelastic constants can either be derived from theoretical models such as the point-charge model[24] or, alternatively, enter as adjustable parameters into the theory. In principle some evidence can also be obtained from the shift of CEF excitation spectra under pressure.[25]

As an example for the CEF striction effect we discuss $TmCu_2$.[22] Note in the orthorhombic symmetry there are nine CEF parameters to be considered: $B^0_2 = -0.94$ K, $B^2_2 = -1.23$ K, $B^0_4 = -0.9 \times 10^{-2}$ K, $B^2_4 = -0.39 \times 10^{-2}$ K, $B^4_4 = -0.36 \times 10^{-2}$ K, $B^0_6 = 0.58 \times 10^{-4}$ K, $B^2_6 = 2.47 \times 10^{-4}$ K, $B^4_6 = -0.48 \times 10^{-4}$ K, $B^6_6 = 6.31 \times 10^{-4}$ K.

In order to determine the influence of the CEF, the thermal expansion measured by temperature dependent powder x-ray diffraction is compared to the nonmagnetic isostructural system YCu_2. In Fig. 19 the lattice constants normalized to T=300 K are shown for both compounds.

The CEF contribution to the thermal expansion can be determined by taking the difference of the strains and is shown in Fig. 20. It can be well described by the simple model equation

$$\epsilon^\alpha_{CEF} = A_\alpha \langle O^0_2(\vec{J}) \rangle_T + B_\alpha \langle O^2_2(\vec{J}) \rangle_T \tag{100}$$

with $A_a = 5.9 \times 10^{-5}$, $A_b = -1.1 \times 10^{-4}$, $A_c = 1.1 \times 10^{-5}$, $B_a = 2.1 \times 10^{-4}$, $B_b = -3.9 \times 10^{-5}$, $B_c = -6.6 \times 10^{-5}$.

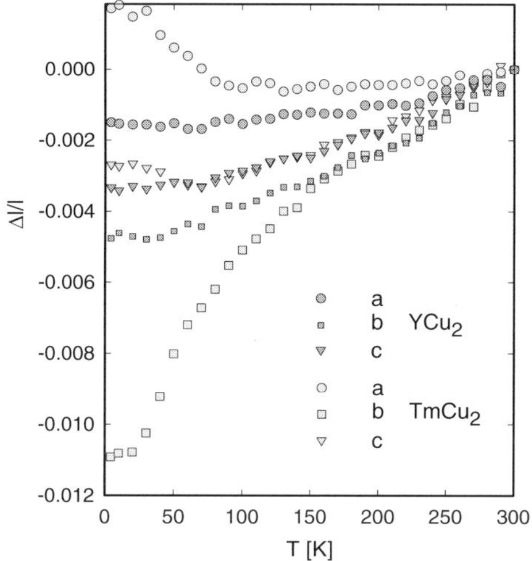

Fig. 19. Lattice constants of $TmCu_2$ and nonmagnetic YCu_2 determined by temperature dependent powder x-ray diffraction and normalized to T=300 K.

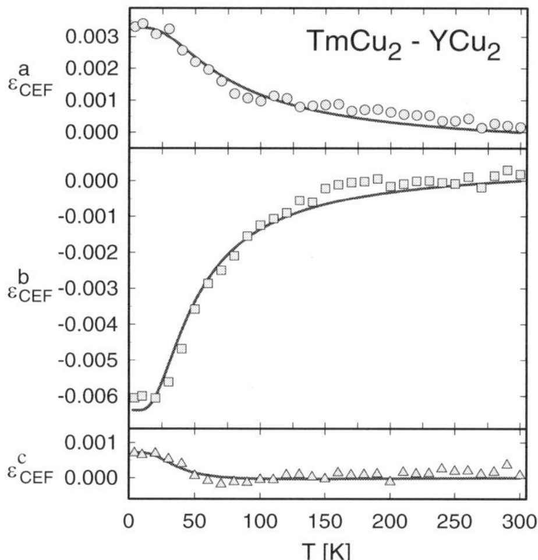

Fig. 20. Difference of normalized lattice constants of $TmCu_2$ and YCu_2 versus temperature. The lines correspond to the results of the CEF model for the thermal expansion.

3. Magnetic Behaviour of Complex Metallic Alloys: Skutterudite $PrFe_4Sb_{12}$

Ternary skutterudites $R_yT_4X_{12}$ with R = Rare Earth, T = Fe, Co, Rh, Ru, ... and X = P, As, Sb, attracted much interest because of a variety of possible ground states and because of their large thermoelectric potential. Depending on the particular Rare Earth element, features like superconductivity as e.g., in $LaRu_4As_{12}$ below $T_c = 10.3$ K or $LaOs_4As_{12}$ below $T_c = 3.2$ K,[26] long range magnetic order in $EuFe_4Sb_{12}$ below $T_{mag} = 84$ K,[27,28] heavy fermion behaviour in $YbFe_4Sb_{12}$,[29] non-Fermi-liquid features in $CeRu_4Sb_{12}$,[30] intermediate and mixed valence behaviour in $Yb(Fe, Co)_4Sb_{12}$ and $Eu(Fe, Co)_4Sb_{12}$[31,32] and hopping conductivity in $YbRh_4Sb_{12}$[31] were already found. Significant interest in this family of compounds, however, stems from the fact that skutterudites are potential candidates for thermoelectric applications. Materials considered for such a use should exhibit values for the *figure of merit* $ZT = S^2T/(\rho\lambda)$ at least of the order of one (T ... temperature, S ... Seebeck coefficient, ρ ... electrical resistivity and λ ... thermal conductivity). Depending on the carrier concentration of a particular skutterudite, Seebeck values above about 100 $\mu V/K$ are frequently observed. Besides, ternary skutterudites are outstanding with respect to their low thermal conductivity which, in some cases, may be near to the theoretical limit. As a matter of fact, the dramatically diminished $\lambda(T)$ values are associated with an exceptionally large thermal displacement parameter of the loosely bound Rare Earth elements, corresponding to a "rattling" (i.e., soft phonon mode) of these atoms in an oversized cage.[33]

Specifically, $PrFe_4Sb_{12}$ is a cubic compound crystallizing in a body-centred cubic lattice; the local point symmetry of Pr is T_h (compare Fig. 21). As already mentioned, Pr is a non-Kramers ion. Then, the cubic symmetry is responsible for a splitting of the $j = 4$ total angular momentum of Pr^{3+} into a Γ_1 singlet, a Γ_3 doublet and in the Γ_4 and Γ_5 triplets. For non-Kramers ions like Pr, nonmagnetic ground states, i.e., Γ_1 and Γ_3, are possible.

Very recently, however, Takegahara et al.[34] have shown that the standard CEF Hamiltonian for simple cubic materials cannot be used for skutterudites. This implies that the Stevens operators ($O_6^2 - O_6^6$) are non-vanishing for the cubic point groups T_h and T due to the lack of Umklappung and fourfold symmetry axis of the point group O_h. Hence, the CEF Hamiltonian reads:

$$H_{cub}^* = B_4^0(O_4^0 + 5O_4^4) + B_6^0(O_6^0 - 21O_6^4) + B_6^2(O_6^2 - O_6^6) \qquad (101)$$

While, in general, the Hamiltonian Eq. (101) leaves the degeneracy of the CEF states unchanged (in comparison to the standard CEF Hamiltonian for cubic symmetry, Eq. (48)), the wave functions, and thus matrix elements

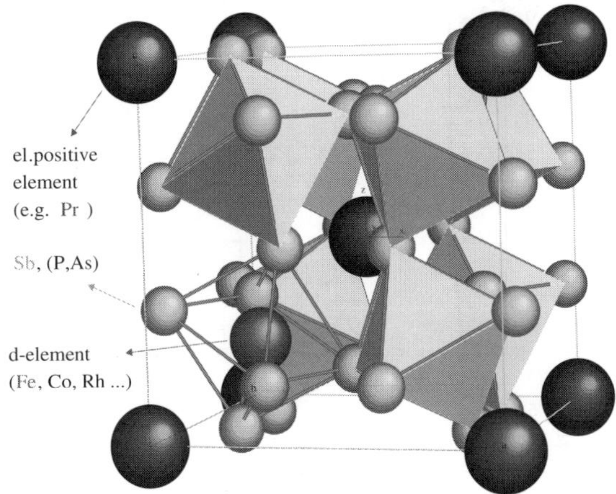

Fig. 21.   Crystal structure of $Pr_{0.73}Fe_4Sb_{12}$.

and magnetic moments are modified, depending on the magnitude of the additional sixth-order terms. Specifically, the relative energy scheme of Γ_1 and Γ_3 is the same, but that of Γ_5 and Γ_5 is dominated by the value of B_6^2.

Realistic CEF parameters B_l^m of a certain system are, in general, obtained from neutron inelastic scattering studies. Exposing at low temperatures a certain material to incoming neutrons, using triple-axis or time-of-flight spectrometers, allows to obtain excitations from the ground state CEF level to excited CEF levels. Such excitations are allowed only, if the matrix elements for magnetic dipole transitions between two distinct states have finite values. Once, the temperature rises, the first, second, etc. level becomes thermally populated and new transitions between excited states become possible with non-vanishing matrix elements. Note that such inelastic transitions may evolve in two different directions: neutrons may absorb energy from the system (at higher temperatures), or loose energy after a collision with the magnetic moment under consideration.

An evaluation of inelastic neutron scattering (INS) data helps in general to define the CEF levels and their splitting from the ground state. Using least squares fitting, the CEF parameters B_l^m may be derived.

In the following subsection we will discuss the CEF scheme $PrFe_4Sb_{12}$ as derived from INS studies at the time-of-flight instrument HET at ISIS, U.K., and will use the B_l^m data derived to describe a number of physical properties which are distinctly governed by CEF effects.

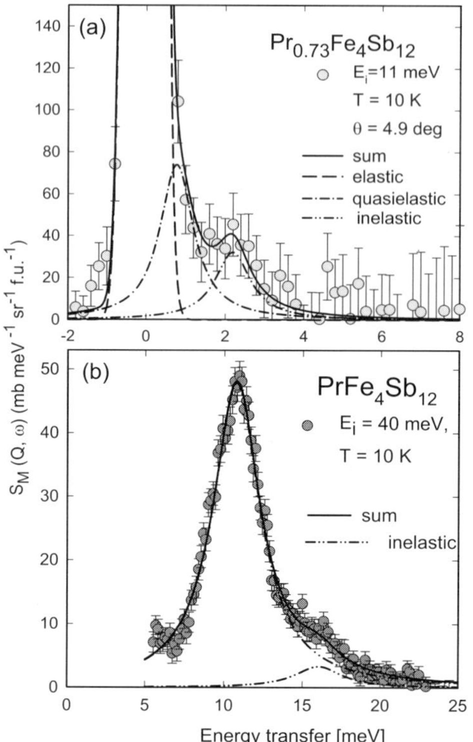

Fig. 22. Energy dependent magnetic scattering intensity S_M of $Pr_{0.73}Fe_4Sb_{12}$. (a): $T = 10$ K, $E_i = 11$ meV; (b): $T = 10$ K, $E_i = 40$ meV. The solid, dashed, dashed-dot and dashed-dot-dot lines describe the sum, elastic, quasielastic, and inelastic contributions, respectively.

The magnetic parts of the inelastic neutron spectra S_M collected for $Pr_{0.73}Fe_4Sb_{12}$ at $T = 10$ K, $E_i = 11$ meV and $E_i = 40$ meV are shown in Figs. 22(a) and (b). S_M is derived by comparison with the respective data of $LaFe_4Sb_{12}$.[35] The low energy excitations are decomposed into a quasielastic line and an inelastic line centred at $E_1 = 2.2$ meV. The spectrum at higher incident energy reveals additional inelastic lines at $E_2 = 10.7$ meV and $E_3 = 16.8$ meV.

Considering the experimentally observed values of inelastic CEF excitations in the context of Eq. (101), a set of CEF parameters is obtained: $B_4^0 = 0.03712$ K, $B_6^0 = 0.00125$ K and $B_6^2 = 0.001$ K, matching the estimated parameters $B_4^0 = 0.04$ K and $B_6^0 = 0.00133$ K derived, when only a simple cubic Hamiltonian is considered.[36] These values determine the wave functions of the various eigenstates and cause the triplet $\Gamma_4^{(2)}$ to be the ground state of the system. $\Gamma_4^{(2)}$, which in the simple version of the Hamiltonian corresponds to the triplet Γ_5, possesses a magnetic moment and thus can give rise to a magnetically ordered ground state. In fact, various bulk properties derived for $Pr_{0.73}Fe_4Sb_{12}$ indicate magnetic ordering below about 4.6 K.[36,37] The CEF singlet Γ_1 is situated 25.8 K above the ground state.

Differences between the appropriate Hamiltonian for skutterudites (Eq. (101)) and the standard Hamiltonian for cubic systems (compare e.g. Eq. (48)) are small particularly for thermodynamic quantities like the specific heat, but have severe consequences for properties like the inelastic neutron scattering, where the intensity of a distinct transition depends on the square of the respective matrix element, i.e. $|\langle\Gamma_i|J_z|\Gamma_j\rangle|^2$; Γ_i and Γ_j represent the initial and the final CEF state, respectively. If this matrix element vanishes, a transition will likely not occur, and no neutron inelastic scattering intensity (in general a Lorentzian line) will be found at an energy $\Delta = \Delta(\Gamma_j) - \Delta(\Gamma_i)$.

The fact that the singlet state Γ_1 is observed roughly 25 K above the magnetic ground state triplet in $Pr_{0.73}Fe_4Sb_{12}$ from INS, cannot be explained by the Hamiltonian Eq. (48), since $|\langle\Gamma_5|J_z|\Gamma_1\rangle|^2 = 0$. The latter follows from the dipole selection rules with $\Delta m_j = \pm 1$. On the contrary, the Hamiltonian for skutterudites, Eq. (101), yields a finite matrix element, i.e., $|\langle\Gamma_4^2|J_z|\Gamma_1\rangle|^2 \neq 0$, thus intensity may be observed, as it is obvious from Fig. 22. Differences regarding the resulting CEF schemes of both CEF Hamiltonians applied to $Pr_{0.73}Fe_4Sb_{12}$ are sketched in Fig. 23.

Results of magnetic susceptibility measurements performed on $Pr_{0.73}Fe_4Sb_{12}$ are shown in Figs. 24(a) and (b). The Curie-Weiss like behaviour at elevated temperatures ($T > 50$ K) reveals an effective magnetic moment $\mu_{\text{eff}} = 4.19\ \mu_B$, as well as a paramagnetic Curie temperature $\theta_p \approx 0.5$ K. In order to match the theoretical Rare Earth moments associated with a 3+ state of the Pr ion, a significant contribution to μ_{eff} of the [Fe_4Sb_{12}] sublattice is required. Assuming that both the Rare Earth and the [Fe_4Sb_{12}] contribution to μ_{eff} are simply additive, i.e.,

$$\mu_{eff}^{meas.} = \sqrt{y\left(\mu_{eff}^{Pr}\right)^2 + \left(\mu_{eff}^{[Fe_4Sb_{12}]}\right)^2} \tag{102}$$

(y is the void filling factor) yields an effective magnetic moment for [Fe_4Sb_{12}] of 2.7 μ_B.

A closer inspection of the low temperature data and of the a.c. measurement (Fig. 24(b)) indicates an onset of long range magnetic order. Defin-

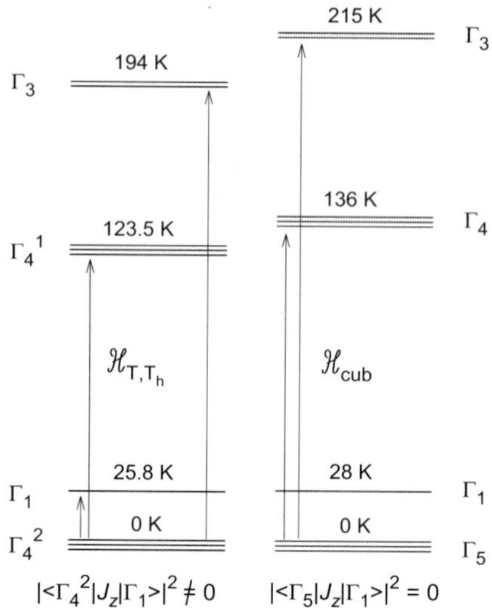

Fig. 23. CEF scheme of $Pr_{0.73}Fe_4Sb_{12}$ derived from both Hamiltonians Eq. (101) (left) and Eq. (48) (right).

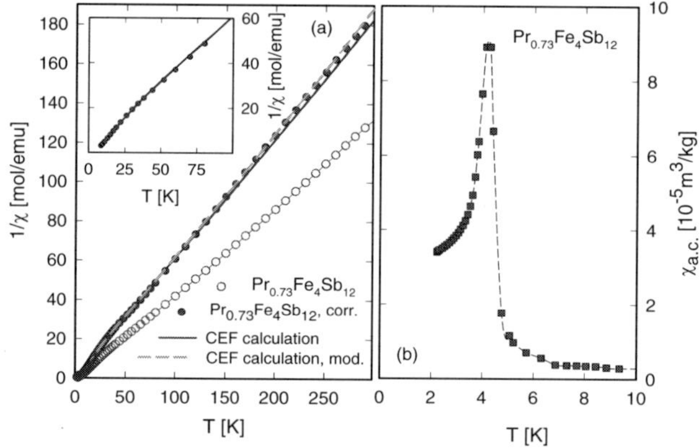

Fig. 24. (a): Temperature dependent magnetic susceptibility χ of $Pr_{0.73}Fe_4Sb_{12}$ plotted as χ^{-1} vs. T. The solid circles represent the Pr-related susceptibility and the solid and the dashed lines are least squares fits according to Eq. (103). The inset shows the low temperature behaviour in more detail. (b): $\chi_{a.c.}(T)$ of $Pr_{0.73}Fe_4Sb_{12}$.

ing the transition temperature of $Pr_{0.73}Fe_4Sb_{12}$ in a standard manner, i.e., by taking the extremum in $d\chi_{dc}T/dT$ as well as the temperature at half height of the a.c. susceptibility anomaly on the paramagnetic side reveals $T_{mag} \approx 4.6$ K. The sharp transition and the overall behaviour of $\chi_{a.c.}(T)$ does not exclude a ferromagnetic ground state.

To fix the contribution originating from by Pr, the susceptibility associated with $[Fe_4Sb_{12}]$ is subtracted from the total measured set of data assuming a simple Curie behaviour and taking into account $\mu_{eff}([Fe_4Sb_{12}]) = 2.7\mu_B$. The thus modified $1/\chi(T)$ curve is added in Fig. 24(a).

The susceptibility $\chi(T)$ related to Pr in $Pr_{0.73}Fe_4Sb_{12}$ is analysed in terms of

$$1/\chi = 1/\chi_{CEF} - \lambda \tag{103}$$

where χ_{CEF} is the susceptibility due to CEF effects and λ is the molecular field parameter caused by exchange interactions between the Pr ions. Note a substantially large value of λ corresponds to large two-ion interactions, which may drive a magnetic instability leading to a long range magnetic ordered ground state.

Taking into account the CEF parameters B_l^m derived from the INS study, in the context of Eq. (83) allows the calculation of $\chi_{CEF}(T)$ as well as of $\chi(T)$ by adjusting λ. A reasonable fit of the experimental data is obtained for $\lambda = 6$ mol/emu. Results of the calculation are shown in Fig. 24(a) as solid line. In particular, the pronounced curvature in $1/\chi(T)$ around 25 K is well reproduced (compare inset, Fig. 24 (a)) and the overall susceptibility behaviour matches fairly well the experiment.

Associated with the Γ_4^2 triplet ground state is a magnetic moment of $\mu = 2\ \mu_B$. This finding is consistent with the occurrence of magnetic order and moreover explains the magnitude of the isothermal magnetization at $T = 2$ K.

Magnetization measurements performed up to 15 T are an additional possibility to consider CEF effects and to corroborate the set of the CEF parameters chosen. Plotted in Fig. 25 is the temperature dependent magnetization for various values of applied magnetic fields. The solid lines are calculations of the magnetization based on Eq. (101) with $g_j = 4/5$, $j = 4$, and the above indicated CEF parameters, revealing reasonable agreement with the experimental data. Note, there are no further adjustable parameters used for this calculation.

Shown in Fig. 26 is the temperature dependent electrical resistivity $\rho(T)$ of $Pr_{0.73}Fe_4Sb_{12}$. The compound behaves metallic, i.e., $\rho(T)$ increases with rising temperature. The particular temperature dependence, however, deviates significantly from a simple metal and is characteristic for materials having a reduced number of charge carriers. At low temperatures, the weak anomaly in $Pr_{0.73}Fe_4Sb_{12}$ indicates the onset of long range magnetic order at 4.6 K.

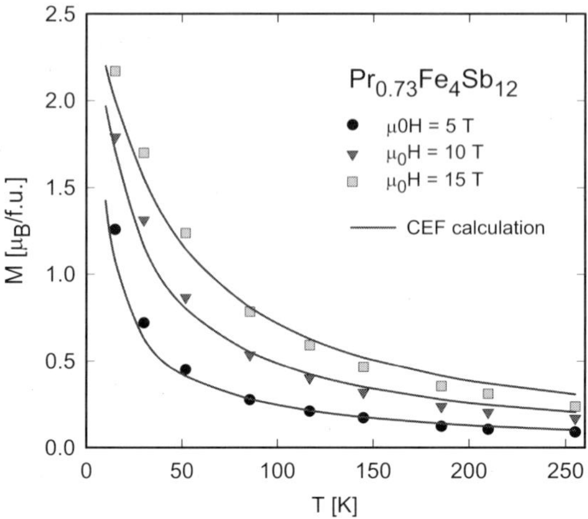

Fig. 25. Temperature dependent magnetization of $Pr_{0.73}Fe_4Sb_{12}$ for various values of externally applied magnetic fields. The solid lines are theoretical values (see text).

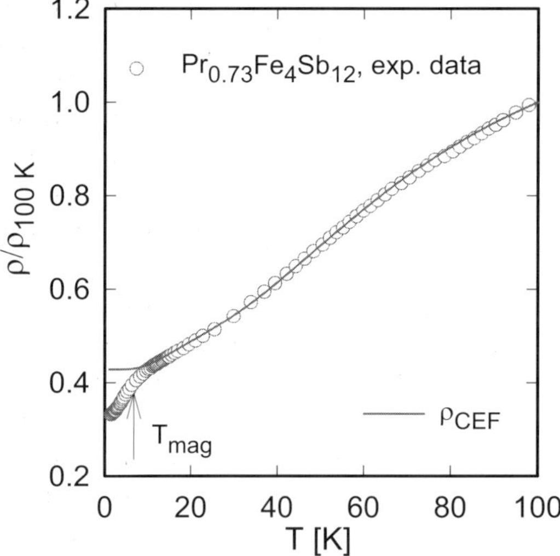

Fig. 26. Normalized resistivity of $Pr_{0.73}Fe_4Sb_{12}$. The solid line represents the CEF derived spindisorder resistivity as derived from the CEF parameters of $Pr_{0.73}Fe_4Sb_{12}$.

To account for the strongly curved $\rho(T)$ behaviour of $Pr_{0.73}Fe_4Sb_{12}$, at least in the lower temperature range ($T < 100$ K), the spin disorder resistivity ρ_{spd} is considered, i.e., scattering of conduction electrons on disordered magnetic moments, in combination with CEF effects, compare Eq. (91).

Using the already derived CEF parameters for the simple cubic model, the normalized spin disorder contribution to the electrical resistivity can be calculated without any further free parameters. Results of this calculation are shown in Fig. 26, together with the resistivity data of $Pr_{0.73}Fe_4Sb_{12}$ normalized to 100 K. Except at low temperatures, where $Pr_{0.73}Fe_4Sb_{12}$ exhibits a phase transition ($T_{mag} = 4.6$ K), the peculiar structure of $\rho(T)$ is well reproduced by this choice of CEF parameters.

4. Outlook

Subjects which could not be touched in this chapter should be briefly mentioned. There is on one hand the extension of the theory to include higher multiplets, such as necessary for example in the case of Sm and Eu.[38] Moreover, in some peculiar cases the CEF may be influenced significantly by optical phonons propagating through the crystal. Such a situation is described by a large CEF phonon interaction,[39] which produces usually a strong damping and consequently increased linewidth of the CEF excitations spectra. Similarly, there may be a strong coupling to conduction electrons resulting also in increased linewidth effects.[40] Moreover, such couplings may result in a temperature dependence of the CEF excitation energies.[41] In very special cases a strong CEF phonon coupling may lead to bound states, which have partly CEF and partly phonon character with a characteristic anticrossing in the dispersion relation.[42]

If the CEF ground state is degenerate it may couple strongly to some lattice strain and below a characteristic temperature the system may lower its lattice symmetry thus lifting the degeneracy of the CEF ground state. Such a Jahn Teller effect is often observed in absence of significant two ion interactions.[43] However, in most Rare Earth systems, two ion interactions are dominant, leading to long range magnetic order. The two ion interactions in the simplest case are isotropic and bilinear, such as for instance *direct* (Heisenberg) and the *indirect exchange interactions* (RKKY, superexchange etc.). In many cases there is strong evidence for anisotropic bilinear interactions,[44] such as the classical dipolar interaction, anisotropic exchange or the completely antisymmetric Dzyaloshinski-Moriya interaction. In many cases the bilinear interactions scale with the deGennes factor. When this factor is small, as for example in the case of Pr and Tm, other types of two ion interactions may become dominant, such as for example the quadrupolar interactions.[43] The Jahn Teller effect in this view is driven by such an effective quadrupolar interaction. However, there exist also "true" two ion quadrupolar or orbital interactions, which are not coupled to a strong lattice distortion. The static behaviour of the ordered states can be extremely complex and is usually treated in mean field theory.[45]

Any two ion interactions will influence not only the static behaviour of the system, but also modify the excitation spectrum. If only the CEF interaction is present, the magnetic modes do not show any dispersion and the main effect of two ion interaction is to produce dispersion in the magnetic excitations. This dispersion can be calculated from the Fourier transform of the two ion interactions within the random phase approximation.[45,46] If the CEF interaction is weak compared to the bilinear (quadrupolar, orbital) interaction the corresponding dispersive excitations are usually called magnons (orbitons).

Computing programs for CEF and two ion interactions

For a quantitative analysis of the CEF problems, software programs have been developed. For users having no or little experience in CEF physics, the program *amos* developed by the group of Prof. Radwankski and Dr. Michalski, Krakow, is recommended, distinguished by an almost self-explaining user interface. The program is free for download at *http://www.css-physics.edu.pl/down.htm*. A very popular and mighty program is *cfield*, which has been incorporated into the *McPhase* package[47] [d]. *McPhase* is a modelling suite for complex problems in magnetism, which allows the quantitative evaluation of most of the effects described above, i.e. long range order, dispersion of excitations[44] etc. As a result of such a modelling exchange and CEF parameters can be quantitatively determined. Moreover, by a quantitative analysis in some cases clear discrepancies between the predictions of the standard model of Rare Earth magnetism and the experimental data have been identified. As examples we mention (i) the discovery of additional magnetic excitations in $CeCu_2$, which are not allowed by symmetry;[48] (ii) the discovery of the magnetoelastic paradox in Gd based compounds.[49,50] These examples demonstrate, that the numerical evaluation of models in magnetism allows to confront theoretical concepts with experimental observations. Discrepancies between experiment and theory, which become evident in this way, will help to improve our understanding of magnetism.

---

[d] www.mcphase.de

Acknowledgment

We thank Prof. M. Loewenhaupt, T.U. Dresden, for critically reading the manuscript.

References

1. see e.g., J. Crangle, *Solid State Magnetism*, Edward Arnold, London (1991).
2. M. T. Hutchings, in: *Solid State Physics* Vol. 16, 227, ed. F. Seitz and D. Thurnbull, Academic Press, New York and London (1964) .
3. P. Fulde, in *Handbook on the Physics and Chemistry of Rare Earths*, ch. 17 p. 295 ed. K. A. Gschneidner and L. Eyring, North Holland Publishing Company (1978).
4. P. Fulde and M. Loewenhaupt, Adv. Physics **34**, (1985) 589.
5. A.J. Freeman and R.E. Watson, Phys. Rev. **127**, (1962) 2058.
6. W. B. Lewis, J. B. Mann, D. A. Liberman, D. T. Cromer, J. Chem. Phys., **53**, (1970) 809.
7. K.W.H. Stevens, Proc. Phys. Soc. A **65**, (1952) 209.
8. A.J. Freeman, *Magnetic Properties of Rare Earth Metals*, ed. by R.J. Elliot, Plenum, New York, (1972).
9. C. Cohen-Tannoudji, Claude, *Quantum Mechanics*, Wiley John + Sons, (1977).
10. J. P. Elliott, P. G. Dawber, *Symmetry in physics* McMillan Press LTD, (1979).
11. U. Walter, Phys. Rev. B **36**, (1987) 2504.
12. J.X. Boucherle, D. Givord and J. Schweizer, J. Phys. (France) **43**, (1982) C7-199
13. J.X. Boucherle and J. Schweizer, Physica B **130**, (1985) 337
14. R. Schedler, U. Witte, M. Rotter, M. Doerr, M. Loewenhaupt and J. Kulda, Physica B **350**, (2004) E107.
15. J. Sereni, in *Handbook on the Physics and Chemistry of Rare Earths*, ch. 98 p. 295 ed. K. A. Gschneidner and L. Eyring, Elsevier Science Publishers B.V., Amsterdam, (1991).
16. V.M. Barthem, D. Gignoux, A. Nait-Saada, D. Schmitt, G. Creuzet, Phys. Rev. B **37**, 1733 (1988).
17. T. Murao and T. Matsubara, Prog. Theor. Phys. **18**, 215 (1957).
18. V.U.S. Rao and W.E. Wallace, Phys. Rev. B **2**, 4613 (1970).
19. J.M. Ziman, *Electrons and phonons: the theory of transport phenomena in solids*, Oxford, The Clarendon Press, (1960).
20. M. Doerr, M. Rotter and A. Lindbaum, Adv. Phys. **54**, (2005) 1.
21. E. Gratz, M. Rotter, A. Lindbaum, H. Müller, E. Bauer and H. Kirchmayr, J. Phys. Cond. Mat. **5**, (1993) 567.
22. E. Gratz, A. Lindbaum and M. Rotter, J. Phys. Cond. Mat. **5**, (1993) 7955.
23. E. Gratz, M. Rotter, A. Lindbaum, H. Müller, E. Bauer and H. Kirchmayr, Materials Science Forum **133-136**, (1993) 461.

24. K. Hense, PhD thesis Technische Universität Wien (2002).

25. M. Rotter, M. Doerr, M. Loewenhaupt, W. Kockelmann, R.I. Bewley, R.S. Eccleston, A. Schneidewind and G. Behr, J. Magn. Magn. Mat. **269**, (2004) 372.

26. I. Shirotani, K. Ohno, C. Sekine, T. Yagi, T. Kawakami, T. Nakanishi, H. Takahashi, J. Tang, A. Matsushita, and T. Matsumoto, Physica B **281-282**, (2000) 1021.

27. M.E. Danebrock, C.B.H. Evers, and W. Jeitschko, J. Phys. Chem. Solids, **57** (1996) 381.

28. E. Bauer, St. Berger, A. Galatanu, M. Galli, H. Michor, G. Hilscher, Ch. Paul, B. Ni, M.M. Abd-Elmeguid, V.H. Tran, A. Grytsiv, and P. Rogl, Phys. Rev. B **63**, (2001) 224414.

29. N.R. Dilley, E.J. Freeman, E.D. Bauer, and M.B. Maple, Phys. Rev. B **58**, (1998) 6287.

30. N. Takeda and M. Ishikawa, J.Phys.: Cond. Mat. **13**, (2001) 5971.

31. E. Bauer, A. Galatanu, H. Michor, G. Hilscher, P. Rogl, P. Boulet, and H. Noël, Eur. Phys. J. B **14**, (2000) 483.

32. A. Grytsiv, P. Rogl, St. Berger, Ch. Paul, E. Bauer, C. Godart, B. Ni, M.M. Abd-Elmeguid, A. Saccone, R. Ferro, and D. Kaczorowski, Phys. Rev. B **66**, (2002) 094411.

33. G.S. Nolas, D.T. Morelli, and T.M. Tritt, Annu. Rev. Mater. Sci **29**, (1999) 89, and references therein.

34. K. Takegahara and H. Harima, J. Phys. Soc. Japan J. Phys. Soc. Jpn. **70**, (2001) 1190.

35. D. T. Adroja, J.-G. Park, K. A. McEwen, N. Takeda, M. Ishikawa and J.-Y. So Phys. Rev. B **68** (2003), 094425.

36. E. Bauer, St. Berger, Ch. Paul, M. Della Mea, G. Hilscher, H. Michor, M. Reissner, W. Steiner A. Grytsiv, P. Rogl and E. W. Scheidt, Phys. Rev. B **66**, (2002) 214421.

37. N.P. Butch, W.M. Yuhasz, P.-C. Ho, J.R. Jeffries, N. Frederick, T.A. Sayles, X.G. Zheng, M.B. Maple, F.M. Woodward, J.W. Lynn, P. Rogl and G. Giester Phys. Rev. B **71**, (2005) 214417.

38. M. Loewenhaupt, Physica B **130**, (1985) 347.

39. M. Loewenhaupt and U. Witte, J. Phys.: Cond. Mat. **15**, (2003) S519.

40. K.W. Becker, P. Fulde and J. Keller, Z. Phys. B **28**, (1977) 9.

41. K. Hense, U. Witte, M. Rotter, R. Schedler, E. Gratz, H. Nowotny and M. Loewenhaupt, J. Magn. Mag. Mat. **272-276**, (2004) 387.

42. P. Thalmeier and P. Fulde, Phys. Rev. Lett. **49**, (1982) 1588.

43. P. Morin and D. Schmitt, in: *Ferromagnetic Materials* Vol. 5, p1, ed. K. H. J. Buschow and E. P. Wohlfarth, Elsevier Sci. Pub. Amsterdam, The Netherlands (1990).

44. M. Rotter, M. Loewenhaupt, S. Kramp, T. Reif, N. M. Pyka, W. Schmidt and R.v.d. Kamp, Europ. Phys. J. **B 14**, (2000) 29.

45. J. Jensen and A. R. Mackintosh, *Rare Earth Magnetism*, Clarendon Press, Oxford, (1991).

46. M. Rotter J. of Comp. Mat. Sci. **38**, (2006) 404.

47. M. Rotter, J. Magn. Mag. Mat. **272-276**, (2004) e481.

48. M. Loewenhaupt, R. Schedler, U. Witte, M. Rotter and W. Schmidt, Physica B **378-380**, (2006), 775.

49. M. Rotter, A. Lindbaum, M. Doerr, M. Loewenhaupt, A. Barcza, M. el Massalami and B. Beuneu, Europhys. Lett. **75**, (2006) 160.

50. M. Rotter, M. Doerr, M. Zschintzsch, A. Lindbaum, H. Sassik and G. Behr, J. Magn. Magn. Mat., **310**, (2007) 1383.

A. Stevens Operators

The appendix summarises all Stevens operators and the tesseral harmonics.

$$X = J(J+1)$$

$$O_0^0 = 1$$

$$O_1^{-1} = \frac{-i}{2}[J_+ - J_-] = J_y$$

$$O_1^0 = J_z$$

$$O_1^1 = \frac{1}{2}[J_+ + J_-] = J_x$$

$$O_2^{-2} = \frac{-i}{2}[J_+^2 - J_-^2] = J_x J_y + J_y J_x = 2P_{xy}$$

$$O_2^{-1} = \frac{-i}{4}[J_z(J_+ - J_-) + (J_+ - J_-)J_z] = \frac{1}{2}[J_y J_z + J_z J_y] = P_{yz}$$

$$O_2^0 = [3J_z^2 - X]$$

$$O_2^1 = \frac{1}{4}[J_z(J_+ + J_-) + (J_+ + J_-)J_z] = \frac{1}{2}[J_x J_z + J_z J_x] = P_{xz}$$

$$O_2^2 = \frac{1}{2}[J_+^2 + J_-^2] = J_x^2 - J_y^2$$

$$O_3^{-3} = \frac{-i}{2}[J_+^3 - J_-^3]$$

$$O_3^{-2} = \frac{-i}{4}[(J_+^2 - J_-^2)J_z + J_z(J_+^2 - J_-^2)]$$

$$O_3^{-1} = \frac{-i}{4}[(J_+ - J_-)(5J_z^2 - X - 1/2) + (J_+ - J_-)(5J_z^2 - X - 1/2)(J_+ - J_-)]$$

$$O_3^0 = [5J_z^3 - (3X - 1)J_z]$$

$$O_3^1 = \frac{1}{4}[(J_+ + J_-)(10J_z^2 - 2X - 1) + (5J_z^2 - X - 1/2)(J_+ + J_-)]$$

$$O_3^2 = \frac{1}{4}[(J_+^2 + J_-^2)J_z + J_z(J_+^2 + J_-^2)]$$

$$O_3^3 = \frac{1}{2}[J_+^3 + J_-^3]$$

$$O_4^{-4} = \frac{-i}{2}[(J_+^4 - J_-^4]$$

$$O_4^{-3} = \frac{-i}{4}[(J_+^3 - J_-^3)J_z + J_z(J_+^3 - J_-^3)]$$

$$O_4^{-2} = \frac{-i}{4}[(J_+^2 - J_-^2)(7J_z^2 - X - 5) + (7J_z^2 - X - 5)(J_+^2 - J_-^2)]$$

$$O_4^{-1} = \frac{-i}{4}[(J_+ - J_-)(7J_z^3 - (3X + 1)J_z) + (7J_z^3 - (3X + 1)J_z)(J_+ - J_-)]$$

$$O_4^{0} = [35J_z^4 - (30X - 25)J_z^2 + 3X^2 - 6X]$$

$$O_4^{1} = \frac{1}{4}[(J_+ + J_-)(7J_z^3 - (3X + 1)J_z) + (7J_z^3 - (3X + 1)J_z)(J_+ + J_-)]$$

$$O_4^{2} = \frac{1}{4}[(J_+^2 + J_-^2)(7J_z^2 - X - 5) + (7J_z^2 - X - 5)(J_+^2 + J_-^2)]$$

$$O_4^{3} = \frac{1}{4}[(J_+^3 + J_-^3)J_z + J_z(J_+^3 + J_-^3)]$$

$$O_4^{4} = \frac{1}{2}[(J_+^4 + J_-^4]$$

$$O_5^{-5} = \frac{-i}{2}[J_+^5 - J_-^5]$$

$$O_5^{-4} = \frac{-i}{4}[(J_+^4 - J_-^4)J_z + J_z(J_+^4 - J_-^4)]$$

$$O_5^{-3} = \frac{-i}{4}[(J_+^3 - J_-^3)(9J_z^2 - X - 33/2) + (9J_z^2 - X - 33/2)(J_+^3 - J_-^3)]$$

$$O_5^{-2} = \frac{-i}{4}[(J_+^2 - J_-^2)(3J_z^3 - (X + 6)J_z) + (3J_z^3 - (X + 6)J_z)(J_+^2 - J_-^2)]$$

$$O_5^{-1} = \frac{-i}{4}[(J_+ - J_-)\{21J_z^4 - 14J_z^2X + X^2 - X + 3/2\} + \{\ldots\}(J_+ - J_-)]$$

$$O_5^{0} = [63J_z^5 - (70X - 105)J_z^3 + (15X^2 - 50X + 12)J_z]$$

$$O_5^{1} = \frac{1}{4}[(J_+ + J_-)(21J_z^4 - 14J_z^2X + X^2 - X + 3/2)$$
$$+ (21J_z^4 - 14J_z^2X + X^2 - X + 3/2)(J_+ + J_-)]$$

$$O_5^{2} = \frac{1}{4}[(J_+^2 + J_-^2)(3J_z^3 - (X + 6)J_z) + (3J_z^3 - (X + 6)J_z)(J_+^2 + J_-^2)]$$

$$O_5^{3} = \frac{1}{4}[(J_+^3 + J_-^3)(9J_z^2 - X - 33/2) + (9J_z^2 - X - 33/2)(J_+^3 + J_-^3)]$$

$$O_5^{4} = \frac{1}{4}[(J_+^4 + J_-^4)J_z + J_z(J_+^4 + J_-^4)]$$

$$O_5^{5} = \frac{1}{2}[J_+^5 + J_-^5]$$

$$O_6^{-6} = \frac{-i}{2}[J_+^6 - J_-^6]$$

$$O_6^{-5} = \frac{-i}{4}[(J_+^5 - J_-^5)J_z + J_z(J_+^5 - J_-^5)]$$

$$O_6^{-4} = \frac{-i}{4}[(J_+^4 - J_-^4)(11J_z^2 - X - 38) + (11J_z^2 - X - 38)(J_+^4 - J_-^4)]$$

$$O_6^{-3} = \frac{-i}{4}[(J_+^3 - J_-^3)(11J_z^3 - (3X+59)J_z) + (J_+^3 + J_-^3)(11J_z^3 - (3X+59)J_z)(J_+^3 - J_-^3)]$$

$$O_6^{-2} = \frac{-i}{4}[(J_+^2 - J_-^2)\{33J_z^4 - (18X + 123)J_z^2 + X^2 + 10X + 102\} + \{\ldots\}(J_+^2 - J_-^2)]$$

$$O_6^{-1} = \frac{-i}{4}[(J_+ - J_-)\{33J_z^5 - (30X - 15)J_z^3 + (5X^2 - 10X + 12)J_z\} + \{\ldots\}(J_+ - J_-)]$$

$$O_6^{0} = [231J_z^6 - (315X - 735)J_z^4 + (105X^2 - 525X + 294)J_z^2 - 5X^3 + 40X^2 - 60X]$$

$$O_6^{1} = \frac{1}{4}[(J_+ + J_-)\{33J_z^5 - (30X - 15)J_z^3 + (5X^2 - 10X + 12)J_z\} + \{\ldots\}(J_+ + J_-)]$$

$$O_6^{2} = \frac{1}{4}[(J_+^2 + J_-^2)\{33J_z^4 - (18X + 123)J_z^2 + X^2 + 10X + 102\} + \{\ldots\}(J_+^2 + J_-^2)]$$

$$O_6^{3} = \frac{1}{4}[(J_+^3 + J_-^3)(11J_z^3 - (3X+59)J_z) + (J_+^3 + J_-^3)(11J_z^3 - (3X+59)J_z)(J_+^3 + J_-^3)]$$

$$O_6^{4} = \frac{1}{4}[(J_+^4 + J_-^4)(11J_z^2 - X - 38) + (11J_z^2 - X - 38)(J_+^4 + J_-^4)]$$

$$O_6^{5} = \frac{1}{4}[(J_+^5 + J_-^5)J_z + J_z(J_+^5 + J_-^5)]$$

$$O_6^{6} = \frac{1}{2}[J_+^6 + J_-^6]$$

B. Tesseral Harmonics

$$Z_0^0 = \frac{1}{\sqrt{4\pi}}$$

$$Z_1^{-1} = \sqrt{\frac{3}{4\pi}}[y/r]$$

$$Z_1^0 = \sqrt{\frac{3}{4\pi}}[z/r]$$

$$Z_1^{+1} = \sqrt{\frac{3}{4\pi}}[x/r]$$

$$Z_2^{-2} = \frac{1}{4}\sqrt{\frac{15}{\pi}}[2xy/r^2]$$

$$Z_2^{-1} = \frac{1}{2}\sqrt{\frac{15}{\pi}}[yz/r^2]$$

$$Z_2^{0} = \frac{1}{4}\sqrt{\frac{5}{\pi}}[(3z^2 - r^2)/r^2]$$

$$Z_2^{+1} = \frac{1}{2}\sqrt{\frac{15}{\pi}}[xz/r^2]$$

$$Z_2^{+2} = \frac{1}{4}\sqrt{\frac{15}{\pi}}[(x^2 - y^2)/r^2]$$

$$Z_3^{-3} = \sqrt{\frac{35}{32\pi}}[(3x^2y - y^3)/r^3]$$

$$Z_3^{-2} = \sqrt{\frac{105}{16\pi}}[2xyz/r^3]$$

$$Z_3^{-1} = \sqrt{\frac{21}{32\pi}}[y(5z^2 - r^2)/r^3]$$

$$Z_3^{0} = \sqrt{\frac{7}{16\pi}}[z(5z^2 - 3r^2)/r^3]$$

$$Z_3^{+1} = \sqrt{\frac{21}{32\pi}}[x(5z^2 - r^2)/r^3]$$

$$Z_3^{+2} = \sqrt{\frac{105}{16\pi}}[(x^2 - y^2)z/r^3]$$

$$Z_3^{+3} = \sqrt{\frac{35}{32\pi}}[(x^3 - 3xy^2)/r^3]$$

$$Z_4^{-4} = \frac{3}{16}\sqrt{\frac{35}{\pi}}[4(x^3y - xy^3)/r^4]$$

$$Z_4^{-3} = \frac{3}{8}\sqrt{\frac{70}{\pi}}[(3x^2y - y^3)z/r^4]$$

$$Z_4^{-2} = \frac{3}{8}\sqrt{\frac{5}{\pi}}[2xy(7z^2 - r^2)/r^4]$$

$$Z_4^{-1} = \frac{3}{4}\sqrt{\frac{5}{2\pi}}[yz(7z^2 - 3r^2)/r^4]$$

$$Z_4^0 = \frac{3}{16}\frac{1}{\sqrt{\pi}}[35z^4 - 30z^2r^2 + 3r^4)/r^4]$$

$$Z_4^{+1} = \frac{3}{4}\sqrt{\frac{5}{2\pi}}[xz(7z^2 - 3r^2)/r^4]$$

$$Z_4^{+2} = \frac{3}{8}\sqrt{\frac{5}{\pi}}[(x^2 - y^2)(7z^2 - r^2)/r^4]$$

$$Z_4^{+3} = \frac{3}{8}\sqrt{\frac{70}{\pi}}[(x^3 - 3xy^2)z/r^4]$$

$$Z_4^{+4} = \frac{3}{16}\sqrt{\frac{35}{\pi}}[(x^4 - 6x^2y^2 + y^4)/r^4]$$

$$Z_5^{-5} = \sqrt{\frac{693}{512\pi}}[(5x^4y - 10x^2y^3 + y^5)/r^5]$$

$$Z_5^{-4} = \sqrt{\frac{3465}{256\pi}}[4(x^3y - xy^3)z/r^5]$$

$$Z_5^{-3} = \sqrt{\frac{385}{512\pi}}[(3x^2y - y^3)(9z^2 - r^2)/r^5]$$

$$Z_5^{-2} = \sqrt{\frac{1155}{64\pi}}[2xy(3z^3 - zr^2)/r^5]$$

$$Z_5^{-1} = \sqrt{\frac{165}{256\pi}}[y(21z^4 - 14z^2r^2 + r^4)/r^5]$$

$$Z_5^0 = \sqrt{\frac{11}{256\pi}}[(63z^5 - 70z^3r^2 + 15zr^4)/r^5]$$

$$Z_5^{+1} = \sqrt{\frac{165}{256\pi}}[x(21z^4 - 14z^2r^2 + r^4)/r^5]$$

$$Z_5^{+2} = \sqrt{\frac{1155}{64\pi}}[(x^2 - y^2)(3z^3 - zr^2)/r^5]$$

$$Z_5^{+3} = \sqrt{\frac{385}{512\pi}}[(x^3 - 3xy^2)(9z^2 - r^2)/r^5]$$

$$Z_5^{+4} = \sqrt{\frac{3465}{256\pi}}[(x^4 - 6x^2y^2 + y^4)z/r^5]$$

$$Z_5^{+5} = \sqrt{\frac{693}{512\pi}}[(x^5 - 10x^3y^2 + 5xy^4)/r^5]$$

$$Z_6^{-6} = \frac{231}{64}\sqrt{\frac{26}{231\pi}}[(6x^5y - 20x^3y^3 + 6xy^5)/r^6]$$

$$Z_6^{-5} = \sqrt{\frac{9009}{512\pi}}[(5x^4y - 10x^2y^3 + y^5)z/r^6]$$

$$Z_6^{-4} = \frac{21}{32}\sqrt{\frac{13}{7\pi}}[4(x^3y - xy^3)(11z^2 - r^2)/r^6]$$

$$Z_6^{-3} \doteq \frac{1}{32}\sqrt{\frac{2730}{\pi}}[(3x^2y - y^3)(11z^3 - 3zr^2)/r^6]$$

$$Z_6^{-2} = \frac{1}{64}\sqrt{\frac{2730}{\pi}}[2xy(33z^4 - 18z^2r^2 + r^4)/r^6]$$

$$Z_6^{-1} = \frac{1}{8}\sqrt{\frac{273}{4\pi}}[yz(33z^4 - 30z^2r^2 + 5r^4)/r^6]$$

$$Z_6^0 = \frac{1}{32}\sqrt{\frac{13}{\pi}}[(231z^6 - 315z^4r^2 + 105z^2r^4 - 5r^6)/r^6]$$

$$Z_6^{+1} = \frac{1}{8}\sqrt{\frac{273}{4\pi}}[xz(33z^4 - 30z^2r^2 + 5r^4)/r^6]$$

$$Z_6^{+2} = \frac{1}{64}\sqrt{\frac{2730}{\pi}}[(x^2 - y^2)(33z^4 - 18z^2r^2 + r^4)/r^6]$$

$$Z_6^{+3} = \frac{1}{32}\sqrt{\frac{2730}{\pi}}[(x^3 - 3xy^2)(11z^3 - 3zr^2)/r^6]$$

$$Z_6^{+4} = \frac{21}{32}\sqrt{\frac{13}{7\pi}}[(x^4 - 6x^2y^2 + y^4)(11z^2 - r^2)/r^6]$$

$$Z_6^{+5} = \sqrt{\frac{9009}{512\pi}}[(x^5 - 10x^3y^2 + 5xy^4)z/r^6]$$

$$Z_6^{+6} = \frac{231}{64}\sqrt{\frac{26}{231\pi}}[(x^6 - 15x^4y^2 + 15x^2y^4 - y^6)/r^6]$$

CHAPTER 6

ELECTRONIC STRUCTURE OF QUASICRYSTAL-RELATED COMPOUNDS INVESTIGATED BY ULTRA HIGH RESOLUTION PHOTOEMISSION SPECTROSCOPY

Ryuji Tamura

Department of Materials Science and Technology
Tokyo University of Science, Chiba, Japan
E-mail: tamura@rs.noda.tus.ac.jp

The relation between the density of states and the transport properties of metals is briefly examined with the electrical conductivity as a typical example. The principles of photoemission spectroscopy and the reached energy resolution are detailed and exemplified with quasicrystal-related materials. For Cd-based compounds only the thermodynamically stable binary quasicrystal and its approximants are considered. Characteristic features of their valence band spectrum such as a sharp Fermi cutoff edge and a depression in the density of states at the Fermi level are highlighted. Comparison between quasicrystals and approximants is made in terms of long-range order and local structure. The dipole selection rule for the X-ray absorption process is briefly considered and the principles of Resonnant Photoemission Spectroscopy are described to interpret the Cd-based compounds spectra. Finally, their stabilisation mechanism is discussed from a comparison between experimental partial densities of states and theoretical calculations.

1. Introduction

The electronic transport of metals is determined by the electronic structure near the Fermi energy. In order to clarify this point, the relation between the density of states (DOS) and the transport properties of metals is briefly reviewed for the electrical conductivity as a typical example, showing why high resolution is necessary for the spectroscopic

249

studies. Then, the principles of the photoemission spectroscopy (PES) are presented and the energy resolution reached is described in detail.

1.1. *Relation between the density of states and the electronic transport*

The electronic properties of metals are determined by the density of states (DOS) $N(E)$ near the Fermi energy (E_F) roughly within an energy range of $2k_BT$ across E_F, i.e., $E_F - k_BT < E < E_F + k_BT$. For instance, the electrical conductivity of the cubic system is expressed

$$\sigma = e^2 \int_{-\infty}^{\infty} D(E)N(E)\left(-\frac{df(E)}{dE}\right)dE \qquad (1)$$

under the relaxation time approximation, where $D(E)[1/3 v(E)^2 \tau(e)$, $v(E)$, $\tau(E)$ and $f(E)$ $(=1/(1+ exp(E\text{-}E_F)/k_BT)$ are the diffusivity, the average velocity and the relaxation time of the conduction electrons having energy E, and the Fermi-Dirac distribution function, respectively. The integral contains the derivative of the Fermi-Dirac distribution function, which is given by

$$-\frac{df(E)}{dE} = \frac{1}{k_BT}\frac{e^{(E-E_F)/k_BT}}{\left(1+e^{(E-E_F)/k_BT}\right)^2} \qquad (2)$$

Its shape is shown in Fig. 1 for a certain value of k_BT, i.e., $k_BT = 0.5$.

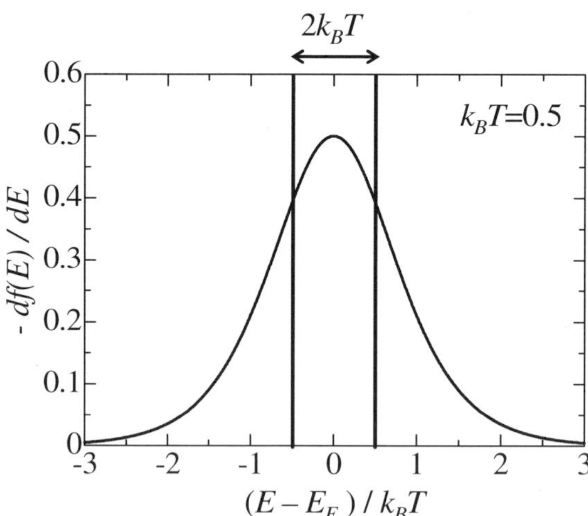

Fig. 1. -*df(E)/dE* as a function of *E-E_F* for *k_BT*=0.5.

Approximately, this component in the integral acts as an energy slit of the width, of roughly $2k_BT$ (unshadowed part in Fig. 1), which means that only the DOS within k_BT from E_F can contribute to the conductivity. The value of k_BT is dependent on temperature, e.g., k_BT = 8.6 meV and 26 meV at 100 K and 300 K, respectively. In an extreme case of T = 0 K, it follows that only the DOS at E_F contribute to the electronic properties of metals. Such consideration leads to the conclusion that the DOS around E_F of less than 20-30 meV is of a particular importance when discussing the electron transport of metals below room temperature.

1.2. *Principles of photoemission spectroscopy*

Photoemission spectroscopy (PES) is principally based on the photoelectric effect, which is related with emission of electrons from a material caused by irradiation with lights such as ultraviolet (UV) or X-ray. The emitted electrons are called photoelectrons. The work function ϕ is defined as an energy barrier for an electron at E_F to escape from a material. Hence, the condition for the photoemission process to occur is expressed by

$$h\nu > \phi \tag{3}$$

where $h\nu$ is the energy of incident photon. The work function ϕ differs among the elements, ranging from 2 to 6 eV as shown in Fig. 2.

Therefore, incident photon with $h\nu > 6$ eV is required for the photoemission spectroscopy and the lights which satisfy the condition are in the range of either ultraviolet (UV) ray or X-ray. When UV ray is used as the photon source, it is called ultraviolet photoemission spectroscopy (UPS), on the other hand, when X-ray is used as a photon source, it is called X-ray photoemission spectroscopy (XPS).

For the photoemission spectroscopy with the incident photon energy $h\nu$, the kinetic energy E'_{kin} of a photoelectron is given by

$$E'_{kin} = h\nu - E_B - \phi \tag{4}$$

where E_{kin} is measured from the vacuum level and E_B is the binding energy of the electron measured from E_F, which is schematically shown

Ryuji Tamura

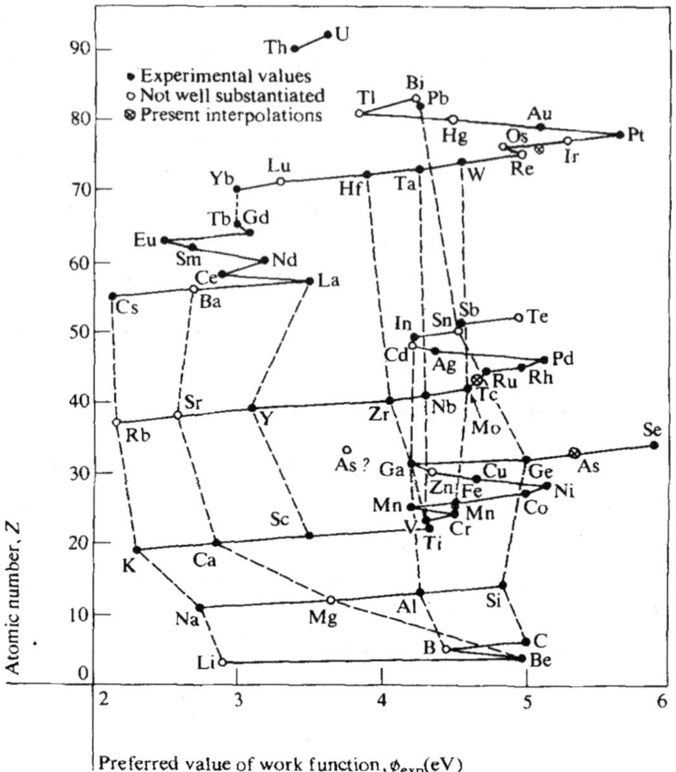

Fig. 2. Work function of the elements (*J. Appl. Phys. 48 (1977) 4729.*)

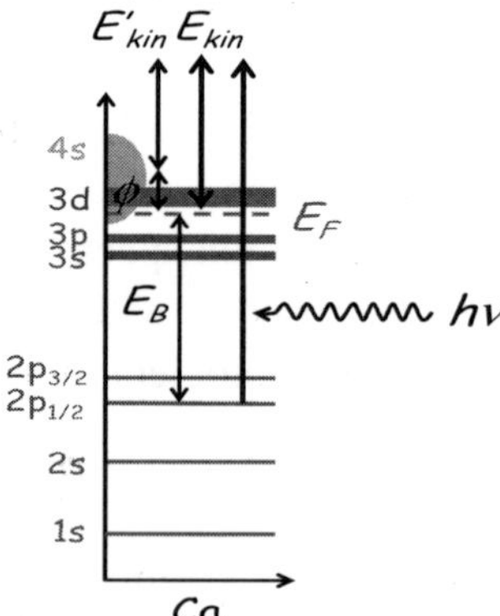

Fig. 3. Schematic illustration of the energy levels of atomic Ca.

the above relation is expressed by

$$E_{kin} = h\nu - E_B,\tag{5}$$

with E_{kin} the kinetic energy of the photoelectron measured from E_F.

In order to obtain the valence band spectrum of metals, the number of photoelectrons at each kinetic energy E_{kin} should be measured with a constant incident photon energy $h\nu$. From Eq. (5), it is clear that uncertainty of the photon energy $h\nu$ should be small in order to measure the binding energy with high resolution.

The above expression of the kinetic energy E_{kin} needs to be modified when the photoelectrons are generated deep inside the bulk of metals since such photoelectrons are likely to lose a part of their kinetic energy before reaching the surface of the metal. In such a circumstance, the kinetic energy of the photoelectrons escaping from the metal is then modified as

$$E_{kin} = h\nu - E_B - \Delta E\tag{6}$$

where ΔE stands for the energy loss by inelastic scattering events. If such inelastic invents occur, the information on the binding energy E_B will be lost. For this reason, the information on the binding energy obtained by PES is restricted only to the region from where the photoelectrons escape without suffering inelastic scattering. This is the reason why the PES is regarded as a surface sensitive technique. For quantitative description of the region probed by PES, inelastic mean free path (IMFP) of a photoelectron should be introduced. The photoelectron IMFP is defined as the average moving distance of photoelectrons without suffering inelastic scattering and it is strongly dependent on the kinetic energy of a photoelectron.

Figure 4 presents IMFP as a function of the kinetic energy (so-called "universal curve"). The IMFP has a minimum < 1nm at around several tens of eV, and increases as the kinetic energy moves towards both higher and lower energies.

The average depth of the signal, defined as the escape depth (*ED*), can be evaluated by using IMFP of electrons in the following manner. *ED* is described by a following expression using the probability

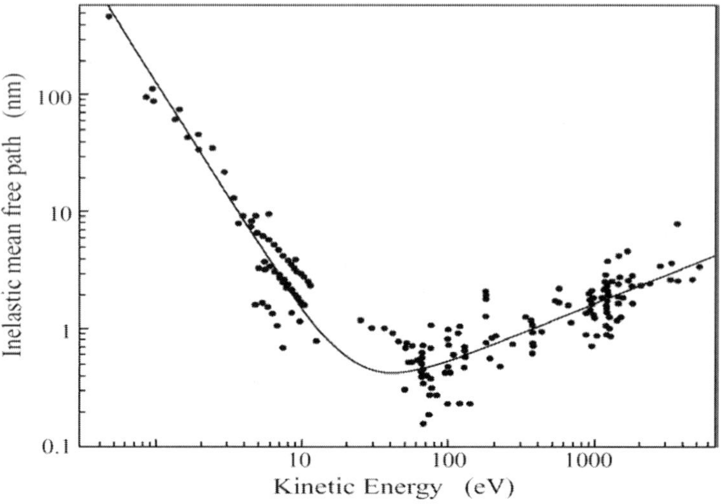

Fig. 4. Inelastic mean free path of a photoelectron as a function of the kinetic energy.

function $P(z)$ that a photoelectron generated at a distance z from the surface reaches the surface without suffering inelastic scattering.

$$ED = \frac{\int_0^\infty zP(z)dz}{\int_0^\infty P(z)dz} \tag{7}$$

The probability function $P(z)$ can be evaluated by taking into account that the scattering probability of a photoelectron while moving a distance dz is given by dz/λ. Then, the reduction of photoelectrons $dN(z)$ after moving from the position z to $z + dz$ is estimated as

$$dN(z) = -\frac{dz}{\lambda}N(z) \tag{8}$$

where $N(z)$ is the number of photoelectrons after moving a distance z from the position that the photoelectrons are generated. $N(z)$ is readily obtained by integrating Eq. 8 as

$$N(z) = N_0 e^{-\frac{z}{\lambda}} \tag{9}$$

where N_0 is the initial number of the photoelectrons at $z = 0$. Then, the probability $P(z)$ that a photoelectron generated at the depth of z reaches the surface without suffering an inelastic scattering is given by

$$P(z) \equiv \frac{N}{N_0} = e^{-\frac{z}{\lambda}} \tag{10}$$

By substituting Eq. 10 into Eq. 7, the escape depth (*ED*) is obtained as

$$ED = \frac{\int_0^\infty z e^{-\frac{z}{\lambda}} dz}{\int_0^\infty e^{-\frac{z}{\lambda}} dz} = \lambda \tag{11}$$

Equation 11 means that the average depth of the signal is equal to the inelastic mean free path of a photoelectron. For this reason, the IMFP is of a particular importance when we interpret PE spectra.

Based on the above estimation, it is clear that the signal comes from a region within 1 nm below the surface when the kinetic energy is several tens of eV from Fig. 4 and hence the PES is a very surface sensitive technique. Since the *ED* is the average depth of the signal, a PE spectrum may contain signal from deep regions of the metal. The ratio of the intensity from a certain region to the total intensity is obtained by using the probability function as follows,

$$\frac{I_x}{I} = \frac{\int_0^x P(z)dz}{\int_0^\infty P(z)dz} = 1 - e^{-\frac{x}{\lambda}} \tag{12}$$

where I_x represents the intensity of the photoelectrons coming from the region $0 < z < x$ below the surface. For instance, 90% (95%) of the signal comes from the region less than $2.3\,\lambda$ ($3\,\lambda$) from the surface.

In XPS, frequently used photon sources are Al Kα (1486.6 eV) and Mg Kα (1253.6 eV). When investigating the valence band profile near

E_F, i.e., E_B ~0 eV, the kinetic energy of the photoelectrons E_{kin} is nearly the same as the incident photon energy hv according to Eq. 5, therefore, the escape depth (*ED*) is estimated to be about 2 nm for both the photon sources from Fig. 4. For UPS, He Iα (20.218 eV) and He IIα (40.814 eV) lines are often used and in this case *ED* is about 0.5 nm, much shorter than that for XPS. Thus, UPS is more sensitive to the surface than XPS.

Although the sensitiveness of the UPS to the surface will be disadvantage when investigate the bulk valence band spectra, on the other hand, the resolution of the UPS is much higher than that of the XPS as explained below. The energy resolution ΔE of the PES is determined by two components. One is the line width ΔE_l of the photon source and the other is the energy resolution ΔE_a of the analyzer. The total resolution ΔE is then given by these two components as

$$\Delta E^2 = \Delta E_l^{\,2} + \Delta E_a^{\,2} \tag{13}$$

The line widths ΔE_l of Al Kα and Mg Kα lines are 0.85 and 0.65 eV, respectively, whereas that of He Iα is about 1 meV. Thus, high resolution PES is realized only by using UV ray as a photon source according to Eq. 13. However, as mentioned above, <u>there is a trade-off between the resolution and the bulk sensitivity.</u>

The energy resolution of the analyzer ΔE_a is evaluated by considering the geometry of the hemispherical analyzer which is schematically shown in Fig. 5.

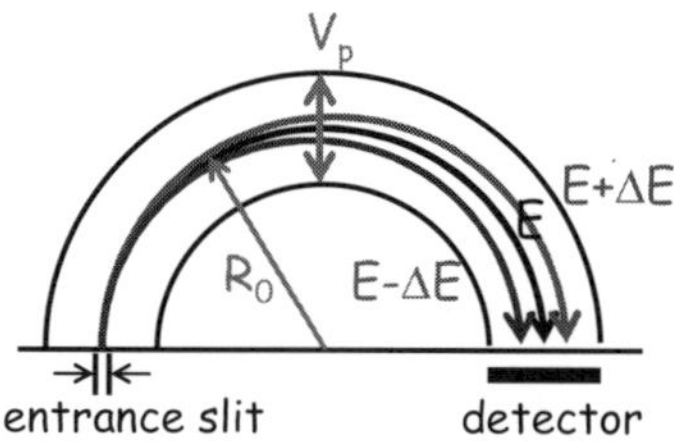

Fig. 5. Cross-sectional view of the hemispherical analyzer.

Here, the principle of the hemispherical analyzer is briefly described in order to explain its instrumental resolution. The photoelectrons emitted from metals are retarded by electrical field before entering the entrance slit of the analyzer. Then, the photoelectrons are guided along the hemispherical pass by applied electrical field Vp (pass energy: $E_0 = \alpha Vp$). Since the photoelectrons having higher kinetic energies are reflected less and vice versa, only photoelectrons within a certain kinetic energy range can reach the detector for a given pass energy E_0. The kinetic energy E_{kin} of the photoelectrons is measured by the position that they arrive at the detector and high resolution is achieved by high dispersion of photoelectrons with respect to the position at the detector. There are three components which affect the resolution of the analyzer; pass energy E_0, the radius of the hemisphere R_0 and the width w of the slit for incoming photoelectrons. Since the kinetic energy of photoelectrons is measured by their positions at the detector, the slit width w should be small, i.e., ideally zero, for better energy separation at the detector position and the radius R_0 should be large for better dispersion at the detector position. On the other hand, as for the pass energy E_0, an increase in E_0 suppresses the dispersion at the detector resulting in the reduction of the resolution. Hence, the energy resolution of the analyzer is given by:

$$\Delta E_a = \frac{w}{2R_0} E_0 \tag{14}$$

Experimentally, the highest resolution has been achieved by a typical condition of $R_0 = 200\text{mm}$, $w = 0.1\text{mm}$ and $E_0 = 1\text{eV}$. In this case, the resolution of the analyzer is estimated to be $\Delta E_a = 250\mu\text{eV}$.

Based on the above evaluation of the energy resolutions of the photon source and the analyzer, the current best resolution can be obtained by Eq. 13 using $\Delta E_l \sim 1$ meV of He I line and $\Delta E_a \sim 0.25$ meV for the analyzer, resulting in $\Delta E \sim 1.0$ meV. Hence, <u>the resolution limit of UPS is determined by the line width of He I line and in order to achieve higher resolution less than 1 meV, photon sources having a shorter line width are severely required.</u>

1.3. *Fermi edge of simple metals*

The intensity $I(E_B)$ of a PE spectrum is given by a convolution of DOS, FD function and instrumental resolution:

$$I(E_B) = \int N(x)\left(\frac{1}{1+e^{-x/k_BT}}\right)\exp\left[-\frac{(x-E_B)^2}{2s^2}\right]dx \qquad (15)$$

where $N(x)$ represents the DOS and the instrumental energy resolution is approximated by a Gaussian function.

Now, let us consider a simple case where DOS is constant in the vicinity of E_F, which corresponds to the most simplified picture of normal metals. In the high resolution limit ($s \to 0$), Eq. 15 is approximated as:

$$I(E_B) = N \int\left(\frac{1}{1+e^{-x/k_BT}}\right)\delta(x-E_B)dx = \frac{N}{1+e^{-E_B/k_BT}} \qquad (16)$$

where the following property of the delta function was utilized.

$$\delta(x-x_0) = \lim_{s\to 0}\exp\left[-\frac{(x-x_0)^2}{2s^2}\right] \qquad (17)$$

From Eq. 16, it is understood that the shape of the Fermi edge is given by FD function for normal metals having smooth and featureless DOS, which is presented in Fig. 6 for three different temperatures.

With increasing temperature, the Fermi edge is broadened according to the temperature dependence of the FD function and low temperature is

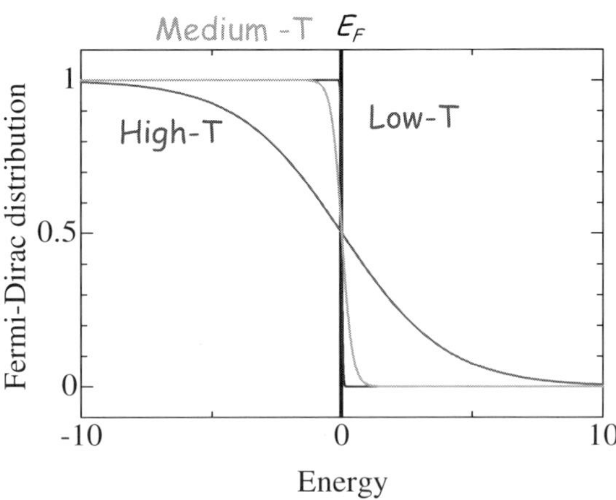

Fig. 6. Shape of the Fermi-Dirac distribution function.

required for obtaining a sharp Fermi cutoff. Low temperature is severely required for high resolution PES since the DOS is smoothed by thermal broadening of the FD function of the energy scale of roughly $k_B T$ at finite temperatures.

2. High Resolution Photoemission Spectroscopy in Quasicrystals and Approximants

Ultrahigh resolution photoemission spectroscopies on quasicrystal-related materials are presented in this chapter. For the quasicrystal-related materials, we have studied Cd-based quasicrystals (QCs) $Cd_{5.7}M(M=Yb,Ca)$, paying attention only into binary icosahedral QCs that are thermodynamically stable, as well as their approximants $Cd_6M(M-Yb,Ca)$. Characteristic features such as a sharp Fermi cutoff and a depression of the density of states at E_F, i.e., occurrence of a pseudogap, in the valence band spectrum of these quasicrystal-related materials are demonstrated. Then, comparisons between the QCs and the approximants are made in detail in terms of the long-range order and the local structure.

2.1. *On the stabilization mechanism of the quasicrystals*

The existence of a pseudogap is one of the characteristic properties of the quasicrystals $(QCs)^{1}$ and the pseudogap has been directly probed by photoemission (PE) spectroscopy with ultrahigh energy resolution of several meV at very low temperatures near 10 K.[2] Owing to the low temperatures, the Fermi cutoff has been clearly observed, indicating the existence of a finite DOS at E_F for QCs. The pseudogap feature, i.e., a marked intensity decrease towards E_F, has also been noticed, well separated from the Fermi-Dirac distribution function. However, no other structure with a finer energy scale reflecting the complex band structures of QCs has been observed even by ultrahigh resolution PE spectroscopy of several meV resolution. The absence of structures on a finer scale has often been attributed to the presence of topological or chemical disorder, which will be discussed later in this text.

The existence of the pseudogap in QCs is explained by Fermi-sphere-quasi-Brillouin zone interaction. In QCs, the wave vector k is not a good quantum number since the QCs do not have lattice periodicity. Therefore, the Brillouin zone cannot be defined in a rigorous sense. However, strong reflections of X-ray occur in QCs and a quasi Brillouin zone can be constructed by a set of the wave vectors corresponding to the strong reflections. Since the quasicrystals have the highest symmetry among ordered materials, i.e., icosahedral symmetry, the shape of the quasi-Brillouin zone is nearly spherical compared with crystalline materials. It results in most effective interaction of the quasi-Brillouin zone with the Fermi sphere and an energy gap opens along various directions. As a result, the electronic states near E_F are repelled from E_F, forming a deep depression of DOS at E_F (pseudogap), and an energetic gain is obtained by occupation of the electronic states below E_F (Hume-Rothery stabilization mechanism). In such a case, the position of E_F is of particular importance. In this scenario, since the position of E_F is determined by average electrons-per-atom (*e/a*) ratio, any change of the *e/a* ratio will destabilize the quasicrystalline structure by moving E_F away from the pseudogap. Soon after the discovery of the ternary stable QCs, it was found that the icosahedral QCs are formed at a specific *e/a* ratio.[3] Subsequent works have shown that the formation regions of the QCs are narrow in a direction perpendicular to a constant *e/a* line and extended along the constant *e/a* line, suggesting that the position of E_F, i.e., the *e/a* ratio, plays an essential role in the stability of the quasicrystals. On the other hand, the elongated shape of the single phase region means that ternary quasicrystals are not stoichiometric compounds in a rigorous sense, but can be regarded as a kind of alloys containing chemical disorder. The absence of fine structures in PE spectra of QCs has often been attributed to this chemical disorder since fine structures of the DOS would be smeared out in the presence of such a chemical disorder.

Another important aspect concerning the stability of quasicrystals is the hybridization effect between s,p and d states for the quasicrystals containing transition metal elements.[4] Hybridization between sp and d states has also been known to repel states at E_F, forming a pseudogap.[5]

Hence, the Hume-Rothery mechanism and the hybridization effect have been regarded as the two major sources of the stability of the QCs, however, it has not been establish clearly which plays the more significant role in the stability. The role of the hybridization will be examined in the Cd-based binary quasicrystals which are free of chemical disorder.

2.2. *Binary stable quasicrystals: Cd-Ca and Cd-Yb QCs*

The Cd-Ca and Cd-Yb QCs are the only two binary icosahedral QCs that are thermodynamically stable.[6] They are of a particular interest for spectroscopic studies of the valence band of the QC for the following reasons. First, they are composed of only two elements, therefore, they are free from chemical disorder. This is also suggested from the structural analysis on the crystalline approximant Cd_6Yb[7], which reveals that there are no statistical sites occupied by both Cd and Yb atoms. Second, the interpretation of experimental spectra is simplified because they are binary systems. Third, as mentioned above, there exist crystalline approximants Cd_6Yb[8] and Cd_6Ca[9] in the close vicinity of the compositions of the QCs, which have the same local structure as the QCs. Since the approximants are periodic crystals, the existence of the approximants makes it possible to compare experimental spectra with theoretical DOS and also to extract the influence of the quasiperiodicity on the electronic structure. In addition, there is another interesting aspect inherited to the Cd-Yb QC. This system allows us to investigate changes of the electronic structure in the surface region since the well localized 4f states produce sharp final-state multiplet structures in the PE spectrum. In the next section, the valence band of the Cd-Yb QC is described in comparison with that of the Cd_6Yb approximant.

2.3. *Near E_F valence band of the Cd-Yb quasicrystal and its approximant*[10]

Figure 7 presents PE spectra of the $Cd_{5.7}Yb$ QC and the $1/1$-Cd_6Yb measured at 5.7 K. It is seen that two distinct peaks appear at the binding energies of about 0.7 and 2.0 eV in both the spectra. The two peaks are assigned to the $4f_{7/2}$ and $4f_{5/2}$ derived states since their energy separation (1.27 eV) is in good agreement with the atomic spin-orbit energy 1.27 eV as illustrated in Fig. 8.

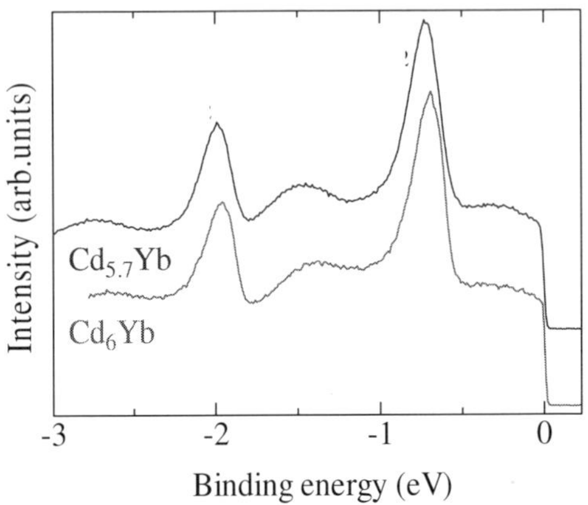

Fig. 7. PE spectra of Cd$_{5.7}$Yb QC and 1/1-Cd$_6$Yb approximant.

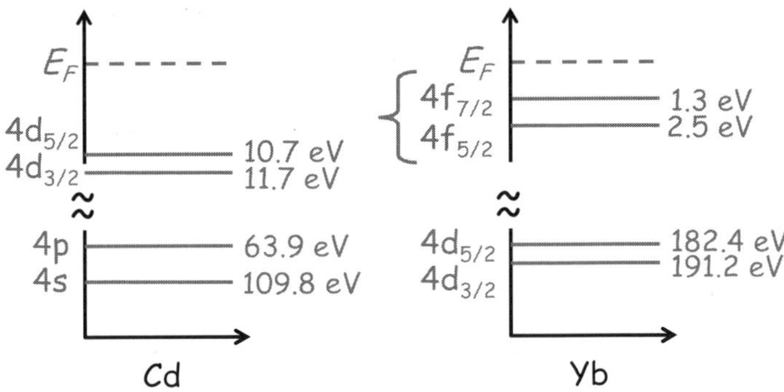

Fig. 8. Electronic levels and their binding energies of atomic Cd and Yb.

From Fig. 7, it is noticed that the peak positions of the Yb 4f doublet are slightly deeper for the QC than for the approximant, by about 25 meV, which implies that the electronic system around Yb atoms is more stable in the QC than in the approximant. A striking feature in the PE spectra in Fig. 7 is the existence of two humps on the higher binding-energy side of the 4f doublet, at about 1.4 and 2.7 eV for the approximant, while they are mixed with the 4f doublet for the QC. The location of the humps of the $1/1$-Cd_6Yb was determined to be 1.40 and 2.68 eV from the intensity maxima of the humps. In the case of the QC, a deconvolution of the spectrum by using Gaussian functions as well as by assuming a linear background has been performed in order to determine their locations. The positions were determined to be 1.08 and 2.37 eV from the least-squares fits. Thus, it is understood that the energy separation of the humps is about 1.3 eV for both compounds. The values are in good agreement with the 4f splitting energy of the bulk states, indicating that the humps are due to Yb 4f-derived states.

In rare-earth-metal compounds, such extra humps in the vicinity of well-localized states such as the 4f states are well known for core-level shifts at the surface, i.e., the so-called surface core-level shift (SCS).[11] The SCS reflects changes of the electronic structure in the surface region. The observation of the surface peaks are of importance because <u>the appearance of the surface peaks verifies that the main doublet is the bulk 4f spin-orbit doublet.</u> The binding energy of the 4f states is higher in the surface region, which indicates that the binding energies of surface atoms are higher than those of bulk atoms. The surface components are shifted to higher binding energies by 0.68 and 0.36 eV for $1/1$-Cd_6Yb and $Cd_{5.7}Yb$ QC, respectively. In addition, the surface peaks are broader than the bulk peaks in both phases, which is attributed to contributions from more than one surface layer or surfaces of different crystallographic orientations. The occurrence of the SCS clearly demonstrates that the electronic structure at the surface is different from that in the bulk in the QC as well as in the approximant. Although there have been a number of indications which could be interpreted as evidence that the electronic structure at the surface differs from that of the bulk in QCs, the SCS thus gives a direct evidence of the surface states in the QC. Next, the valence band structure in the close vicinity of E_F is focused at. As seen from Fig. 9, the spectral intensity decreases towards E_F for both the QC and the approximant, which is well distinguished from the Fermi cut-off.

The decreasing trend becomes enhanced as E_F is approached. A comparison of the two spectra reveals that the intensity depression at E_F is stronger for the QC than the approximant, which indicates that the dip at E_F is deeper in the QC than in the approximant. The stronger intensity decrease near E_F, namely less occupied states in the vicinity of E_F in the QC, means a larger energy gain due to the dip formation in the QC than in the approximant.

Mori *et al.*[12] introduced the following simple phenomenological model in order to fit near-E_F PE spectra of QCs, where the pseudogap profile is approximated as the inverse of a Lorentz function centered at E_F. The expression of the PE spectrum is then given by $\left[1 - C\Gamma^2 / \left\{ (E - E_F)^2 + \Gamma^2 \right\} \right]$, where C and Γ denote the depth and the width of the pseudogap, multiplied by the Fermi-Dirac function and convoluted with a Gaussian function describing the instrumental broadening. Although the positions of E_F are not necessarily located at the bottom of the pseudogap, the depression of the DOS at E_F can be evaluated by using the model. The results of the fitting of the PE spectra of the $Cd_{5.7}Yb$ QC and the Cd_6Yb approximant are shown in Fig. 9, which gives $C = 18.8\%$ and $\Gamma = 96 meV$ for the $Cd_{5.7}Yb$ QC, and $C = 11.1\%$ and $\Gamma = 94 meV$ for the Cd_6Yb approximant.

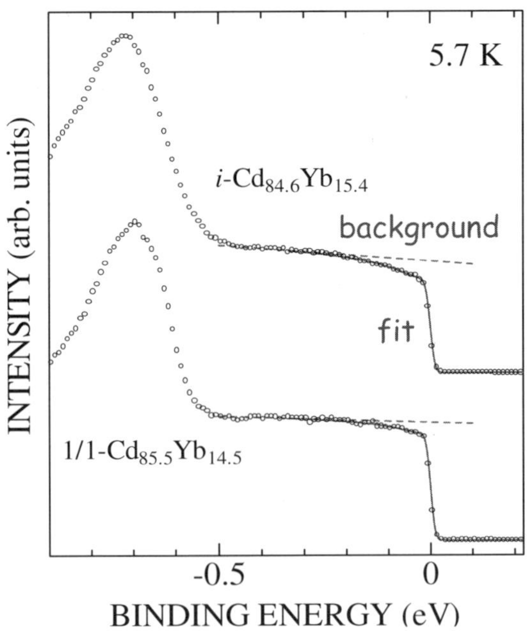

Fig. 9. Valence band spectra of $Cd_{5.7}Yb$ QC and $1/1$-Cd_6Yb.

The shapes of the DOS obtained by the fits are plotted in Fig. 10 and it is clear that the depression of the DOS at E_F is deeper in the QC by a factor of 1.7. Similar results were reported in several ternary systems.

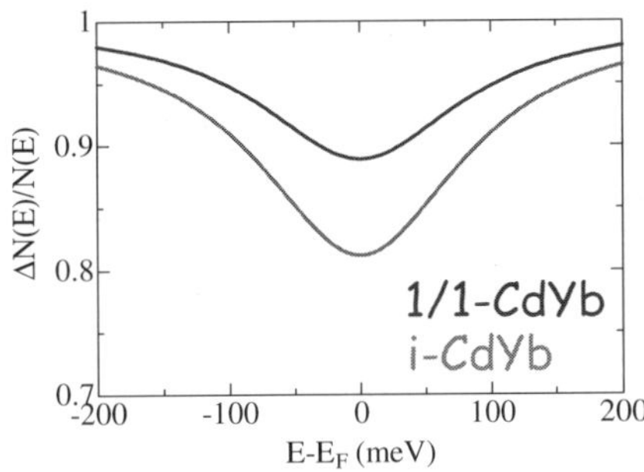

Fig. 10. Shape of the pseudogap obtained from the fits in Fig. 9.

2.4. *Near E_F valence band of the Cd-Ca quasicrystal and its approximant[13]*

In this section, the electronic structure of Cd_6Ca binary QC is described in detail. The electronic structures of Cd_6Ca and Cd_6Yb have been theoretically investigated by Ishii and Fujiwara.[14] As in the cases of the ternary quasicrystals, a pseudogap is observed in the vicinity of E_F for the binary approximants and, in addition, it has been demonstrated that unoccupied d states are essential for the pseudogap formation. In atomic Cd and Yb, the d states are not occupied and a part of the d states are descended below E_F by the formation of the approximant, which has been attributed to hybridization between Cd d states and Yb/Ca sp states. This sp-d hybridization produces bonding orbitals near E_F, forming a pseudogap between the bonding and the antibonding orbitals. According to the calculations, E_F is located at the shoulder of the occupied band. Electron occupation of the bonding states stabilizes the approximant structure, which can account for the cohesion of Cd_6Ca and Cd_6Yb. Such importance of sp-d hybridization was originally pointed out for Al-based approximants by Trambly de Laissardiere *et al.*[4], who have shown based on calculations that both the Hume-Rothery mechanism and the sp-d hybridization work in the pseudogap formation and the role of the sp-d hybridization exists in the enhancement of the pseudogap via resonances of the conduction electrons by strong potentials of transition metal atoms.

Figure 11 presents near-E_F photoemission (PE) spectra of Cd_6Ca at photon energies of 21.2 and 350 eV measured at 5.7 and 20 K, respectively. The inset shows near-E_F spectra of both the Cd_6Ca approximant and the $Cd_{5.7}Ca$ QC at 5.7 K for comparison.

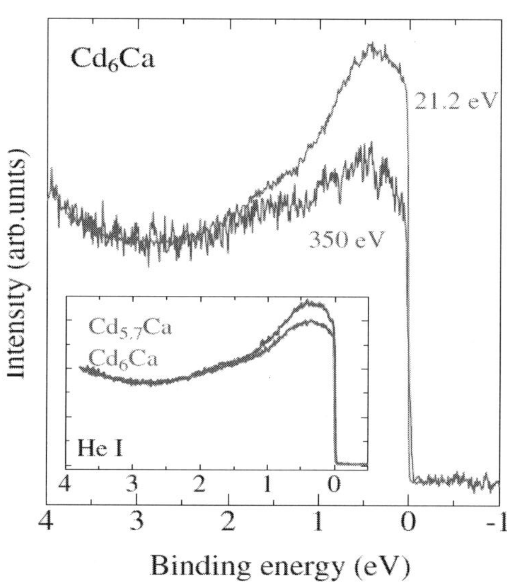

Fig. 11. Valence band spectra of 1/1-Cd_6Ca with the photon energy $h\nu$=21.2 and 350 eV. The inset shows a comparison of He I spectra between $Cd_{5.7}Ca$ QC and 1/1-Cd_6Ca.

Figure 12 presents a valence band spectrum of the Cd_6Ca approximant in a wider binding energy range measured with $h\nu$=350eV. As seen in Fig. 12, two narrow peaks are observed at 11.3 and 10.4 eV, which are assigned to Cd $4d_{3/2}$ and $4d_{5/2}$-derived states, respectively, since the binding energies as well as their splitting energy agree well with those of Cd $4d_{3/2}$(11.7eV) and Cd $4d_{5/2}$ (10.7eV) of the Cd metal (see Fig. 8). These fully occupied Cd 4d states do not contribute to cohesion since electrons occupy both the bonding and the antibonding states.

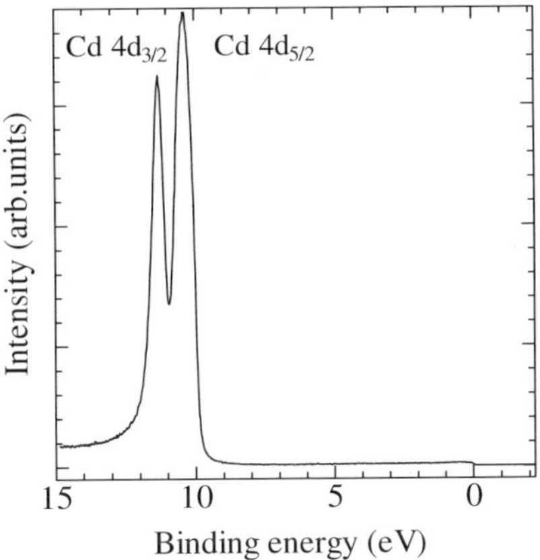

Fig. 12. Valence band spectrum of 1/1-Cd6Ca with the photon energy $h\nu$=350 eV.

On the other hand, as seen in Fig. 11, a peak structure with a width of about 1 eV is observed just below E_F, and E_F is located at the shoulder of the peak. This suggests that the E_F is situated at the edge of the pseudogap. A similar peak structure is also noticed in the near-E_F PE spectrum of the $Cd_{5.7}Ca$ QC, and the positions of the peak maxima are exactly the same, i.e., 0.36 $\pm$ 0.01 eV, for the two compounds although they have different long-range orders. <u>This indicates that the peaks are due to the same origin, i.e., the common local structure, not to the long-range order.</u> From the figure, it is also seen that the peak structure is more pronounced in the QC than in the approximant. Another remarkable feature in the PE spectra is the variation of the peak intensity with the incident photon energy, i.e., the peak is enhanced at the photon energy of 21.2 eV compared with that at 350 eV.

The possible states which constitute the valence band of the Cd_6Ca approximant are Cd 5s, 5p and Ca 3d states. The PE spectrum is given by a sum of the contributions from all the states weighed by their photoionization cross sections. The photoionization cross section of a state is dependent on the incident photon energy and the energy dependence is different among different states. Therefore, the variation of the intensity with the photon energy is due to the different energy dependence of the photoionization cross section among states, i.e., Cd 5s, 5p, and Ca 3d states. Among the three states, the contributions from Cd

5p and Ca 3d states are more enhanced when the photon energy is reduced from 350 to 21.2 eV than that from Cd 5s state. Therefore, the part of the valence band which is enhanced by reducing the photon energy is composed of Cd 5p and/or Ca 3d states, which means that the observed peak is made of the Cd 5p and/or Ca 3d states. This is the first indication that the peak structure is due to hybridization between Cd 5p and Ca 3d states. The hybridization gives rise to the stability of the local Cd-Ca bonds which are common to both the approximant and the QC.

3. Resonant Photoemission Spectroscopy on Cd-based Quasicrystal and Approximant

Resonant photoemission spectroscopy (RPES) is a powerful tool for investigation of the partial DOS in the valence band. In this chapter, the dipole selection rule for the X-ray absorption process is briefly reviewed and, then, the principles of the RPES are described in order to interpret the RPE spectra of Cd-based quasicrystal (QC) and approximant. Finally, the stability mechanism of the binary QCs and the approximants is discussed based on a comparison between the experimental partial DOS and the theoretical partial DOS.

3.1. *X-ray absorption spectrum*

Resonant PE spectroscopy can provide information on the partial DOS by virtue of the dipole selection rules of the X-ray absorption process. According to the dipole selection rules, electron transitions take place only between the states satisfying $\Delta l = \pm 1$ where l denotes the orbital quantum number: For instance, p electrons are allowed to be excited only to s or d levels, not to p and f levels. Therefore, by tuning the photon energy to a certain absorption edge such as K and L edges, p state and sd states are investigated separately. For instance, in the case of Ca, 2p-3d absorption occurs with the photon energy around 350 eV(Ca $L_{2,3}$ absorption edge) as shown in Fig. 13.

By using photon in this range, 2p-3d and 2p-3s transitions can be selectively excited. Figure 14 shows an X-ray absorption (XAS) spectrum of Cd_6Ca with the incident photon energy in the vicinity of Ca

$L_{2,3}$ absorption edge, where 2p electrons are mostly excited to unoccupied 4s and 3d states.

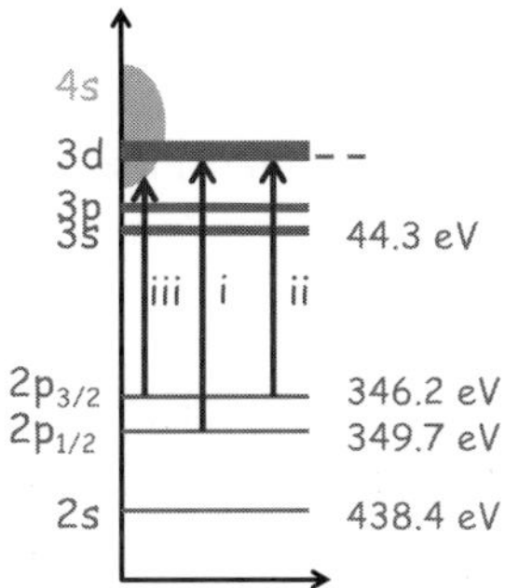

Fig. 13. Schematic illustration of a Ca atom electronic levels and their binding energies.

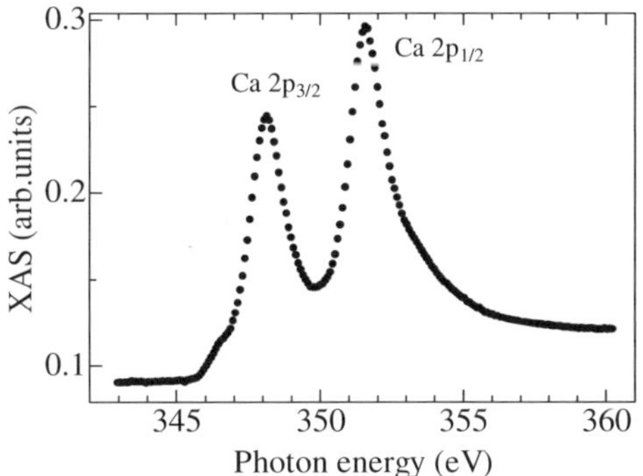

Fig. 14. X-ray absorption spectrum of Cd_6Ca.

Therefore, the XAS spectrum displays the local unoccupied Ca 4s and 3d states due to the dipole selection rule. As seen in the Fig. 14, the XAS spectrum is composed of two peaks at photon energies 348.2 and 351.6 eV with an additional weak structure at the onset. The sharp peaks are ascribed to the Ca 3d level ($2p^6 3d^n \rightarrow 2p^5 3d^{n+1}$) and the occurrence of the peak splitting is attributed to the splitting of the Ca 2p core level due to the spin-orbit interaction. The observed splitting energy is 3.4 eV, which agrees well with the spin-orbit splitting energy (3.5 eV) of the Ca 2p level, i.e., $2p_{3/2}$ and $2p_{1/2}$, as shown in Fig. 13. On the other hand, the weak structure at about 346 eV is assigned to E_F since electron excitation starts to occur above 346 eV where there exist unoccupied states. This energy corresponds to $2p_{3/2}$ (L_3 edge) to E_F transition (346.2 eV) according to Fig. 13.

3.2. *Resonant photoemission spectroscopy on Cd$_{5.7}$Ca QC and 1/1-Cd$_6$Ca[13]*

In order to clarify the states composing the peak structure near E_F, resonant photoemission spectroscopy (RPES) on 1/1-Cd$_6$Ca and Cd$_{5.7}$Ca QC was performed at photon energies ranging from 345 to 352 eV, which is shown in Figs. 15(a) and (b), respectively.

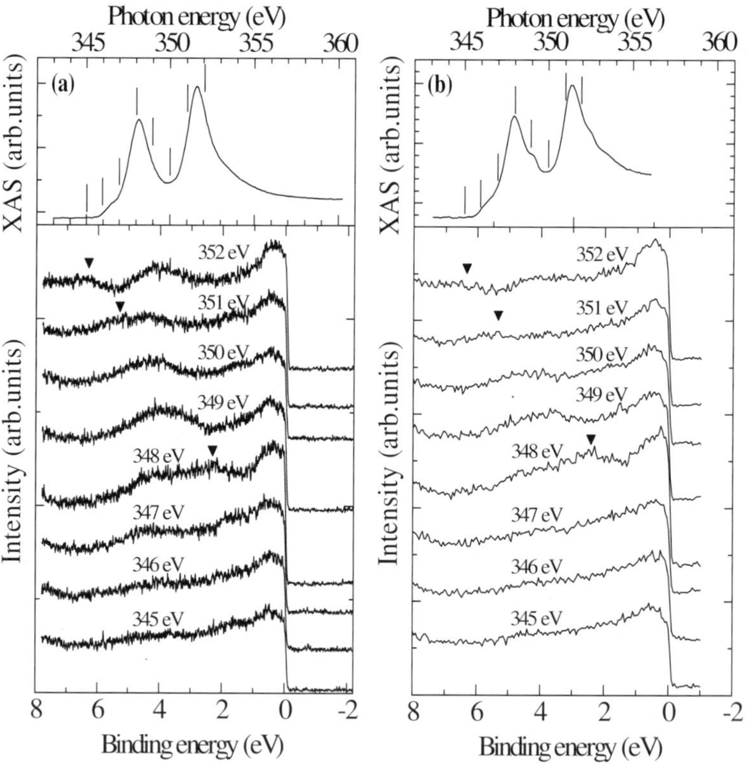

Fig. 15. Resonant photoemission spectra measured in the vicinity of Ca $L_{2,3}$ absorption edge for (a) 1/1 Cd$_6$Ca and (b) Cd$_{5.7}$Ca QC. In the upper panels are shown X-ray absorption spectra in the vicinity of Ca $L_{2,3}$ absorption edge.

There is a relationship between the photon energy (PE), the binding energy (BE) and the kinetic energy (KE), which is given by

$$KE = PE - BE .\tag{18}$$

In the figure, it is noticed that the features marked by a solid triangle have the same kinetic energy of about 345.5 eV. Hence, the features are ascribed to Auger electrons since Auger electrons escape from solids with the help of the transition energy of a nearby electron down to a deep core level as schematically shown in Fig. 16.

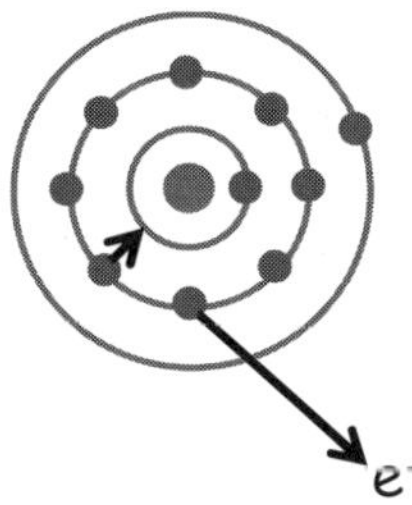

Fig. 16. Emission of an Auger electron.

In the present case, the features are assigned to Ca $L_3M_{4,5}M_{4,5}$ Auger where L_3 and $M_{4,5}$ denote 2p and 3d levels, respectively; the first $M_{4,5}$ stands for an electron transition from the $M_{4,5}$ states to an empty L_3 core level and the second $M_{4,5}$ denotes an emission of the Auger electron form the $M_{4,5}$ states with the help of the transition energy from $M_{4,5}$ to L_3. Therefore, the kinetic energy of the Auger electrons is constant owing to the constant energy difference between $M_{4,5}$ and L_3 core levels, independent on the incident photon energy.

By comparing with the XAS spectrum, the RPES spectra at 348 and 352 eV correspond to on-resonant spectra, where the incident photon energy ($h\nu$) just coincides with the energy difference between the 2p and the 3d levels. In such a case, both the emission of photoelectrons and electron excitations from the 2p to 3d level occur simultaneously. The excited 3d electrons again make a transition to the empty 2p level emitting a nearby 3d electron outside the solid as a photoelectron with energy $h\nu$, which is just as the same as the incident photon energy ($h\nu$). The final states of these two processes are the same having the same electronic configurations. As a result, <u>a strong resonance</u>

occurs between the direct photoelectron emission process $(2p^63d^n + h\nu \rightarrow 2p^63d^{n-1} + e^-)$ and the indirect photoemission process $(2p^63d^n + h\nu \rightarrow 2p^53d^{n+1} \rightarrow 2p^63d^{n-1} + e^-)$, which is a quantum-mechanical interference effect. Such a resonance occurs because the 2p-3d transition energy is just the same as the incident photon energy. This enhancement helps detecting very weal signals which is difficult to obtain by conventional PES. Therefore, if there are 3d states in the valence band, the 3d profile should be enhanced because of the resonance. It is seen that the RPES intensity is enhanced in the close vicinity of E_F for the on-resonant spectra, indicating that Ca 3d states exist at E_F, which is in agreement with the theoretical calculations by Ishii and Fujiwara[14] where a part of 3d DOS exists just below E_F due to hybridization between Cd 5p and Ca 3d states.

Since only the 3d contribution is enhanced in the RPES spectra, 3d partial DOS can be obtained by subtracting an off-resonant spectrum from the on-resonant spectrum. Figure 17(a) shows Ca 3d partial DOS determined by subtracting the spectra at 346 and 352 eV respectively.

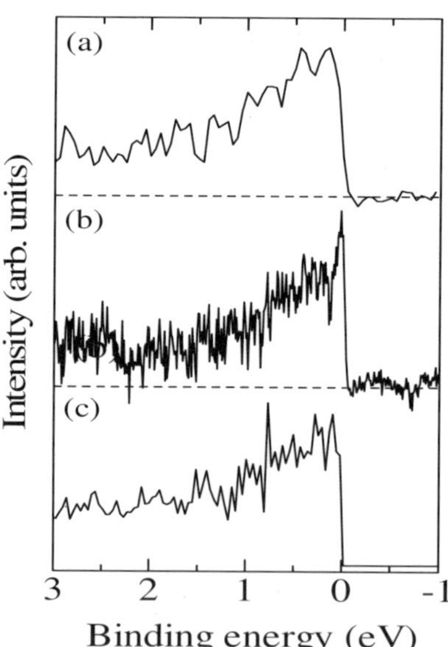

Fig. 17. Experimental 3d partial DOS of (a) $Cd_{5.7}Ca$ QC and (b) 1/1 Cd_6Ca. (c) theoretical 3d partial DOS of Cd_6Ca. In (c), only DOS below E_F is shown.

When subtracting two spectra at different photon energies, a care should be taken since the energy dependence of the photoionization cross section differs among the different states, which might result in an artifact not due to the 3d states. It is confirmed that the photoionization cross sections of the Cd 5s, 5p, and Ca 3d states show a smooth variation without any singularities for the energy range around 350 eV and subtraction of the two spectra with small energy difference of only several eV does not cause any artifact. In Fig. 17(a), the contribution from the d states near E_F, especially in the close vicinity of E_F, is clearly noticed. This experimental d DOS shows that the Ca 3d states, which are unoccupied in the Ca metal, are lowered below E_F by the formation of the crystalline approximant.

Figure 17(b) presents theoretical d DOS of Cd_6Ca calculated with the tight-binding linear muffin-tin orbitals (LMTO) method in the atomic sphere approximation (ASA) for comparison. The agreement between the experimental d DOS and the theoretical d DOS is excellent, in particular, in the close vicinity of E_F where both the d DOS steeply increases as E_F is approached. Such an agreement between the theory and the experiment supports the theoretical prediction that the Ca 3d states are descended below E_F due to the p-d hybridization, which is schematically shown in Fig. 18.

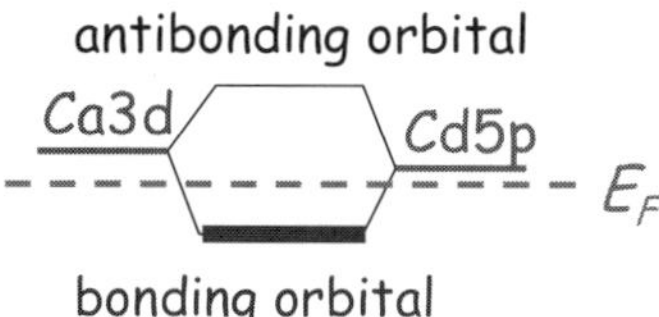

Fig. 18. Schematic illustration of bonding/antibonding orbitals formed below E_F by p-d hybridization.

3.3. *Origin of the p-d hybridization*

Since hybridization occurs between Cd and Ca atoms, the nearest Cd neighbors for Ca atoms should be responsible for the hybridization and the specification of the hybridization bonds is of importance for understanding of the stabilization mechanism of the Cd_6Ca approximant as well as of the $Cd_{5.7}Ca$ QC. A close examination of the structural model of Cd_6Ca reveals that the Ca atoms occupy only a single site, which constitutes the third shell of the 66-atom icosahedral cluster, and there are three types of Cd neighbors around a Ca atom as shown in Fig. 19. One is the second shell Cd atoms (S2 site), another is the fourth shell Cd atoms (S4 site) and the other is so-called glue Cd atoms (S5 site) which bridge the icosahedral clusters. Therefore, some of the three Cd sites are responsible for the p-d hybridization. In order to clarify this point, comparison has been made between the experimental PE spectrum and the theoretical local p DOS at the three different Cd sites.

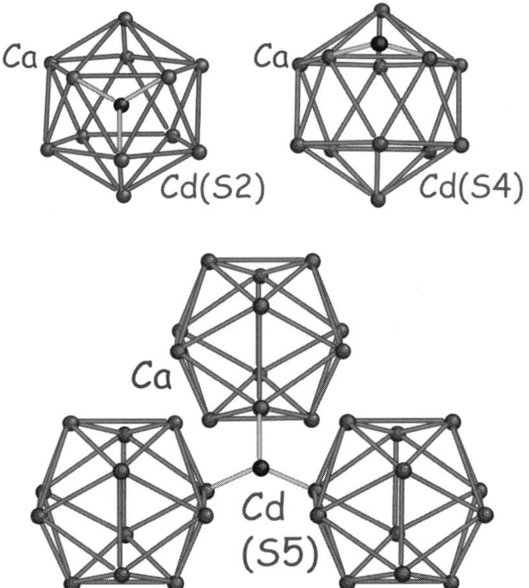

Fig. 19. Three different Cd-Ca bonds found in the structural model of Cd_6Ca.

Figure 20 presents comparison of an experimental PE spectrum with theoretical local p DOS at the S2, S4 and S5 sites. From the figure, it is obvious that the local p DOS at the S2, S4 and S5 sites all exhibit a similar peak structure in the close vicinity of E_F, i.e., within 1eV below E_F. This means that the Cd 5p states at all the three sites contribute to the peak structure observed near E_F in the total DOS, suggesting that they are hybridized with Ca 3d states.

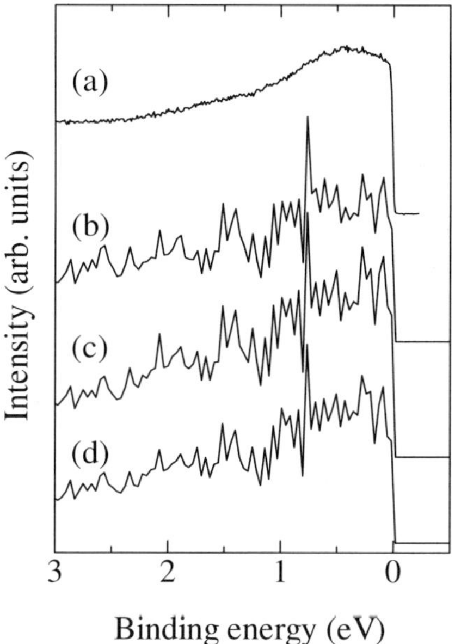

Fig. 20. (a)Experimental valence band spectrum of Cd_6Ca Theoretical local p-DOS at (b)S2, (c)S4 and (d)S5 sites.

A further inspection of the structure of Cd_6Ca reveals that the hybridization between the Ca and Cd atoms at the S2 and S4 sites strengthens the bonding of the Ca atoms into a Ca_{12} icosahedron. Therefore, it gives rise to the stability of the icosahedral cluster. On the other hand, the hybridization between the Ca and the Cd atoms at the S5 site strengthens the connection of the icosahedral clusters along [111] and [100] directions.

Therefore, it gives rise to the stability of the network of the icosahedral clusters. Henceforth, it is concluded that the p-d hybridization stabilizes both the icosahedral cluster and the cluster network in the approximant structure.

A similar peak is also observed in the PE spectrum of the $Cd_{5.7}Ca$ icosahedral QC at the same binding energy, indicating that the origin of the peak is the same as that of the approximant. Therefore, the peak observed in the QC is also understood as a consequence of the p-d hybridization between Cd and Ca atoms, which means that the local

<u>structure of the icosahedral QC is stabilized by the hybridization between Cd and Ca atoms.</u>

4. Summary and Outlook

First, the principles of the photoemission spectroscopy (PES) have been reviewed focusing on the energy resolution obtained by the PES. Second, ultrahigh resolution PES and resonant PES on binary Cd-based quasicrystals and their approximants, as examples of the quasicrystal-related materials, were presented. Characteristic features such as a sharp Fermi cutoff, a pseudogap, a hybridization peak and surface core level shifts have been clearly observed by the ultrahigh resolution PES and, in particular, the important role of the sp-d hybridization effect in the stability of the binary approximants as well as the binary quasicrystals has been elucidated.

Recently, a substantial progress has been made in the UPS with respect to the energy resolution as well as the bulk sensitivity: Ultrahigh resolution up to 360 μeV and bulk sensitivity of >10 nm have been achieved by using a laser as a photon source, which has a negligibly small line width and photon energy of ~7 eV. New insight into the intrinsic properties of QCs is now being clarified.

References

1. For a review, see T. Fujiwara, in *Physical Properties of Quasicrystals*, Ed. Z. M. Stadnik (Springer-Verlag, Berlin, 1999), p. 169; J. Hafner and M. Krajci, *ibid.*, p. 209.
2. For a review, see Z. M. Stadnik, in Physical Properties of Quasicrystals (Ref. 1), p. 257.
3. For a review, see A. P. Tsai, in Physical Properties of Quasicrystals (Ref. 1), p. 5.
4. G. Trambly de Laissardiere, D. Nguyen Manh, L. Magaud, J. P. Julien, F. Cyrot-Lackmann and D. Mayou, *Phys. Rev.* **B 52**, 7920 (1995).
5. E. Belin, Z. Dankhazi, A. Sadoc, Y. Calvayrac, T. Klein, J. M. Dubois, *J. Phys.: Condens. Matter* **4**, 4459 (1992).
6. J. Q. Guo, E. Abe and A. P. Tsai, *Phys. Rev.* **B 62**, 14605 (2000); A. P. Tsai, J. Q. Guo, E. Abe, H. Takakura and T. J. Sato, *Nature* **408**, 537 (2000).
7. H. Takakura, J. Q. Guo and A. P. Tsai, *Philos. Mag. Lett.* **81**, 411 (2001).
8. A. Palenzona, *J. Less-Common Met.* **25**, 367 (1971).

9. G. Bruzzone, *Gazz. Chim. Ital.* **102**, 234 (1972).

10. R. Tamura, Y. Murao, S. Takeuchi, T. Kiss, T. Yokoya and S. Shin, *Phys. Rev.* **B 65**, 224207 (2002).

11. En-Jin Cho, J. S. Chung, S. J. Oh, S. Suga, M. Taniguchi, A. Kakizaki, A. Fujimori, H. Kato, T. Miyahara, T. Suzuki and T. Kasuya, *Phys. Rev.* **B 47**, 3933 (1993).

12. M. Mori, S. Matsuo, T. Ishimasa, T. Matsuura, K. Kamiya, H. Inokuchi and T. Matsukawa, *J. Phys.: Condens. Matter* **3**, 767 (1991).

13. R. Tamura, T. Takeuchi, C. Aoki, S. Takeuchi, T. Kiss, T. Yokoya and S. Shin, *Phys. Rev. Lett.* **92**, 146402 (2004).

14. Y. Ishii and T. Fujiwara, *Phys. Rev. Lett.* **87**, 206408 (2001).

CHAPTER 7

FIRST-PRINCIPLES CALCULATIONS AND APPLICATIONS FOR MATERIALS DESIGN

Ryoji Asahi

Toyota Central R&D Labs., Inc.
Nagakute, Aichi 480-1192, Japan
E-mail: rasahi@mosk.tytlabs.co.jp

With development of theory along with hardware for computation, quantum materials design—synthesis of materials with desired properties in a controlled way via materials engineering on the atomic scale—becomes a major component of materials research. Computational prediction based on first-principles calculations has successfully provided an efficient way to develop materials, such as a visible-light sensitized photocatalyst as demonstrated. The universality and applicability of the methods lead to physical insight in many applications and serendipity for new finding with close collaboration between theory and experiment.

1. Introduction

Density functional theory[1,2] has successfully established an accurate description to predict ground-state properties of materials based on first-principles quantum calculations. With recent development of theory along with computing power, it extends to predict excitation properties, thermodynamic properties, and nano-scale properties; this allows us to realize quantum materials design-synthesis of materials with desired properties in a controlled way via materials engineering on the atomic scale, becoming a major component of materials research. In fact, we have demonstrated successful development in functional materials based on computational prediction using first-principles calculations as found in visible-light photocatalysts[3] and super elasto-plastic titanium alloys.[4] In these examples, we have seen that the first-principles calculations and

279

experimental works together indeed improve the materials properties to reach a realistic production level in industry beyond scientific achievement.

In this chapter, general aspects of the computational materials design by using the first-principles calculations are first described. We emphasize the points particularly important for materials design in industrial applications. We then present recent development of photocatalysts and thermoelectric materials, enlightening on how the computational materials design plays an important role in collaboration with experimental works.

2. Procedure of the Computational Materials Design

What we discuss here is just one of the possible procedures, which is often employed in industry though, for performing the computational materials design. It consists of the following steps:

(1) Set a target property required for the material to be designed
(2) Choose a benchmark system
(3) Construct a computational model
(4) Compute the bench mark system
(5) Predict a new system

In step (1), usually we set the target property so that it can satisfy the industrial or social requirements. No materials structure is assumed in this stage. However, prerequisites (or restrictions) for materials constitution are often enforced, e.g., to be non-toxic, environmentally benign, or sustainable. The materials price is often to be considered after all. In step (2), we may take a material system that shows the best performance so far reported for the target property as an instructive benchmark. For this purpose, we can choose the one which does not satisfy the above prerequisites. In step (3), since the real material system is often too complicated to simulate directly, one may construct a simple model capable of computations in a reasonable time but without loosing any physical essence. This procedure requires considerable insight into the materials science, computational physics, and engineering aspects of the problem, and therefore, close discussion with experimentalists is indispensable. In practice, one of the crucial discussions often made in industry is to judge efficiency and meaning of the computation compared to experiment. However, one should keep in mind that the computation can access an extreme condition, such as no impurity or under very high pressure, which is not easily prepared in experiment, and may show possible maximum properties of materials.

Computing the bench mark system in step (4), we carefully make sure of computational accuracy and validity of the model by comparing with experimental results available. Without this knowledge the deduced consequences can be wrong. The most important and powerful aspect of the computation is to understand microscopic mechanism and to derive a materials trend for the target property. It is followed by prediction of a new system that may improve the target property over the conventional benchmark system as in step (5). To this end, systematic modifications of the material by elementary substitution or doping can be often used. We can check stability and/or process conditions to synthesize the candidates, which can convince people to proceed to the experiment.

If we know a suitable benchmark system in advance, we can find some of the candidates smoothly from the above procedure. If not, however, one must introduce a new concept and choose a system, which may or may not realize the target properties. Usually it is quite a difficult issue.

3. Materials Design of Visible-Light Sensitized Photocatalysts

Since photo-induced decomposition of water on TiO_2 electrodes was discovered[5], semiconductor-based photocatalysis has attracted extensive interest.[6] Most of investigation as well as development of applications have been focused on TiO_2, which shows relatively high reactivity and chemical stability under ultraviolet (UV) light (wavelengths $\lambda<387$ nm) whose energy exceeds the band gap of 3.2 eV in the anatase crystalline phase. On the other hand, photocatalysts that can yield high reactivity under visible light ($\lambda>380$ nm) have been desired as they should allow the main part of the solar spectrum and even poor illumination of interior lighting to be utilized. To this end, the following two ways to modify TiO_2 have been mainly employed; one is to dope transition metals into TiO_2[7,8,9], and another is to form reduced TiO_x photocatalysts.[10,11] Both approaches introduce impurity/defect states in the band gap of TiO_2, which lead TiO_2 to absorb visible light. However, subtle issues reside in the nature of these impurity states: first, reducing TiO_2 introduces localized oxygen vacancy states located at 0.75-1.18 eV below the conduction band minimum (CBM) of TiO_2,[11] which may sacrifice photo reduction activity because a redox potential of the hydrogen evolution (H_2/H_2O) locates just below the CBM of TiO_2.[12] Second, the transition metal doping, where quite localized d states appear deep in the band gap of the host semiconductor, often results in an increase of the carrier recombination[8] as described by the Shockley-Read-Hall model.[13] Therefore, if the doping produce localized deep levels (E_i, much larger than the thermal excitation energy k_BT) in the band gap, the lifetime of the mobile carriers may become shorter, giving a low photocatalytic activity, as typically observed in a system with the transition metal doping.

We started to consider whether visible-light activity could be introduced in TiO_2 by doping, and set the following requirements: (i) doping should produce states in the band gap of TiO_2 that absorb visible light; (ii) the CBM, including subsequent impurity states, should be as high as that of TiO_2 or higher than the H_2/H_2O level to ensure its photo-reduction activity; and (iii) the states in the gap should overlap sufficiently with the band states of TiO_2 to transfer photo-excited carriers to reactive sites at the catalyst surface within their lifetime. These conditions lead to an idea that anionic species can be suitable doping for the visible light sensitization rather than the cationic metals.

We calculated densities of states (DOS) of the substitutional doping of C, N, F, P, or S for O in the anatase TiO_2 crystal, by the full-potential linearized augmented plane wave (FLAPW) formalism[14,15] in the framework of the local density approximation (LDA)[16] as shown in Fig. 1 (A). The substitutional doping of N was the most effective because its p states contribute to the band-gap narrowing by mixing with O $2p$. Although doping with S shows a similar band-gap narrowing, it would be difficult to incorporate it into the TiO_2 crystal because of its large ionic radius, as evidenced by a much larger formation energy required for the substitution of S than that required for the substitution of N. The states introduced by C and P are too deep in the gap to satisfy the above condition (iii). The calculated imaginary parts of the dielectric functions of $TiO_{2-x}N_x$ indeed show a shift of the absorption edge to a lower energy by the N doping (Fig. 1 (B)). Dominant transitions at the absorption edge have been identified with those from N $2p_\pi$ to Ti d_{xy}, instead of from O $2p_\pi$ as in TiO_2.[17]

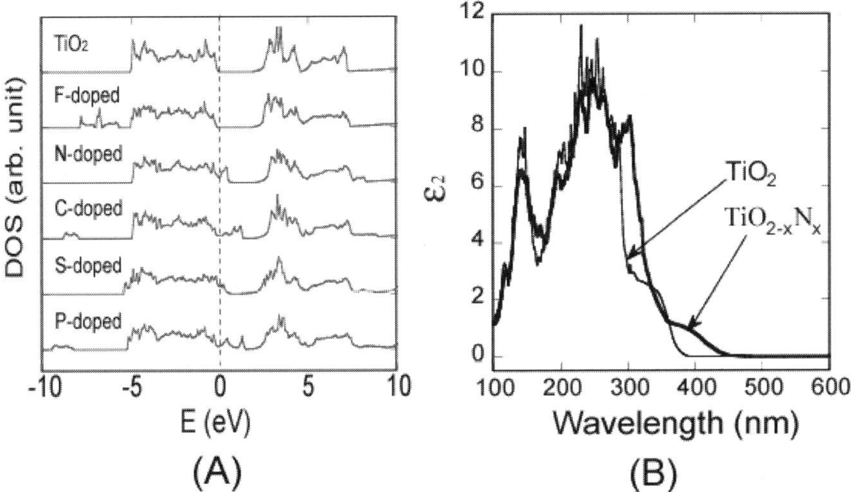

Fig. 1. (A) Calculated total density of states (DOS) of doped TiO_2. The dopants F, N, C, S, and P were located at a substitutional site for an O atom in the anatase TiO_2 crystal (the eight-TiO_2 units per cell). The energy is measured from the top of the valence bands of TiO_2, and the DOS for doped TiO_2 are shifted so that the peaks of the O $2s$ states (at the farthest site from the dopant) are aligned with each other. (B) Calculated imaginary parts of the dielectric functions (ε_2), which are averaged over three (x, y, and z) polarization vectors, of $TiO_{2-x}N_x$ (thick lines) compared with TiO_2 (thin lines).

Effectiveness of the substitutional N doping was verified by experiment. Film samples prepared by sputtering the TiO_2 target in N_2(40%)/Ar gas mixture with post-annealing at 550°C in N_2 gas, and powder samples prepared by treating anatase TiO_2 powder in the NH_3(67%)/Ar atmosphere at 600°C, showed significant photocatalytic activity under visible light, while both revealed similar activity to TiO_2 under UV light as shown in Fig. 2. In fact, optical absorption spectra show that the $TiO_{2-x}N_x$ films noticeably absorb the light at less than 500 nm, whereas the TiO_2 films do not. N $1s$ core levels were investigated by the x-ray photoemission spectroscopy (XPS). The N-doped samples showed that the peak intensity at the binding energy of 396 eV, assigned the substitutional N for the O site, was well correlated with the photocatalytic activity. Thus the active site of N for photocatalysis under visible light was concluded to be the substitutional one.

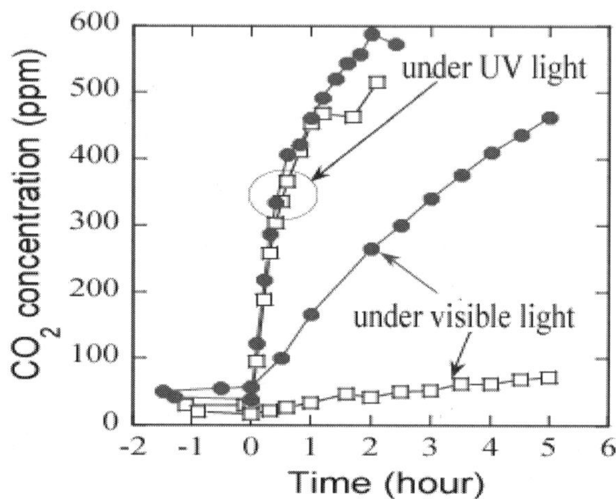

Fig. 2. Photocatalytic properties of $TiO_{2-x}N_x$ samples (solid circles) compared with TiO_2 samples (open squares). CO_2 evolution as a function of irradiation time (light-on at zero) during the photo-degradation of acetaldehyde gas [with an initial concentration of 485 parts per million (ppm)] under UV irradiation (BL with a peak at 351 nm and the light power of 5.4 mWcm^{-2}) and visible irradiation [fluorescent light cut by the optical high-path filter (SC42, Fuji Photo Film), with a peak intensity at 436 nm and a light power of 0.9 mWcm^{-2}].

Since Asahi *et al.* reported[3], a large number of studies have reported on the photocatalytic properties of N-doped TiO_2 under visible light, showing a promising extension of applications where the conventional TiO_2 is not used. However, arguments of the chemical nature of N and the mechanism of the visible-light response have not been reconciled to date. Hydrolysis of $TiCl_4$ with NH_4OH followed by calcination at 400°C also lead to the formation of yellowish TiO_2 powder showing the visible light absorption.[18] Such an N species was assigned to be NO_x type complex, which was consistent with the results of electron paramagnetic resonance (EPR) measurement on TiO_2 samples treated with ammonia and calcined at high temperature.[19] Chen and Burda suggested that the NO_x type complex species with a binding energy of 400 eV is responsible for the visible light sensitization.[20] These experimental results suggest that the detailed N states in TiO_2 would vary with a nitrification process employed and assignment of their binding energies remains unsolved.

Considering a variety of nitrification processes we introduced several kinds of N complex species including N, NO, and NO_2 into a substitutional site for O (denoted as $(NO)_O$ etc.) or an interstitial site (denoted as $(NO)_i$ etc.) in the anatase TiO_2 crystal. Note that the N_2 complex may be controlled by kinetics rather than thermodynamics because two N atoms from different source points must arrive simultaneously at the same site, and thus the formation can be suppressed when the N concentration is as small as that observed in experiment.[21] A supercell of 16 TiO_2 units (i. e., 48 atoms of the TiO_2 anatase matrix) was employed to simulate each kinds of doping which is thus typically 3.1% N for O. We obtained the stable geometry by performing first-principles calculations using the projector augmented wave method[22] implemented in Vienna *ab initio* simulation package (PAW-VASP) code.[23,24] The generalized gradient approximation[25] was used as the exchange and correlation potential. To calculate band structures we employed FLAPW method which treats all electrons including the core electrons self-consistently and relativistically, thus best suitable for comparing with the experimental XPS results. Note that an absolute value of the binding energy (a core level with respect to the Fermi energy) is not accurately evaluated from an eigenvalue of the single electron within the present theoretical framework. Instead, we calculated an energy difference ($\Delta_{N\text{-}Ti}$) between the N $1s$ core level and an averaged value of Ti $2p_{3/2}$ over the unit cell. We then obtained the binding energy (BE), $BE = \Delta_{N\text{-}Ti} + \Delta_0$, where a rigid offset Δ_0 is rather arbitrary but here we set it so that the BE of N $1s$ for the $(N)_O$ species agrees with one of the experimental values, 395.7 eV.[26] An error of this evaluation is roughly given by the standard deviation of the Ti $2p_{3/2}$ levels within the unit cell because the considered unit cell would not be large enough if the error is large.

We list in Table 1 the N $1s$ binding energies calculated by the self-consistent core levels with errors estimated by the Ti $2p_{3/2}$ levels within the unit cell. The error is at most 0.2 eV, which is small enough to distinguish the N $1s$ binding energies among the N species. The $(NO)_i$ and $(NO_2)_O$ species go with a large distortion of the TiO_2 lattice so we expect these species may be found mostly at the surface or in the voids of the solid as suggested in reference 21. The $(N)_i$ and $(NO)_O$ can be interpreted as an intermediate state from weak bonding with an isolated molecule (BE=400 eV) to an atomic N chemisorption with the lattice Ti (BE=396 eV).

Table 1. Calculated N $1s$ binding energies, BE in eV, and their errors in eV estimated by the Ti $2p_{3/2}$ levels within the unit cell. The BE for $(N)_O$ is assumed to be an experimental value.[26]

Dopant	BE	error
$(N)_O$	395.70	0.17
$(N)_i$	397.66	0.11
$(NO)_O$	398.10	0.12
$(NO_2)_O$	399.81	0.03
$(NO)_i$	399.85	0.10

We examined experimental XPS spectra of the samples prepared by different methods. A powder sample was obtained by treating anatase TiO_2 powder (ST01, Ishihara Sangyo Kaisha Ltd.) in NH_3(67%)/Ar atmosphere for 3 hours at 600°C (denoted as Powder1). We also prepared another powder sample (denoted as Powder2) by hydrolysis of $TiCl_4$ with NH_4OH, followed by calcination at 400°C, as reported by Sato.[18] The film sample was prepared by sputtering the TiO_2 target in N_2(40%)/Ar gas mixture and with post-annealing at 550°C in N_2 gas for 4 hours (denoted as Film). These samples showed fairly good photocatalytic activity under visible light.[3,18]

In Fig. 3, we show the XPS spectra of Powder1, Powder2, and Film along with the BEs obtained by the FLAPW calculations. Powder1 and Film samples show peaks around 396 eV and 398 eV in addition to peak at 400 eV while only the peak at 400 eV is observed in Powder2, suggesting that the N species in Powder2 differ from those in Powder1 or Film. In fact, chemical stability among these samples are different; the photocatalytic activity under visible light for Powder2 disappeared after the sample was heated in air at 500°C, while Powder1 still exhibited significant visible-light photocatalysis even after annealing in air at 550°C.[27]

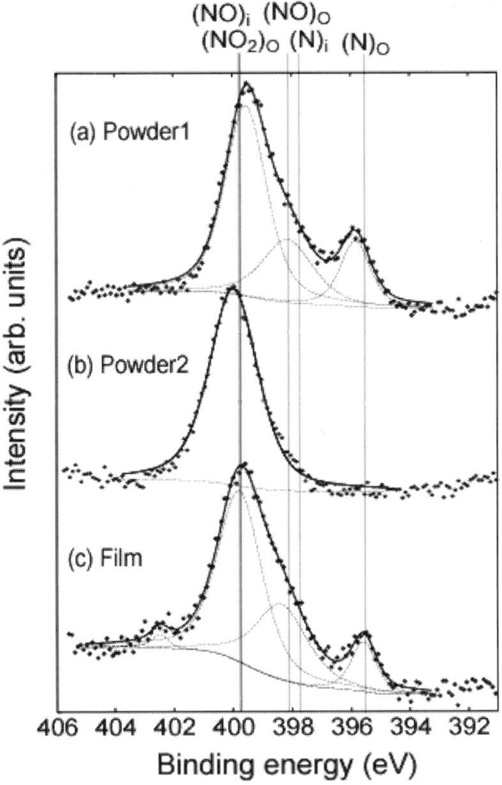

Fig. 3. The measured X-ray photoemission spectra of the N 1s states for (a) Powder1, (b) Powder2, and (c) Film. The peak decompositions (dashed lines) were performed to find the detailed peak positions. The binding energies evaluated by the FLAPW calculations (Table 1) are indicated at the top and thin vertical lines.

These results indicate that the states with the lower BEs, namely, 396 eV and 398 eV, are chemically more stable. The decomposed XPS peaks coincide with the calculated BEs very well. Besides the peak at 396 eV assumed to be $(N)_O$, the peaks at 398 eV and 400 eV are assigned to be $(N)_i$ or $(NO)_O$ and $(NO)_i$ or $(NO_2)_O$, respectively. In the Powder2, the

$(NO_2)_O$ or $(NO)_i$ is suggested to dominate among the N species. This is considered to be a consequence of the hydrolysis process. In fact, the detailed studies of the formation energies show that $(NO_2)_O$ and $(NO)_i$ are stabilized in the oxygen rich condition, and when going to the Ti-rich condition, $(N)_i$ and $(NO)_O$ then $(N)_O$ become stable.[28] This explains why Powder1, nitrified under a rather reduced condition, gives the peaks at 398 eV and 396 eV, and shows higher thermal stability than Powder2. Note that the reduced condition may introduce oxygen defects. More efficient and less oxygen-defect nitrification thus can be realized by using an oxygen-containing nitrogen gas source such as NO.[28]

We performed first-principles calculations for materials design of the visible-light sensitized photocatalyst and comprehensive studies regarding possible N complex species introduced in TiO_2. We also investigated effects of different nitrification conditions. The substitutional N for oxygen is stable under a reduced process condition while NO_x species may appear under an O-rich process condition. The calculated results show good agreement with the experimental XPS spectra and consistency with the trend of the process conditions. It is clear that to realize the N species more in a controlled way via process conditions is crucial to achieve the optimized photocatalytic performance.

The active wavelength of $TiO_{2-x}N_x$, of less than 500 nm, promises a wide range of applications. In fact, recently, various kinds of products using this photocatalyst (V-CAT, Toyota Tsusho Co.), including wallpaper, cloth, coating for glasses, etc., have been commercialized. Their photocatalytic functions, such as anti-bacteria, deodorizing, and purifying the air, proved to be significant under interior lighting over the conventional TiO_2.

4. Conclusions

As demonstrated in development of the visible-light sensitized photocatalysts we particularly emphasized the materials designs on the atomic scale with computation. The universality and applicability of these methods provide exciting insight to many applications, and even lead to serendipity for new findings. Since the density functional theory was born in 1964, the first-principles calculation has now becomed a practical tool in materials research even in industry. The essentials for success will be the close collaboration between theory and experiment, meaningful models to be pursued, and understanding the limitations and efficiency of both theory and experiment.

Acknowledgments

I am pleased to acknowledge and thank T. Morikawa, T. Ohwaki, K. Suzuki, Y. Taga, S. Yamamoto, M. Suzuki, and N. Takahashi for fruitful discussion and collaboration.

References

1. P. Hohenberg and W. Kohn, *Phys. Rev.* **136**, B864 (1964).
2. W. Kohn and L.J. Sham, *Phys. Rev.* **140**, A1133 (1965).
3. R. Asahi, T. Morikawa, T. Ohwaki, K. Aoki and Y. Taga, *Science* **293**, 269 (2001).
4. T. Saito, T. Furuta, J.H. Hwang, K. Kuramoto, K. Nishino, N. Suzuki, R. Chen, A. Yamada, K. Ito, Y. Seno, T. Nonaka, H. Ikehata, N. Ngakasako, C. Iwamoto, Y. Ikuhara and T. Sakuma, *Science* **300**, 464 (2003).
5. A. Fujishima and K. Honda, *Nature* **238**, 37 (1972).
6. *Photocatalytic Purification and Treatment of Water and Air,* Eds D.F. Ollis and H. Al-Ekabi, (Amsterdam: Elsevier, 1993).
7. A.K. Ghosh and H.P. Maruska, *J. Electrochem. Soc.* **124**, 1516 (1997).
8. W. Choi, A. Termin and M.R. Hoffmann, *J. Phys. Chem.* **98**, 13669 (1994).
9. M. Anpo, *Catal. Surv. Jpn.* **1**, 169 (1997).
10. R.G. Breckenridge and W.R. Hosler, *Phys. Rev.* **91**, 793 (1953).
11. D.C. Cronemeyer, *Phys. Rev.* **113**, 1222 (1959).
12. T. Sakata and T. Kawai, in *Energy Resources Through Photochemistry and Catalysis*, Ed. M. Grätzel (New York: Academic Press, 1983) p. 332.
13. W. Shockley and W.T. Read, Jr., *Phys. Rev.* **87**, 835 (1952).
14. E. Wimmer, H. Krakauer, M. Weinert and A.J. Freeman, *Phys. Rev.* **B24**, 864 (1981).
15. H.J.F. Jansen and A.J. Freeman, *Phys. Rev.* **B30**, 561 (1984).
16. L. Hedin and B.I. Lundqvist, *J. Phys.* **C4**, 2064 (1971).
17. R. Asahi, Y. Taga, W. Mannstadt and A.J. Freeman, *Phys. Rev.* B **61**, 7459 (2000).
18. S. Sato, *Chem. Phys. Lett.* **123**, 126 (1986).
19. O. Diwald, T.L. Thompson, T. Zubkov, E.G. Goralski, S.D. Walck and J.T. Yates, Jr., *J. Phys. Chem.* **B108**, 6004 (2004).
20. X. Chen and C. Burda, *J. Phys. Chem.* **B108**, 15446 (2004).
21. S. Livraghi, A. Votta, M.C. Paganini and E. Giamello, *Chem. Commun.* 498 (2005).
22. P.E. Blochl, *Phys. Rev.* B **50**, 17953 (1994).
23. G. Kresse and J. Furthmuller, *Comput. Mater. Sci.* **6**, 15 (1996).

24. G. Kresse and J. Furthmuller, *Phys. Rev.* B **54**, 11169 (1996).
25. J.P. Perdew, K. Burke and M. Ernzerhof, *Phys. Rev. Lett.* **77**, 3865 (1996).
26. N.C. Saha and H.G. Tompkins, *J. Appl. Phys.* **72**, 3072 (1992).
27. R. Asahi, T. Morikawa, T. Ohwaki, K. Aoki and Y. Taga, *Science* **295**, 627 (2002).
28. R. Asahi and T. Morikawa, *Chem. Phys.* **339**, 57 (2007).

CHAPTER 8

SIMULATING STRUCTURE AND PHYSICAL PROPERTIES OF COMPLEX METALLIC ALLOYS

Hans-Rainer Trebin, Peter Brommer, Michael Engel, Franz Gähler,
Stephen Hocker, Frohmut Rösch and Johannes Roth

Institut für Theoretische und Angewandte Physik
Universität Stuttgart, Germany
Email: trebin@itap.uni-stuttgart.de

An introduction is presented to numerical methods, by which the behaviour of complex metallic alloys can be simulated. We primarily consider the molecular dynamics (MD) technique as implemented in our software package IMD, where Newton's equations of motion are solved for all atoms in a solid. After a short discourse on integration algorithms, some possible types of interactions are addressed. Already simple model potentials, as for example the Lennard-Jones-Gauss potential, can give rise to complex structures, where the characteristic length scales of the order by far exceed the range of the pair interaction. Realistic interactions are modelled by highly parametrized effective potentials, like the EAM (Embedded Atom Method) potential. Our program *potfit* allows to fit the parameters such that data from experiment or from ab-initio calculations are well reproduced. Several applications of the methods are outlined, notably the simulation of aluminium diffusion in quasicrystalline d-Al-Ni-Co, the computation of the phonon dispersion via the dynamical structure factor of $MgZn_2$, the propagation of cracks in $NbCr_2$, and an order-disorder phase transition in $CaCd_6$.

1. Numerical Simulation of Matter

The basic equation for computing the structure of atomistic matter is known since the invention of quantum theory: It is the many-body Schrödinger equation. Powerful analytical tools have been developed in the sequel to make it solvable. For periodic crystals, the Bloch theorem and the Kohn- Sham density functional theory allowed to determine ground state structures, cohesion energies and electronic band structures. Within the Born-Oppenheimer approximation, phonon dispersion relations, dynamical and even and non-equilibrium properties can be evaluated.

Explicit realistic data were gained, however, only with the advent of computers and efficient algorithms. Yet, when dealing with complex metallic alloys, where the number of atoms in the unit cell is large, and when exploring phenomena, where periodic boundary conditions are hardly applicable, like defect motion or fracture, the power of even massively parallel computer systems is not sufficient. The same is valid when long simulation times are required, as for example for diffusion processes. For such cases one employs an approximation denoted molecular dynamics (MD). The atoms are considered as point particles of classical mechanics, which move according to Newton's equations of motion under their mutual (but possibly also external) forces. The atom-atom interactions are the central quantities that tune the resulting physical properties. In principle, these interaction could even be computed quantum-mechanically, e.g. in density functional theory under the Born-Oppenheimer approximation, where it is assumed that the electronic degrees of freedom are always in the instantaneous ground state or a given set of nuclear coordinates. For large systems, however, this is far too slow, and it is more promising to use a classical effective potential approximating the quantum-mechanical interaction. As classical potentials are much cheaper to compute, far larger systems and longer time scales can be reached in this way.

Many efforts have been made to understand the response of the particle systems to different types of interaction potentials. A simple isotropic pair potential like the Lennard-Jones potential with a single minimum typically yields densely packed simple structures as found in many hexagonal or face centered cubic monoatomic metals. Even with such simple potentials, it is possible to explain e.g. the formation and motion of partial dislocations. With binary systems or modified potentials, e.g. with an additional bump or minimum, more complicated structures may be reached, whose period lengths can by far exceed the range of the potential and which we hence denote as *complex*. For realistic simulations of particular materials, however, it is necessary to go beyond such model potentials. For various types of materials there exist highly parameterized potential models tailored for the specific situation, e.g. for metallic, covalent, or ionic systems. The parameters are fitted such that either experimental data (melting temperature, elastic constants, cohesion energies) or data from quantum mechanical calculations of small systems (cohesion energies, stresses, forces on individual atoms) are optimally reproduced. With such realistic potentials and efficient algorithms on supercomputers it is possible to routinely simulate even the most demanding situations with high reliability, whether they require large system sizes (crack propagation, shock waves, dislocation entanglement) with up to several hundred million atoms, or long simulation times (atomic diffusion). But also equilibrium material parameters (elastic constants, heat conductivities, toughness) can be extracted and used as input for continuum theories. By forwarding data from quantum mechanical ab-initio calculations to molecular dynamics simulations and further to continuum theories (e.g. finite element methods) one can establish a multi-scale algorithm by which matter can be analysed up to mesoscopic length scales.

A problem remaining is the span of real time achievable in MD simulations, which reaches up to the order of nanoseconds, but not much beyond. Stochastic methods like the Monte Carlo algorithm may then be used to accelerate equilibration processes, where it is not necessary to track precisely the physical trajectories of the particles. Rather, the displacements of the atoms are chosen probabilistically, with weights depending on the energy gain achieved. By eliminating irrelevant degrees of freedom from the set of proposed moves, a considerable speed-up can be achieved, compared to MD. There are many other accelerators on the market like dissipative particle dynamics (PDP) or stochastic rotation dynamics (SRD), which average out short-time motions, but we will not dwell upon them further.

In this article we will proceed as follows. In Sec. 2 the molecular dynamics method is described. In particular, we sketch some popular integration methods and show how modified equations of motion can be used to simulate different kinds thermodynamic ensembles, e.g. ones which allow to control temperature, pressure, or shear. In Sec. 3 several model potentials are introduced, and some of the possible complex structures they generate are described. In Sec. 4 we show examples of realistic potentials, and discuss their range of validity. In Sec. 5 it is then demonstrated how the optimal potential parameters can be determined from results of ab-initio simulations, thus making quantum-mechanical information available also to classical simulations. Several potentials of embedded-atom-method (EAM) type have been generated by this method and were used in simulations of various physical properties and processes, such as atomic diffusion, phonon dispersion, crack propagation and a phase transition. These applications are described in Sec. 6, before we summarize in Sec. 7.

2.   Molecular Dynamics Simulations

In molecular dynamics (MD) simulations[1-3] the trajectories of particles in a many-body system are calculated in classical approximation. Thus Newton's equations

$$\frac{dp_i}{dt} = F_i = -\nabla_i V(r_1, \ldots, r_N) \tag{1}$$

$$\frac{dr_i}{dt} = \frac{p_i}{m_i} \tag{2}$$

have to be solved under a specific interaction potential V. Here $r_1, \ldots, r_N$ are the coordinate vectors of the particles and $p_1, \ldots, p_N$ their momenta.

In a first step one must establish an initial structure in the form of particle coordinates for the different atomic species. This can be done by either distributing the particles randomly, or using a model structure or a realistic structure derived from experimental (e.g. diffraction) data.

In a second step the interaction potentials have to be determined. These can be two-, three- or many-body potentials. They can be model potentials, addressed in Sec. 3, as for example the Lennard-Jones pair potential and modifications thereof. Or one chooses realistic effective potentials, fitted to experimental or ab-initio data, as discussed in Sec. 4 and 5.

In a third step the boundary conditions must be fixed. One can apply open boundaries or enclose the system in a hard box. In most instances periodic boundary conditions are chosen. They allow to simulate an infinite system by packing it into a periodically repeated array of unit cell. Thus one can avoid the influence of surfaces. But even for the calculation of surfaces frequently periodic boundary conditions are used by simulating a periodic layer structure. For specific situations a variety of other boundary conditions have been invented. In Lees-Edward boundary conditions, for example, the periodically stacked boxes are displaced relative to each other with constant velocity to mimic a shear deformation. Or twisted boundaries can model chiral structures, as e.g. cholesteric liquid crystals.

Newton's equations form a $3N$-dimensional coupled set of second order differential equation for the particle trajectories $r_i(t)$. In a fourth step they must be solved by numerical integrators subject to requirements like high precision, stability, time-reversal symmetry, energy and momentum conservation. The conditions are met by the relatively simple expressions of the Verlet and leap frog algorithms.

In the Verlet algorithm the Taylor expansions of $r_i(t+\delta t)$ and $r_i(t-\delta t)$,

$$r_i(t + \delta t) = r_i(t) + \delta t \dot{r}_i(t) + \frac{1}{2}(\delta t)^2 \ddot{r}_i(t) + \dots \tag{3}$$

$$r_i(t - \delta t) = r_i(t) - \delta t \dot{r}_i(t) + \frac{1}{2}(\delta t)^2 \ddot{r}_i(t) - \dots \tag{4}$$

are added such that the odd powers of δt vanish and the velocities do not appear explicitely:

$$r_i(t + \delta t) = 2r_i(t) - r_i(t - \delta t) + (\delta t)^2 \ddot{r}_i(t) + \dots \tag{5}$$

with

$$\ddot{r}_i(t) =: a_i(t) = -\frac{1}{m_i} \nabla_i V(r_1, \dots, r_N) \tag{6}$$

The algorithm is correct to order $(\delta t)^4$ and time-reversible as it is centered.

For short time-steps the term of order $(\delta t)^2$ is small compared to the $O(1)$-terms which leads to numerical imprecision. A modification avoiding this deficiency is the half-step leap-frog algorithm:

$$r_i(t + \delta t) = r_i(t) + \delta t\, \dot{r}_i(t + \frac{1}{2}\delta t) \tag{7}$$

$$\dot{r}_i(t + \frac{1}{2}\delta t) = \dot{r}_i(t - \frac{1}{2}\delta t) + \delta t\, a_i(t) \tag{8}$$

Here, only terms of $O(\delta t)$ appear. Moreover, the second order equation (Eq. 5) is replaced by a first order one, in which also the velocities have become independent degrees of freedom. Inserting Eq. 7 and Eq. 8 into the centering procedure (Eq. 3)-(Eq. 5) shows that the leap-frog scheme is algebraically equivalent to Verlet. With the leap-frog integrator we can simulate the microcanonical NVE ensemble, where particle number N, volume V and energy E are kept constant. For other ensembles, like the canonical NVT - ensemble, or the NpT - ensemble one must apply numerical thermostats and barostats.

Temperature and pressure control can be achieved with suitable modifications of the equations of motion, Eq. 1 and Eq. 2. Specifically, for a thermostat a (positive or negative) friction term is added, and for a barostat the box and all coordinates are rescaled. For a thermostat and barostat of Nosé-Hoover type, the modified equations of motion then read as follows:

$$\dot{p}_i = F_i - (\eta + \xi)\, p_i \tag{9}$$

$$\dot{r}_i = \frac{p_i}{m_i} + \xi r_i \tag{10}$$

$$\dot{\eta} = \frac{1}{\tau_T^2} \frac{T - T_{ext}}{T_{ext}} \tag{11}$$

$$\dot{\xi} = \frac{1}{\tau_p^2} \frac{V(p - p_{ext})}{N k_B T_{ext}} \tag{12}$$

$$\dot{h} = \xi h \tag{13}$$

Here, η and ξ are two extra dynamical degrees of freedom. η is a friction parameter, and ξ the rate of rescaling of the coordinates and the matrix of the box vectors, h. The amount of friction and rescaling is determined by the deviation of the actual temperature and pressure from the desired values, Text and pext. The actual temperature and pressure are computed from the coordinates and momenta:

$$T = \frac{1}{3Nk_B} \sum_i \frac{|p_i|^2}{m_i} \tag{14}$$

$$pV = \frac{1}{3}\left(\sum_i \frac{|p_i|^2}{m_i} + \frac{1}{2}\sum_{i,j} r_{ij} \cdot F_{ij} \right) \tag{15}$$

where $r_{ij} = r_i - r_j$ and F_{ij} is the force atom j exerts on atom i. The time evolution of η and ξ is governed by time constants τ_T and τ_p, which must be chosen suitably to ensure the best possible coupling of η and ξ to the other degrees of freedom. If only a thermostat is needed, ξ can be chosen identically zero (and thus $\tau_p = \infty$). By using tensor quantities for p, p_{ext}, and ξ, it is also possible to simulate an arbitrary external stress tensor, and with similar types of equations even a constant shear flow can be modelled.

In a fifth step the usually large amount of data must be analyzed. It is possible to extract global quantities like cohesion energies or melting temperatures or elastic constants. Also static and time dependent correlation functions can be calculated, like density-density or velocity-velocity correlations. From them structural characteristics can be gained like the pair correlation function or transport coefficients such as diffusion constants or viscosities. By Fourier transformation of the density-density correlation the diffraction images can be obtained.

But frequently the local situation is of interest. Here the primary tool is the visualization of the structure. The atoms can be plotted directly as spheres. Additionally they can be coloured, e.g. according to their energy content. Thus places of enhanced energy are marked, which usually are close to defects like vacancies or interstitials or dislocations. Or they are coloured according to their number of next neighbours to image surfaces, e.g. cracks. Observables like energy or temperature can be averaged over small spatial regions and then depict propagation of phonons. Animations of atomic dynamics give insight into mechanisms like energy transfer from elastic dilation into the crack tip where bonds are broken. Thus numerical simulations combined with visualization bring us back from the diagrammatic representation of physics to a way manifest to the visual senses.

All algorithms mentioned in this and the following sections are implemented in our own MD code IMD.[4] The results discussed in Sec. 1.6 have been obtained with IMD, which supports a large variety of integrators, potential models and other features, making it an excellent choice for the simulation of complex metals. Simple metals and even covalent solids are also supported, of course. IMD is efficiently parallelized and shows excellent performance and scaling on a large variety of hardware, from simple Linux PCs to massively parallel supercomputers.[5]

3. Model Potentials

3.1. *Lennard-Jones potentials*

The Lennard-Jones potential (see Fig. 1)

$$v(r) = \epsilon\left(\left(\frac{\sigma}{r}\right)^{12} - 2\left(\frac{\sigma}{r}\right)^{6}\right) \tag{16}$$

is an interaction widely applied in statistical physics. It is particularly suited for noble gases. Frequently it has been used also for the simulation of metals as long as computing power was too small to treat more complicated interactions and better descriptions like EAM potentials had not been developed. The potential was invented by Sir John Edward Lennard-Jones (1894-1954) in 1924 for quantum chemistry applications. The mathematical form was chosen such that it is possible to evaluate integrals easily.

The stable solid phase of the Lennard-Jones potential in two dimensions is the hexagonal close packing. Thermodynamical properties and the phase diagram have been given by Abraham.[6]

In three dimensions the energetic ground state is the hexagonal close packing (hcp) which is slightly more stable than fcc[7]. Reference 7 also includes many citations of Lennard-Jones system properties like virial coefficients and thermodynamic and transport properties. The thermodynamic phase diagram has been determined to high precision.[2,8,9]

In statistical physics the Lennard-Jones potential is frequently used to model binary liquids and glasses. In general there are six parameters ε_{AB} and σ_{AB} for the bond energies and the bond distances and an additional parameter for the composition. Usually two parameters are eliminated via the approximations $\varepsilon_{AB} = (\varepsilon_{AA} + \varepsilon_{BB})/2$ and $\sigma_{AB} = \sqrt{\sigma_{AA}\sigma_{BB}}$ (Lorentz-Berthelot rule), but this is not always possible for solid phases.[10–12]

The reason is that the bond lengths in solids are determined by geometric distances, and the Lorentz-Berthelot rule would lead to internal stresses that could distort or destroy the solid.

For glasses there are a number of special choices for the potential parameters which have been shown to avoid crystallization under ordinary conditions.[13–15] For some of them, however, a crystalline ground states has resulted.[16–20]

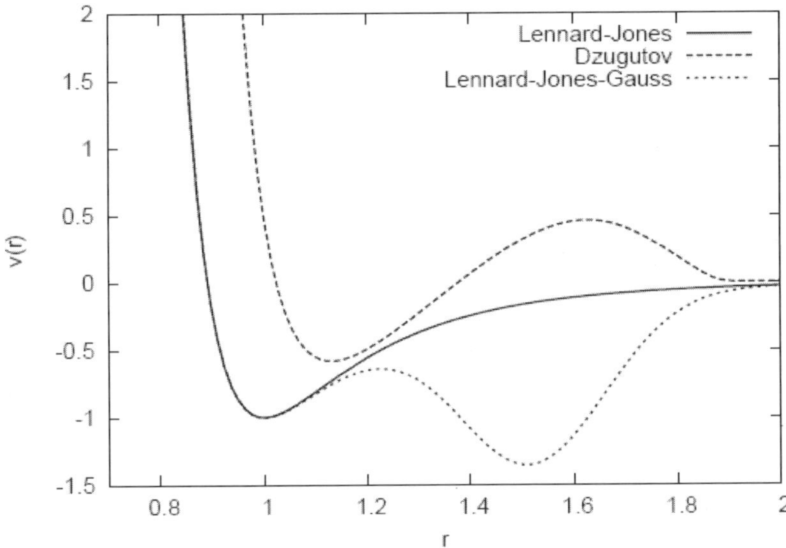

Fig. 1. The Lennard-Jones, Dzugutov and Lennard-Jones-Gauss potential for $\alpha=1.2$, $\beta^2=0.02$ and $r_0=1.52$.

3.2. *Dzugutov potentials*

The Dzugutov potential has been invented by M. Dzugutov for the description of monatomic metallic glasses (in contrast to covalent glasses).[21,22] The problem with single-minima-interactions like Lennard-Jones or Morse potentials ($v(r) = \mathrm{D}(1 - e^{-a(r-r_0)})^2$) is that if a sample is cooled from the liquid state to the solid state it will *always* freeze into a close-packed structure, either fcc or hcp or a mixture of both. Dzugutov's idea was to prevent the crystalline order by introducing a maximum at the distance of second nearest neighbours in close packed structures.

The Dzugutov pair potential is defined by the following equations (see Fig. 1):

$$\Phi(r) = \Phi_1 + \Phi_2 \tag{17}$$

$$\Phi_1 = A(r^{-m} - B)\exp\left(\frac{c}{r-a}\right), r < a \tag{18}$$

$$\Phi_1 = 0, r \geq a \tag{19}$$

$$\Phi_2 = B\exp\left(\frac{d}{r-b}\right), r < \ell \tag{20}$$

$$\Phi_2 = 0, r \geq b \tag{21}$$

with the parameters:

m	A	c	a	B	d	b
16	5.82	1.1	1.87	1.28	0.27	1.94

The special feature of the Dzugutov potential is a minimum at 1.13 of depth -0.581 followed by a maximum at 1.63 of height 0.460. The minimum has the same shape as in the case of the Lennard-Jones-potential. Past the maximum the potential goes to zero continously. The potential is cut off at a range of $r_c = 1.94$[a].

As a surprise came the discovery that the potential leads to dodecagonal quasicrystals if cooling is slow enough.[23] An evaluation of the pressure temperature phase diagram of the potential leads to a rich variety of phases: at low temperatures and pressures the bcc phase has the lowest energy, and at high pressures the ground state is fcc.[24] Tetrahedrally close packed phases appear also, especially the σ phase,

---

[a] Potential strength and range can be further scaled to adapt it to specific solids.

which can be considered as a low-order approximant of the quasicrystalline phases. It is assumed that the quasicrystalline phase is stabilized by entropy at low pressures and elevated temperatures.

The Dzugutov potential does not possess separate liquid and gaseous phases due to its maximum and its rather short range.[25] Since 1991 the potential has frequently been applied in the simulation of glasses, and it works quite well for this purpose.[26] Vibrational[27,28] and structural[29–31] properties have been studied extensively. The global ground state of small Dzugutov clusters has been determined.[32,33] In contrast to the Lennard-Jones and Morse potentials the interaction strongly favours polytetrahedral clusters. Diffusion has been studied[34] in monoatomic dodecagonal quasicrystals. Below 60% of the melting temperature only single atomic jumps occur, while above that temperature diffusion is greatly enhanced and flips are found which rearrange the underlying quasicrystal tiling structure. More recently the potential has also been applied in shock wav simulations.[35,36] Phase transitions between crystalline phases and solitons have been observed. The results show similarities to shock waves in iron simulated with EAM potentials although the interaction has not been fitted to any specific material.

3.3. *Lennard-Jones-Gauss potentials*

Electronic energy considerations have shown that the general form of pair potentials in metals resembles of a strongly repulsive core plus a decaying oscillatory Friedel term[37] (see also Sec. 4.1). The shape of the Dzugutov potential reproduces the first oscillation. The next natural step is the addition of a second minimum leading to a double-well potential. One possible choice is the Lennard-Jones-Gauss potential (Fig.1) given by

$$
V(r) = \frac{1}{r^{12}} - \frac{2}{r^6} - \alpha \exp\left(-\frac{(r-r_0)^2}{2\beta^2}\right)
\tag{22}
$$

It consists of a Lennard-Jones minimum of depth 1, positioned at $r = 1$ and a negative Gaussian of depth α and width β at $r = r_0$. Structure formation and dynamics can be studied as a function of the three parameters. At the moment the phase diagram of the system has only been calculated in two dimensions.[38] For T = 0, P = 0, and $\beta^2 = 0.02$ it is shown in Fig. 2 over the α-r_0 parameter space. A surprising variety of structures, four simple crystals (Hex1, Hex2, Sqa, and Rho) and five complex crystals (Pen, Xi, Sig1, Sig2, and Sig3) occur as stable ground states. The largest one (Xi) has 13 atoms per unit cell and a lattice constant about three times r0, which is already remarkably complicated – at least for a monatomic crystal in two dimensions and a simple isotropic pair potential.

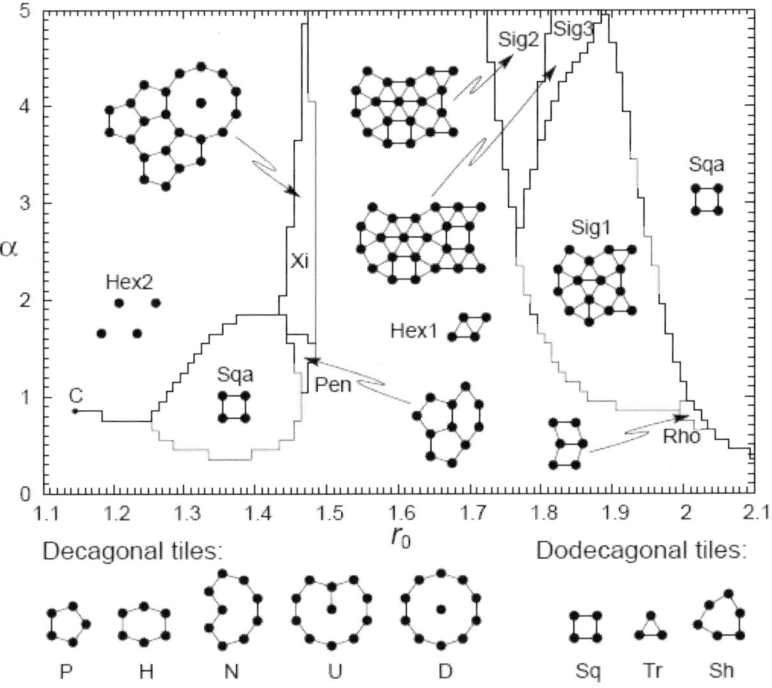

Fig. 2. Phase diagram of the Lennard-Jones-Gauss potential. Unit cell decorations of the crystals are depicted. For $r_0 \approx 1.47$ pentagonal local order and for $r_0 \approx 1.85$ dodecagonal order is found. The tiles are basic building blocks of the complex phases.

Simulations reveal, that the particle dynamics during crystal growth, but also in equilibrium crucially depends on the structural complexity. In addition to the usual vacancy and interstitial diffusion in simple phases, particle jumps (phason flips) are observed in the complex phases. They generate local tile rearrangements, but do not alter the long-range order.

At elevated temperature the resulting entropic contribution to the free energy plays an important role in the thermodynamic stabilization. This is observed for example in the case of a reversible order-disorder transition between the low-temperature periodic crystal Xi and a high temperature decagonal random tiling quasicrystal.[38]

By the choice Eq. 22 for the interaction potential two competing particle distances are introduced into the system. Depending on their ratio various local configurations can be favourable. If the most stable local configuration is not compatible with periodicity, structural complexity is a natural consequence. In fact complexity seems to be closely related to the occurrence of two competing characteristic length scales also for real systems, where additionally the interplay of different atom species can play a role. An example are the binary Laves-phases, which are stabilized in a narrow range of the atom size ratio. Note that the complex phases in Fig. 2 primarily appear in the region, where the second minimum is deeper than the first minimum, i.e. $\alpha > 1$. In realistic pair potentials the first minimum is often either degenerated to a shoulder or deeper than the second one. This might be a reason, why complex phases are not found in elementary metals, at least under ambient external conditions. This is not true, however, for metalloid elements due to stronger angle-dependent multi-body contributions to the interaction.

The phase β-boron with 105 atoms per unit cell, thermodynamically stable at high temperatures, is primarily built from icosahedrons arranged on a rhombohedral lattice.[39] Furthermore, complex phases are found quite frequently in many elemental metals at high pressures.[40] It is conceivable that shoulders in the interaction potential are increasingly significant under compression.

4. Realistic Potentials

More realistic potentials can be obtained in essentially two ways, which are often applied in combination. One way is to derive an effective potential from a more fundamental theory, such as Density Functional Theory (DFT), by making a suitable series expansion, or by applying various physically motivated approximations, which naturally must depend on the nature of the system at hand (whether it is metallic, covalent, etc.). The other way is to choose a parametrized potential family, and to determine the parameters so that the potential correctly reproduces certain reference quantities, which can either be measured or computed ab-initio. The choice of the potential family is again dictated by the nature of the system, and its functional form is often inspired by approximations to DFT.

The first potentials of this kind had relatively few parameters and a functional form closely related to DFT. This often results in relatively complicated analytical formulae which are expensive to evaluate numerically. Moreover, it is difficult to verify the correctness of a program. Examples for such potentials are the original Finnis-Sinclair[41] and Embedded-Atom- Method (EAM) potentials[42–45] for metals, Tersoff potentials[46] for covalent solids, and generalizations of them like bond-order potentials,[47] and the modified EAM (MEAM) potentials.[48–50] With the availability of an almost unlimited amount of ab-initio computed reference quantities which allow to fit a high number of potential parameters, it has become more convenient and also more efficient to use much larger potential families with completely free parameter functions. One example are the modern EAM or glue potentials, whose parameter functions no longer are required to have a functional form prescribed by DFT, but are freely fitted to reference quantities. Similarly, ADP potentials[51] (Angle Dependent Potentials) generalize in a sense the MEAM potentials. They contain similar angle dependent terms, but have a simpler functional form at the expense of a larger number of potential parameters, and are much simpler and more efficient to evaluate.

In the following, we introduce in some more detail the Moriarty-Widom pair potentials directly derived from DFT, and the EAM and ADP potentials, whose parameter functions are determined by fitting to a data base of reference quantities.

4.1. *Moriarty–Widom pair potentials*

By a systematic series expansion of DFT into n-body terms, and by making various approximations, Moriarty and Widom[52, 53] developed a family of potentials for aluminium-rich Al-TM (transition metal) alloys.

Of course, n-body interactions with arbitrarily high n are not suitable for numerical simulations. Fortunately, it turns out that the complete interaction is already well approximated by the one- and two-body terms. At fixed volume, even the (density dependent) one-body term can be neglected. Only the higher-order terms of the TM-TM interactions give a non-negligible contribution, which can be approximated by an extra effective pair term not yet included in the expansion. Such an extra pair term could be fitted empirically to an ab-initio simulation of a decagonal quasicrystal.[54] The resulting effective pair potentials, an example of which is shown in Fig. 3, exhibit characteristic Friedel oscillations, which are believed to be important for the stabilization of complex crystals and quasicrystals.

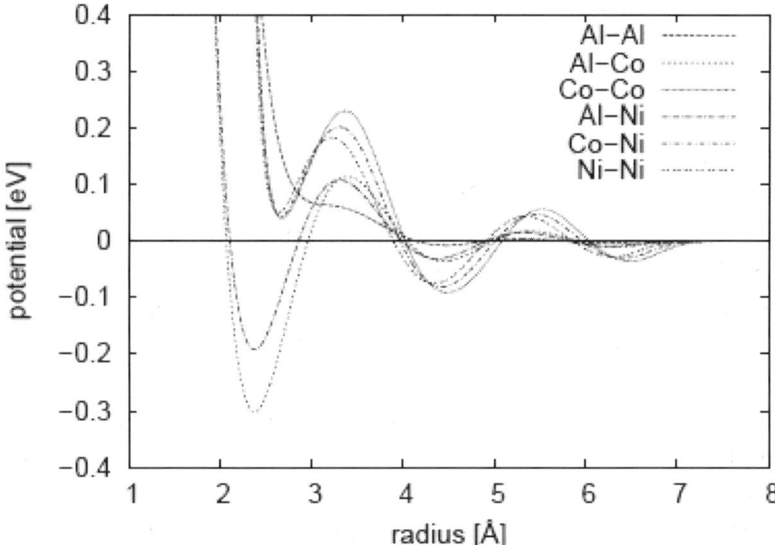

Fig. 3. Effective pair potentials for decagonal AlNiCo according to Moriarty and Widom.[52,53]

The potentials have been used very successfully for different kinds of simulations of decagonal quasicrystals, such as structure optimisation through total energy calculations,[55] or the simulation of aluminium diffusion in decagonal Al-Ni-Co.[56]

4.2. *Embedded atom method potentials*

Pure pair potentials, even with optimally chosen parameter functions, cannot describe the interactions in a metal adequately. This holds true at least for relatively simple, short range potentials (note that the Moriarty-Widom potentials discussed above have long-range Friedel oscillations, which remedy part of the deficiencies). The problem is that in a metal the interactions are density-dependent; one cannot simply add up the contributions of the different neighbours to the potential. In order to better mimic the physics of metallic cohesion, the so-called Embedded Atom Method (EAM) was introduced.[41–43,45] The corresponding potentials contain, in addition to a conventional pair interaction, a density-dependent embedding term:

$$V = \sum_{i,j<i} \phi_{s_i s_j}(r_{ij}) + \sum_i U_{s_i}(n_i) \qquad \text{with} \quad n_i = \sum_{j\neq i} \rho_{s_j}(r_{ij}) \qquad (23)$$

The embedding term consists of a non-linear function $U_{s_i}(n_i)$, whose argument, the local density n_i at the position of atom i, is a sum of contributions from the neighbours j through a transfer function ρ_{s_j}. $U_{s_i}(n_i)$, yields the energy associated with placing atom i at the density n_i. This form was actually inspired by DFT, from where also the form of the embedding function was suggested. In modern versions, however, all parameter functions in Eq. 23 are freely optimized. This more general version is also known as glue potential.[44] EAM or glue potentials have been shown to describe the interactions in a metal much better than pair potentials.[44] Even though they contain many-body terms through the non-linearity of the embedding function, they are still relatively efficient to compute. Compared to pair potentials, a second loop over all neighbour pairs is required (one to compute ni, another to compute the energy and the forces), so that the total effort is approximately doubled.

The EAM potential Eq. 23 has two gauge degrees of freedom, i.e. two sets of parameter changes which do not alter the physics of the potential:

$$\rho_s(r) \rightarrow \kappa \rho_s(r)$$

$$U_{s_i}(n_i) \rightarrow U_{s_i}\left(\frac{n_i}{\kappa}\right) \qquad (24)$$

and

$$\phi_{s_i s_j}(r) \rightarrow \phi_{s_i s_j}(r) + \lambda_{s_i}\rho_{s_j}(r) + \lambda_{s_j}\rho_{s_i}(r)$$
$$U_{s_i}(n_i) \rightarrow U_{s_i}(n_i) - \lambda_{s_i}n_i \tag{25}$$

According to Eq. 24, the units of the density ni can be chosen arbitrarily. The transformation Eq. 25 states that certain energy contributions can be moved freely between the pair and the embedding term. An embedding function U which is linear in the density n can be gauged away completely. This also makes any separate interpretation of the pair potential part and the embedding term void; the two must be judged together. The latter degeneracy is usually lifted by choosing the gradients of the $U_{s_i}(n_i)$ to vanish at the average density for each atom type.

4.3. *Angular-Dependent Potentials (ADP)*

Neither pair nor EAM potentials contain any angle-dependent terms. Therefore they cannot be used for covalent systems. But even for certain metals, such terms cannot be completely neglected. This is likely to be true for metals having only a small number of valence electrons, or which are on the border of being metals. For such cases, Baskes[48,49] has proposed the Modified EAM (MEAM) potentials, which were later modified[50] once more by introducing a screening function. All variants involve relatively complicated analytical formulae, which are difficult to evaluate numerically. This is especially true for molecular dynamics, where not only the potential, but also its derivative is required. For this reason, implementations of MEAM are rarely available, and if they are, they are slow and not well suited for large-scale simulations.

Recently, however, Mishin *et al.*[51] introduced a potential which contains essentially the same angle-dependend terms as the MEAM potential, but with a functionally much simpler form. They termed it ADP for Angle- Dependent Potential. Formally, ADP is an extension of EAM by further embedding terms:

$$V = \frac{1}{2}\sum_{i \neq j}\phi(r_{ij}) + \sum_i U(n_i) + \frac{1}{2}\sum_{i,\alpha}(\mu_i^\alpha)^2 + \frac{1}{2}\sum_{i,\alpha,\beta}(\lambda_i^{\alpha\beta})^2 - \frac{1}{6}\sum_i \nu_i^2 \tag{26}$$

The first two terms are the standard EAM potential (Eq. 23). The third term measures the energy cost of a dipole distortion of an atomic environment, the fourth and the fifth term that of a quadrupole distortion. For this purpose, for each atom a net dipole distortion vector

$$\mu_i^\alpha = \sum_{j \neq i} u_{ij}(r_{ij}) r_{ij}^\alpha \tag{27}$$

and a quadrupole distortion tensor

$$\lambda_i^{\alpha\beta} = \sum_{j \neq i} w_{ij}(r_{ij}) r_{ij}^\alpha r_{ij}^\beta \tag{28}$$

are computed. The last term in Eq. 26 compensates for the trace of

$$\nu_i = \sum_\alpha \lambda_i^{\alpha\alpha} \tag{29}$$

In the equations above, u_{ij} and w_{ij} are parameter functions which must be determined from reference data, and α and β represent Cartesian coordinates x, y, or z. As the angle dependent terms occur only implicitly (in the squares of μ_i^α and $\lambda_i^{\alpha\beta}$) the computation of the ADP potential is algorithmically similar to EAM. It requires about twice as much computation time as EAM, and is thus at least two to three times faster than MEAM. Moreover, the program code is much simpler and more transparent. Also the expressions for the forces are fairly simple.[51]

5.   Potential Development with potfit

The reliability and predictive power of classical MD simulations depend crucially on the quality of the effective potentials employed. In the case of elementary solids, such potentials are usually obtained by adjusting a few potential parameters to optimally reproduce a set of reference data, which typically includes a number of experimental values like lattice constants, cohesive energies, or elastic constants, sometimes supplemented with abinitio cohesive energies and stresses.[45] In the case of more complex systems with a large variety of local environments and many potential parameters to be determined, such an approach cannot help, however: there is simply not enough reference data available.

The force matching method[57] provides a way to construct physically justified potentials even under such circumstances. The idea is to compute forces, energies and stresses from first principles for a suitable selection of small reference systems and to adjust the parameters of the potential to optimally reproduce them.

The force matching algorithm is implemented in the program *potfit*, the details of which have been described previously.[58] By separating the process of optimization from the form of the potential, *potfit* allows for maximal flexibility in the choice of potential model and parametrization.

5.1. *Algorithms*

potfit consists of two separate parts. The first one implements a particular parametrized potential model and calculates from a set of potential parameters ξ_i the target function that quantifies the deviations of the forces, stresses and energies from the reference values. Wrapped around is a second, potential independent part which implements a least square minimization module. As this part is completely independent of the potential model and just deals with the list of parameters ξ_i, it is fairly straightforward to change the parametrization of the potential (tabulated or analytic), or even to switch to a different potential model.

From a mathematical point of view, force matching is a basic optimization problem: There is a set of parameters ξ_i, a set of values $b_k(\xi_i)$ depending on them, and a set of reference values $b_{0,k}$ which the b_k have to match. This leads to the well-known method of least squares, where one tries to minimize the sum of squares of the deviations between the b_k and the $b_{0,k}$. In this case, the reference values can either be the components of the force vector $\sim f_{0,j}$ acting on each individual atom j, or global data $A_{0,k}$ like stresses, energies, or certain external constraints. It is helpful to measure the relative rather than the absolute deviations from the reference data, except for very small reference values. The least squares target function thus becomes

$$Z = Z_F + Z_C \tag{30}$$

with

$$Z_{\mathrm{F}} = \sum_{j=1}^{N_A} \sum_{\alpha=x,y,z} W_j \frac{(f_{j\alpha} - f_{0,j\alpha})^2}{\bar{f}_{0,j}^2 + \varepsilon_j} \tag{31}$$

and

$$Z_{\mathrm{C}} = \sum_{k=1}^{N_c} W_k \frac{(A_k - A_{0,k})^2}{A_{0,k}^2 + \varepsilon_k} \tag{32}$$

where Z_F represents the contributions of the forces, and Z_C that of the global data. The (small and positive) ε_ℓ impose a lower bound on the denominators, thereby avoiding too an accurate fitting of small quantities which are actually not known to such a precision. The W_l are the weights of the different terms. It proves useful for the fitting to give the total stresses and the cohesion energies an increased weight, although in principle they should be reproduced correctly already from the forces. Even if all forces are matched with a small deviation only, those deviations can add up in an unfortunate way when determining stresses, thus leading to potentials giving wrong elastic constants. Including global quantities in the fit with a sufficiently high weight suppresses such undesired behaviour of the fitting process.

As the evaluation of the highly nonlinear target function Eq. 30 is computationally rather expensive, a careful choice of the minimization method has to be made. There are two algorithms in *potfit*. The first is a deterministic algorithm described by Powell,[59] which takes advantage of the form of the target function (which is a sum of squares). By re-using data obtained in previous function calls it arrives at the minimum faster than standard least squares algorithms. It also does not require any knowledge of the gradient of the target function.

The other minimization method implemented is a simulated annealing[60] algorithm proposed by Corana.[61] While the deterministic algorithm mentioned above will always find the closest *local* minimum, simulated annealing samples a larger part of the parameter space and thus has a chance to end up in a better minimum. The price to pay is a computational burdon which can be several orders of magnitude larger.

The basic Monte Carlo move is adding Gaussian-shaped bumps to the potential functions. The bump heights are normally distributed around zero, with a standard deviation adjusted so that on average half of the Monte Carlo steps are accepted. This ensures optimal progress: Neither are too many calculations wasted because the changes are too large to be accepted, nor are the steps too small to make rapid progress.

Generally, potential functions can be specified in various ways. They can either be given in analytic form, using a small number of free parameters, like for a Lennard-Jones potential, or in tabulated form together with an interpolation scheme for distances between the tabulation points. Whereas the parameters of an analytic potential can often be given a physical meaning, such an interpretation is usually not possible for tabulated potentials. On the other hand, an inappropriate form of an analytic potential may severely constrain the optimization, leading to a poor fit. For this reason, the functions are defined by tabulated values and spline interpolation, thus avoiding any bias introduced by an analytic potential. This choice results in a relatively high number of potential parameters, compared to an analytic description of the potentials. This is not too big a problem, however: Force matching provides enough reference data to fit even a large number of parameters. The potential functions only need to be defined at pair distances r between a minimal distance r_{min} and a cutoff radius r_{cut}, where the function should go to zero smoothly.

5.2. *Implementation*

The program is parallelized using the standard Message Passing Interface (MPI[62]), simply by distributing the calculation of forces, energies and stresses in different configurations on several processes. Additionally, neighbour lists and lookup tables are used to speed up spline interpolation.

potfit was designed to cooperate closely with the first-principles code VASP[63,64] and with IMD,[4] our own classical MD code. VASP, which is a plane wave code implementing ultrasoft pseudopotentials and the Projector-Augmented Wave (PAW) method,[65,66] is used to compute the reference data for the force matching, whereas the resulting potentials are intended to be used with IMD. For this reason, *potfit* provides import and export filters for potentials and configurations to communicate with these programs. These filters are implemented as scripts, which can easily be modified to interface with other programs. All *potfit* input and output files are plain text files, so they can be manipulated without problems.

5.3. *Results and validation*

The correctness of potfit was verified by recovering a classical potential from reference data computed with that potential. This test succeeded perfectly for both a monatomic Lennard-Jones potential and a binary EAM potential. One should keep in mind, however, that reference data from ab-initio computations in general cannot be reproduced perfectly by any classical potential.

Although the potentials to be generated are intended for (aperiodic) quasicrystals and crystals with large unit cells, all reference structures have to be periodic crystals with unit cell sizes suitable for the ab-initio computation of the reference data. On the other hand, the reference structures should approximate the quasicrystal in the sense, that all their unit cells together accommodate all relevant structural motifs. To do so, they must be large enough. For instance, the quasicrystalline and related crystalline phases of Ca-Cd and Mg-Zn consist of packings of large icosahedral clusters in different arrangements. Reference structures must be able to accommodate such clusters. A further constraint is, that the unit cell diameter must be larger than the range of potentials. Reference structures with 80–200 atoms represent a good compromise between these requirements.

Starting from a selection of basic reference structures, further ones are obtained by taking snapshots of MD simulations with model potentials at various temperatures and pressures. Also samples which were strained in different ways are included. For all these reference structures, the ab-initio forces, stresses and energies are determined with VASP, and a potential is fitted to reproduce these data. As reference energy, the cohesive energy is used, i.e., the energies of the constituent atoms are subtracted from the VASP energies. Instead of absolute cohesive energies one can also use the energy relative to some reference structure. Once a first version of the fitted potential is available, the MD snapshots can be replaced or complemented with better ones obtained with the new potential, and the procedure is iterated.

During the optimization, the target function Eq. (30) does not converge to zero, which indicates that quantum mechanical reality (taking density functional theory as reality) is not represented perfectly by the potential model used. The forces computed from the optimal potential typically differ by about 10% from the reference forces, which seems acceptable. For the energies and stresses a much higher agreement can be reached. Cohesion energy differences for instance can be reproduced with accuracy better than 1%.

The generated potentials can then be used in molecular dynamics simulations to determine various material properties. Examples are described in Sec. 1.6 below.

It should be kept in mind that force-matched potentials will only work well in situations they have been trained to. Therefore, all local environments that might occur in the simulation should also be present in the set of reference configurations. Otherwise the results may not be reliable. Using a very broad selection of reference configurations will make the potential more transferable, making it usable for many different situations, e.g. for different phases of a given alloy. On the other hand, giving up some transferability may lead to a higher precision in special situations. By carefully constraining the variety of reference structures one may generate a potential that is much more precise in a specific situation than a general purpose potential, which was trained on a broader set of reference structures. The latter potential, on the other hand, will be more versatile, but less accurate on average. Finding sufficiently many suitable reference structures might not always be trivial. For certain complex structure like quasicrystals, there may be only very few (if any) approximating periodic structures with small enough unit cell.

potfit does currently not use experimental data during force matching. The potentials are determined exclusively from ab-initio data, which means they cannot exceed the accuracy of the first principles calculations. While it is possible, in principle, to support also the comparison to experimental values, we decided against such an addition. For once, available experimental values can often not be calculated directly from the potentials, so determining them would considerably slow down the evaluation of the target function Eq. 30. Secondly, experimental values often also depend on the exact structure of the system, which in most cases is not completely known beforehand for complex structures, for instance due to fractional occupancies in the experimentally determined structure model. A better way to use experimental data is to test whether the newly generated potentials lead to structures that under MD simulation show the behaviour known from experiment.

It should be emphasized, however, that constructing potentials is still tedious and time-consuming. Potentials have to be thoroughly tested against quantities not included in the fit. In this process, candidate potentials often need to be rejected or refined. Many iterations of the fitting-validation cycle are usually required. It takes experience and skill to decide when a potential is finished and ready to be used for production, and for which conditions and systems it is suitable. *potfit* is only a tool that assists in this process. Flexibility and easy extensibility was one of the main design goals of *potfit*. While at present only pair and EAM potentials with tabulated potential functions are implemented in *potfit*, it would be easy to complement these by other potential models, or to add support for differently represented potential functions.

6. Simulations of Physical Properties

6.1. *Diffusion in d-Al-Ni-Co*

Aluminium is the majority element in many quasicrystals and expected to be the most mobile element, but due to the lack of suitable radio tracers its diffusion properties are hardly accessible to experiment. Here, we investigate aluminium diffusion in decagonal Al-Ni-Co quasicrystals by molecular dynamics simulations. The calculations were carried out with newly developed EAM potentials[67] (see also Sec. 5).

The model structures of Ni-rich decagonal Al-Ni-Co ($Al_{72}Ni_{21}Co_9$) consist of an alternate stacking of two different layers which are decorations of the same hexagon-boat-star (HBS) tiling, resulting in a period of about 4Å.[68,69] We use a slightly modified variant determined in relaxation simulations, where it was found that the two innermost Al atoms in the star tiles prefer different positions, and also break the 4Å periodicity locally to an 8Å periodicity.[56] In the quasiperiodic plane the structure contains decagonal clusters which consist of 5, 7 or 8 Al atoms and 5, 3 or 2 Ni atoms.

Significant anisotropic diffusion of Al was observed at temperatures above 0.6 T_{melt}. It was found that the diffusion in the decagonal plane proceeds via mechanisms which are specific to quasicrystals. Of great importance are sites which tend to emit atoms, whereas other sites can absorb atoms. Chain processes occur, where the initial and the final positions are at these sites. As shown in Fig. 4, mobile Al atoms are located in the clusters with 7 or 8 Al atoms and within supertiles.

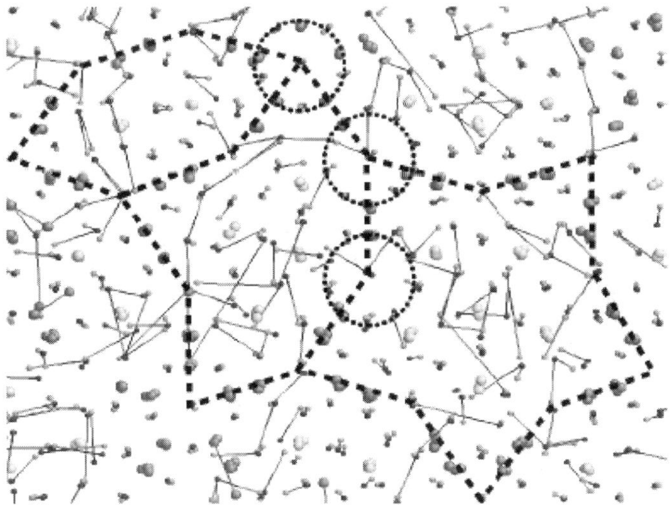

Fig. 4. Diffusion processes in the decagonal plane. Dark grey, large: Ni; light grey, large: Co; dark grey, small: Al initial positions; light grey, small: Al positions after 100ps. Initial and final positions are connected. Dashed lines mark supertiles. Dotted circles mark three types of decagonal clusters, differing in the Ni content.

An important characteristic of this decagonal structure is that some Al positions are not localized, but there are regions of continuous Al density (Fig. 5). In the periodic direction, channels of continuous Al density spread through the structure. The diffusion in this direction runs via such channels. The diffusion mechanism is the direct position exchange of a column of atoms. In each period there are three atoms which are part of this column. With more than three periods the process is usually coupled with a jump process in the decagonal plane. An atom of the decagonal plane jumps into the diffusion channel, whereas another atom leaves the channel. In this case, only the atoms in-between diffuse along the channel. The diffusion channels are located in the centres of supertiles and in the centres of decagonal clusters which contain three atoms per period.

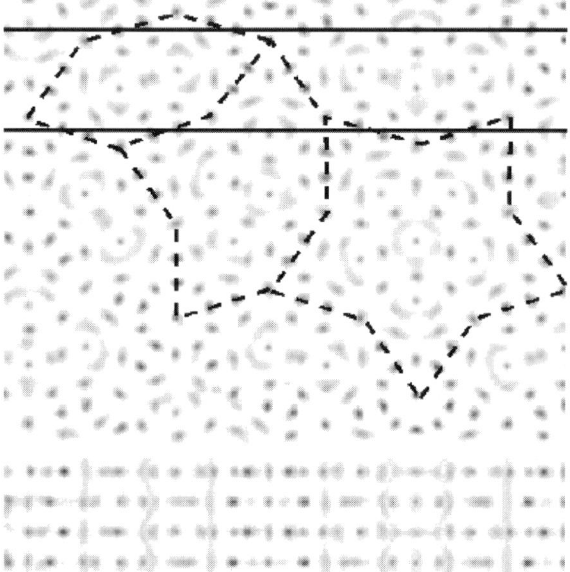

Fig. 5. Atom density map with regions of continous Al density. The lower part of the figure shows the marked stripe perpendicular to the periodic direction.

There are clearly more Al atoms which contribute to the diffusion in the decagonal plane than in the periodic direction, in which the diffusion is limited to the channels. However, since the mobility of Al atoms in the diffusion channels is significantly higher, the diffusion coefficients in the periodic direction are larger than those in the decagonal plane.

Time-averaged atom density maps calculated with ab-initio MD as well as classical MD with EAM potentials look very similar and confirm the existence of the continuous Al density regions. Furthermore, energy barriers calculated for specific diffusion processes using EAM potentials were validated in ab-inito calculations as well. The energy barriers determined in the periodic direction are about 0.3 eV per 8Å-period. For chain processes in the decagonal plane involving 3 atoms, an energy barrier of 0.5-2.5 eV was determined.[70]

6.2. *Dynamical structure factor*

The dynamical structure factor and other vibrational properties of a solid can in principle be derived from its eigenfrequencies and the corresponding eigenmodes. The eigenfrequencies are usually obtained in harmonic approximation: The system is relaxed to its ground state, around which the potential is expanded to second order to compute the dynamical matrix, which then is diagonalized numerically. An important advantage of this approach is that one can work with small systems (one unit cell) by sampling the Brillouin zone using Born-von-Karman boundary conditions. Computationally, it is much easier to diagonalize many small matrices instead of one big matrix for a large system. This makes the method usable even with ab-initio programs. Some MD or ab-initio programs directly provide the required derivatives of the forces. Otherwise it is always possible to take derivatives numerically. Unfortunately, there are also disadvantages. For complex structures, the ground state may not be very well defined, with many shallow minima, which can lead to numerical difficulties like negative eigenvalues of the dynamical matrix. Even more importantly, if there are low-energy structural transformations occurring, harmonic approximation may not be correct or accurate enough at higher temperatures.

Experimentally, phonons are usually studied by inelastic neutron scattering. These experiments measure certain correlation functions, which can be measured also in a molecular dynamics simulation. The advantage of such a procedure is the close resemblance to the experimental setup, so that a direct comparison with experiment becomes possible. In particular, the simulation can be performed at the same temperature as the experiment, and is not restricted to harmonic approximation. In contrast to the dynamical matrix approach, however, it is not possible to use Born-von-Karman boundary conditions, so that large samples have to be simulated over long physical times, which excludes the use of ab-initio methods.

Most correlation functions of interest are actually time-dependent ones. As typical MD codes usually do not keep the whole MD trajectory in memory, they are best computed with some post-processing tool. nMoldyn[71] is such a post-processor. It is freely available and has been written exactly for the purpose of computing correlation functions related to neutron scattering experiments. nMoldyn uses as input the entire MD trajectory of the system, with all atom positions and velocities. For many MD programs, converters to the input format of nMoldyn exist. There is one problem, however: a long trajectory of a large system represents a huge amount of data. Note that if the Brillouin zone of a complex crystal phase with large unit cell shall be sampled with adequate accuracy, then a large system has to be simulated, and in order to obtain a reasonable frequency resolution, long trajectories are required. For one-particle correlation functions (selfcorrelations), where correlations of the properties of a particle with its own history are averaged over the system, one can restrict the trajectory to a representative subset of the particles, e.g. those in one unit cell. This trick can be used e.g. for the velocity auto-correlation, whose Fourier transform is the vibrational density of states.[71]

For collective correlation functions however, such an approach is not possible. Among those, the most interesting is the dynamical structure factor, which consists of a coherent and an incoherent part. The incoherent part is a one-particle correlation function and provides only a relatively smooth background. Interesting structural information is mostly contained in the (collective) coherent part, which is given by the Fourier transform of the coherent intermediate scattering function:

$$\mathcal{F}^c(q,t) = \frac{1}{N} \sum_{i,j} b_i^c b_j^c \langle \exp(-iq \cdot r_i(0)) \exp(iq \cdot r_j(t)) \rangle \tag{33}$$

$$= \frac{1}{N} \langle f(q,0)\overline{f(q,t)} \rangle = \frac{1}{N} \langle f(q,t_0)\overline{f(q,t+t_0)} \rangle_{t_0} \tag{34}$$

$$f(q,t) = \sum_i b_i^c \exp(-iq \cdot r_i(t)) \tag{35}$$

Here, b_i^c is the coherent Neutron scattering amplitude. As we can see, for a fixed wave vector q the coherent intermediate scattering function

$\mathscr{F}^c(q,t)$ is the autocorrelation of the time series $f(q,t)$, which can be measured for a selection of interesting values of q, say those on a suitable path in the Brillouin zone. Writing the time series $f(q,t)$ for a few hundred values of q produces much less data than the whole trajectory of all particles. From that data, it is then easy to compute the Fourier transform of its autocorrelation via the fast Fourier transform method,[71] which is done with a simple post-processor. As an example, we show in Fig. 6 the dynamical structure factor for longitudinal polarisation of the Laves phase of $MgZn_2$, determined with EAM potentials constructed with *potfit*,[58] as described in Sec.5.

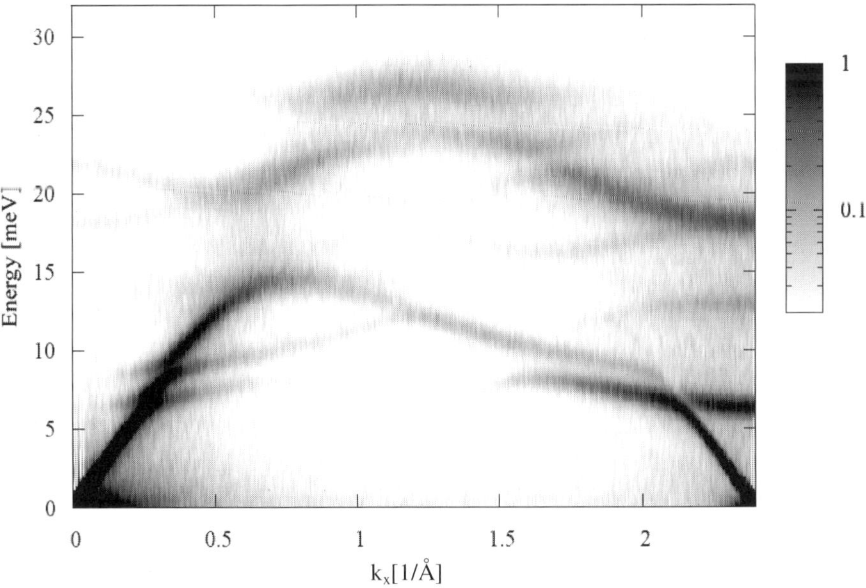

Fig. 6. Coherent part of the dynamical structure factor (longitudinal polarisation) for the Laves phase of $MgZn_2$.

6.3. *Cracks in NbCr2*

By stretching a specimen, elastic energy is stored in the system. When a crack moves through such a strained brittle solid, free surfaces are generated. Thus, crack propagation should be possible, if the elastic energy is sufficient for the generation of new fracture surfaces. It is

                    Hans-Rainer Trebin et al.

evident from this so-called Griffith criterion,[72] that cleavage should probably occur on planes of lowest surface energy. However, this global thermodynamic picture does not take into account the discrete, atomistic nature of matter. Atomic bonds break at the crack tip. Their arrangement and individual strength can vary on different planes. From model calculations[73] it is known that cracks indeed can be stable for loads that lie well above the Griffith value. This is due to the fact that breaking bonds costs a high amount of energy locally. The crack can therefore be trapped by the lattice. To overcome his barrier, the energy for crack propagation now also has to be higher than the value coming only from the generation of flat surfaces. As a consequence, fracture surfaces are not necessarily those of lowest energy or lowest roughness (see e.g. Rösch *et al.*[74,75]). To investigate the influence of the lattice-trapping[73] effect on cleavage at low temperatures, we perform molecular dynamics simulations on the Friauf-Laves phase $NbCr_2$.

The EAM potentials for this compound were generated using the forcematching method as described in Sec. 5. Ab-initio reference data on deformed samples as well as on systems with free surfaces has been included in the fit. The lattice constant, the elastic constants, and the melting temperature obtained with these interactions agree with experimental findings and ab-initio results.[76] The relaxation of atoms at free surfaces and the surface energy are also reproduced well.[76] The Friauf-Laves phases[77–80] form the largest subset of topologically close-packed intermetallic compounds.[81] They frequently combine interesting properties like high melting point, high temperature strength, and low density. However, their applications are limited by an extreme brittleness at low and ambient temperatures. In C15 $NbCr_2$ the Cr atoms sit on a tetrahedral network, whereas the Nb atoms form a diamond lattice. A cubic cell contains 24 basis atoms. On the other hand, the Friauf-Laves phases can be described as a stacking of layers (see e.g. Livingston[82] and Hazzledine *et al.*[83]) along the [111] direction.

We insert an atomically sharp seed crack in a relaxed specimen between (111) planes of lowest surface energy. The samples consist of about five million atoms. Periodic boundary conditions are applied parallel to the crack front. For the other directions, atoms in the outermost boundary layers are held fixed. The system is uniaxially strained perpendicular to the crack plane up to the Griffith load and relaxed. The crack then is further loaded by scaling the resulting displacement field at low temperature. Sound waves emitted by the propagating crack are damped away by the applied molecular dynamics scheme.[84]

From pure energy considerations, the crack propagation direction on one and the same plane should not change fracture behaviour. The required surface energy stays constant. From an atomistic viewpoint, however, the number and orientation of bonds that are approached by a propagating crack front may differ. In fact, this influence can be observed for two different propagation directions on a (111) plane in $NbCr_2$.

Geometrically scanned fracture surfaces for cracks that propagated along a $[2\bar{1}\bar{1}]$ and a $[0\bar{1}1]$ direction are given in Fig. 7. The flat seed crack has been built-in on the left. The energy release rate has been increased to about 170% of the Griffith value to allow both cracks to overcome the lattice-trapped state and to propagate with constant velocity. The crack moving in $[2\bar{1}\bar{1}]$ direction initially climbs up one atomic layer. Brittle cleavage fracture on this plane then is observed. The surplus of energy causes several point defects (see Fig. 7, top). In contrast, the crack propagating along a $[0\bar{1}1]$ axis causes a rougher surface (see Fig. 7, bottom). It seems that the crack could not decide whether to stay on the initial cleavage plane or to move up one atomic layer. The energy of the generated surfaces is in both cases higher than the energy of the initial flat cleavage planes ($[2\bar{1}\bar{1}]$: +13%, $[0\bar{1}1]$: +28%). So, the main part of the energy surplus goes into radiation. Rougher surfaces cost more energy to generate. The associated crack also travels slower ($[2\bar{1}\bar{1}]$: 1.05 km/s, $[0\bar{1}1]$: 0.76 km/s). Thus, macroscopic properties differ for the two propagation directions. This reveals that computer experiments on an atomic scale are necessary to simulate and to understand fracture properly.

The crack travelling in $[2\bar{1}\bar{1}]$ direction changes the initial cleavage plane for diverse applied loads always as shown in Fig. 1.7. Flat cuts at both heights cost the same amount of energy. The height thus could be chosen randomly. However, a crack shifted up one atomic layer does not deviate from this plane (see Fig. 8). The initial lattice trapping for the shifted crack is somewhat lower (an energy surplus of about 32% instead of 44% allows propagation). Thus, the bonds that are approached first at the crack tip even seem to select the cleavage plane. This subtle detail again emphasizes the importance of the lattice-trapping effect. The molecular dynamics simulations show, that fracture is determined by processes on the atomic level. These define whether, where, and how a crack propagates.

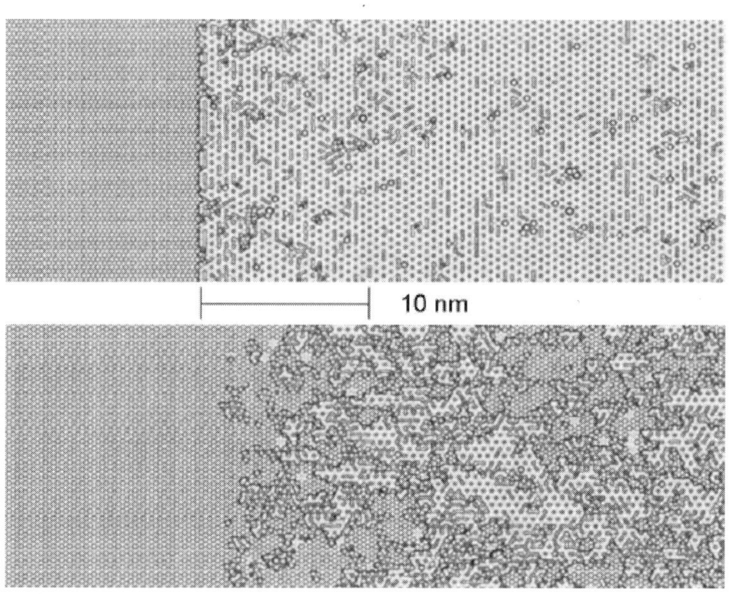

Fig. 7. Sections of (11$\bar{1}$) fracture surfaces geometrically scanned with a Nb atom. The crack propagated in [2$\bar{1}$$\bar{1}$] (top) and in [0$\bar{1}$1] direction (bottom).

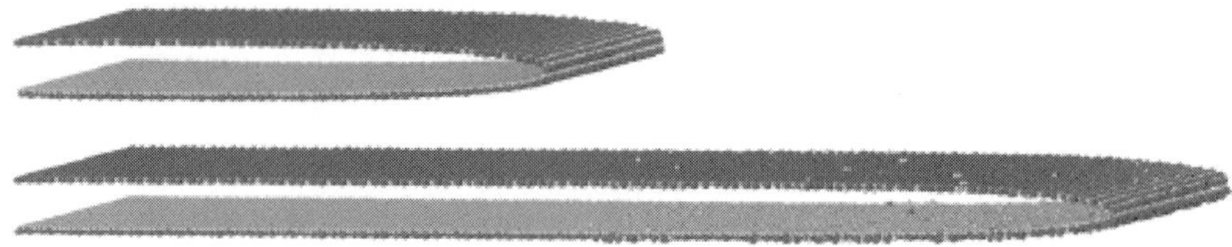

Fig. 8. Atoms forming the fracture surfaces of the shifted crack. Only particles with low coordination number are displayed. The seed crack (top) propagates smoothly on the initial height (bottom: after 0.1 ns).

6.4. *Order-disorder transition in CaCd$_6$*

CaCd$_6$ can be understood as the cubic 1/1-approximant of the thermodynamically stable Ca-Cd icosahedral quasicrystal.[85,86] Both the quasicrystal and the complex periodic phases can be described as packings of essentially identical clusters with some additional glue atoms in-between.[87,88] In the cubic CaCd6 phase, these clusters are arranged on a bcc lattice. The clusters each contain a total of 66 atoms in four shells: A Cd tetrahedron is believed to be the central shell, surrounded by a Cd dodecahedron, a Ca icosahedron and a Cd icosidodecahedron.[89] While the outer three shells are icosahedrally symmetric, this symmetry is broken by the innermost tetrahedron.

The cubic 1/1-approximant CaCd$_6$ (and the isostructural YbCd$_6$) show an order-disorder phase transition at 100K (and 110K, respectively), which is attributed to a reordering or disordering of the orientations of the tetrahedral central shell. At the phase transition temperature, there is a peak in the heat capacity and a discontinuity of the electrical conductivity.[90,91]

From X-ray diffraction, Gómez and Lidin[89] concluded that the central shell is a distorted tetrahedron suffering from two types of disorder. The first is a 90° rotational disorder along the twofold axis of the tetrahedron, which leads to a cube with alternately occupied corners. The second type of disorder is a triple split of the tetrahedron corner atoms that arises when the threefold rotational axis does not coincide with the body diagonal of the cubic unit cell. These two types of disorder combined lead to 24 possible positions for four tetrahedral Cd atoms in each cluster's central shell, as shown in Fig. 9(a). Based on this proposal, Nozawa and Ishii used ab-initio methods to compare the energies of structures with different tetrahedron shapes and orientations.[92] However, the computational demands of these methods make a similar treatment of larger supercells impossible.

Simulations with effective potentials turned out to be much better suited to this task.[93] The *potfit*[58] program (Sec. 5) was used to create EAM potentials for Ca-Cd, with the help of which the precise structure and ordering of the innermost CaCd cluster shell could be analyzed in a level of detail that is inaccessible to ab-initio methods. To study the stability of the Gómez-Lidin tetrahedra[89] as innermost shells, one generates all possible combinations of orientations of the tetrahedra in a cubic cell with two clusters and relaxes these structures to the nearest energy minimum. The resulting relaxed structures have six different energies in total, differing maximally by 0.17 meV/atom. The relaxed tetrahedra are all identical (up to symmetry), the energy difference coming from the different relative orientations of the two tetrahedra. This ideal tetrahedron is not equal to the Gómez-Lidin one, however. Each atom relaxes in a specific way (two atoms by 0.29Å, the other two by 0.17Å) to new tetrahedron corners, without further symmetry breaking. The number of distinct ideal tetrahedral therefore is still twelve, like the number of Gómez-Lidin tetrahedra, but the total number of possible corner positions has doubled. In Fig. 9(b), the new positions of a sample tetrahedron are shown in black, while the medium gray cirles signify the original Gómez-Lidin positions. The light gray positions are occupied by atoms belonging to rotated tetrahedra. The relaxation simulations also show that the surrounding shells of the clusters are strongly deformed by the inner tetrahedron. The stability of the ideal cluster has been verified also with ab-initio relaxations.

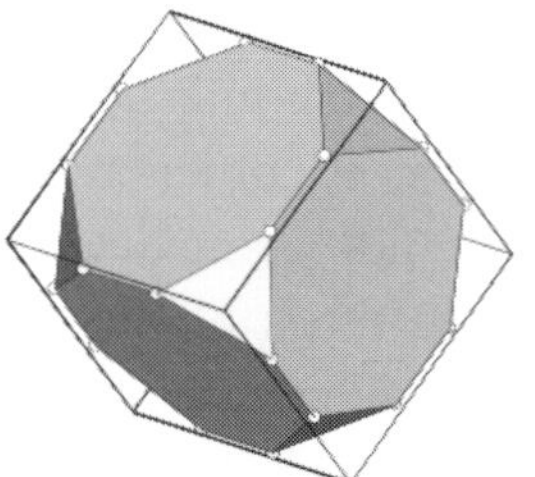

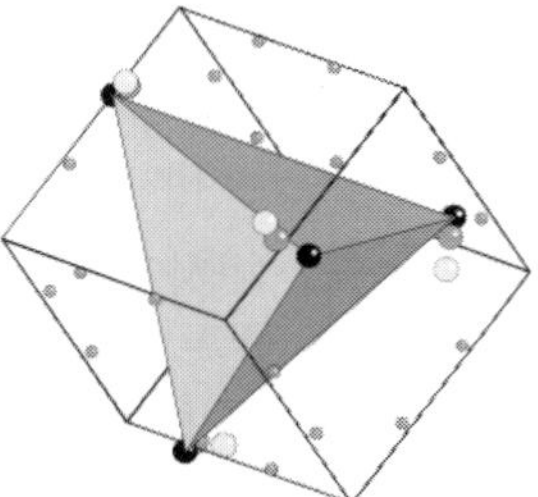

(a) Potential locations of the Cd tetrahedal atoms according to Gomez and Lidin.

(b) Improved tetrahedron. For an explanation, see text.

Fig. 9. Configuration of the Ca-Cd tetrahedron.

In a second step, finite temperature MD simulations were performed with a larger sample, consisting of $5 \times 5 \times 5$ cubic unit cells (250 clusters). These simulations show that, starting from an ordered low-temperature state, the tetrahedron orientations start to change at around 100 K, and even a few non-ideal tetrahedra show up. It was not possible, however, to see any clear sign of the expected phase transition, like a peak in the heat capacity or a jump in the inner energy, probably due to the bad statistics. A further step in our modelling of the interactions is therefore required.

In order to obtain better statistics, the number of degrees of freedom needs to be reduced. All clusters in low temperature states apparently have the same innermost shells in the form of the ideal tetrahedron introduced above, which occurs in twelve different orientations. One then can set up an effective hamiltonian modelling the interactions between neighbouring clusters via induced deformations of the outer shells. Neighbouring clusters can be in contact along two-fold or three-fold directions. Taking into account the different orientations of the inner tetrahedra at both ends of a bond, there are, up to symmetry, 26 types of 2-fold bonds and 16 types of 3-fold bonds. Each bond of type α contributes an energy E_α to the hamiltonian.

The energies E_α can be determined by fitting to energies obtained in MD relaxations of supercells containing up to 64 clusters with randomly oriented tetrahedra. The cluster hamiltonian so derived can now be used in extensive Monte Carlo simulations. With a sample of 4×4×4 unit cells containing 128 clusters, the internal energy as a function of temperature was determined,[93] which shows a sharp jump at about 89K (Fig. 10), not far from the experimental transition temperature of 100K.[91]

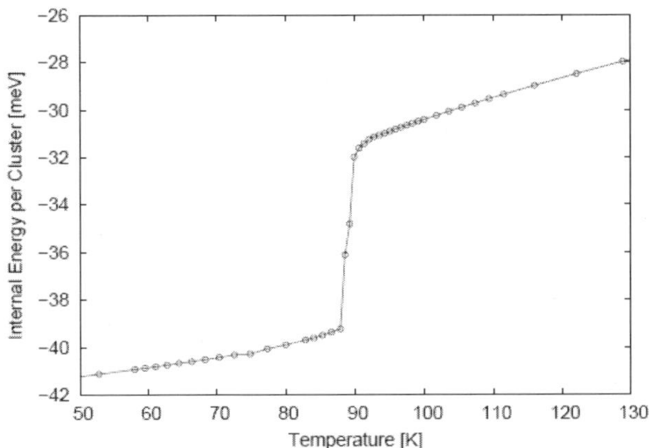

Fig. 10. Internal energy as a function of temperature for the effective cluster hamiltonian. A sharp jump is displayed at about 89K.[93]

This energy jump ΔE corresponds to an entropy jump $\Delta E/k_B T_0$ of roughly $1 k_B$ per cluster, which is about twice the amount estimated in Reference 90.

7. Conclusions

The molecular dynamics method allows to simulate a large variety of material properties and dynamical processes at the atomistic level. For complex materials and inhomogeneous systems with phase or grain boundaries and large defects, such a level of detail is indeed necessary. With classical effective potentials and large parallel computers, it is possible to routinely simulate systems with linear sizes exceeding $100°A$, thus reaching well into the nano-domain. Even though classical potentials may seem to be a gross approximation to the quantum mechanical reality, with a well selected potential model (depending on the system type) and with carefully adjusted potential parameters, it is possible to perform simulations with high accuracy and reliability. This can be achieved by fitting the potential parameters to data obtained in quantum-mechanical calculations, thus making quantum-mechanical information available also to classical simulations.

In many situations, however, system sizes of $100Å$ and time spans of a few nanoseconds may still be far too small. But even in such cases, it is possible to use results of molecular dynamics simulations to determine parameters of higher level theories, such as continuum models, or effective models where part of the degrees of freedom have been integrated out, like models used in Monte Carlo simulations. Molecular dynamics therefore provides the basis also for those higher level theories, and is an essential ingredient for any multi-scale modelling of matter.

References

1. M. P. Allen and D. J. Tildesley, *Computer Simulation of Liquids*. (Clarendon Press, Oxford, 1992).
2. D. Frenkel and B. Smit, Understanding molecular simulation: from algorithms to applications. (San Diego: Academic Press, 2003).
3. D. C. Rapaport, *The Art of Molecular Dynamics Simulation*. Second Edition, (Cambridge University Press, 2004).

4. J. Stadler, R. Mikulla, and H.-R. Trebin, *Int. J. Mod. Phys. C.* **8**, 1131, (1997). http://www.itap.physik.uni-stuttgart.de/~imd/.

5. F. Gähler and K. Benkert, in *Performance Computing on Vector Systems*, Eds. M. Resch, T. Bönisch, K. Benkert, T. Furui, Y. Seo, and W. Bez, High p. 173. Springer, ISBN 3–540–29124–5 (2006).

6. F. F. Abraham, *Adv. Phys.* **35**, 1 (1986).

7. F. H. Stillinger, *J. Chem. Phys.* **115**, 5208 (2001).

8. M. A. Barroso and A. L. Ferreira, *J. Chem. Phys.* **116**, 7145 (2002).

9. M. A. van der Hoef, *J. Chem. Phys.* **113**, 8142 (2000).

10. J. Roth, R. Schilling, and H.-R. Trebin, *Phys. Rev.* **B41**, 2735 (1990).

11. J. W. Roth, R. Schilling, and H.-R. Trebin, *Phys. Rev.* **B51**, 15833 (1995).

12. J. Roth and Ch.Henley, *Phil. Mag.* **A75**, 861 (1997).

13. W. Kob and H. C. Andersen, *Phys. Rev.* **E48**, 4364 (1993).

14. W. Kob and H. Anderson, *Phys. Rev.* **E51**, 4626 (1995).

15. W. Kob and H. C. Anderson, *Phys. Rev.* **E52**, 4134 (1995).

16. M. J. Vlot, H. E. A. Huitema, A. de Vooys, and J. P. van der Eerden, *J. Chem. Phys.* **107**, 4345 (2003).

17. M. Hitchcock and C. K. Hall, *J. Chem. Phys.* **110**, 11433 (1999).

18. T. F. Middleton, J. Hernández-Rojas, P. N. Mortenson, and D. J. Wales, *Phys. Rev.* **B64**, 184201 (2001).

19. J. R. Fernández and P. Harrowell, *Phys. Rev.* **E67**, 11403, (2003).

20. J. R. Fernández and P. Harrowell, *J. Chem. Phys.* **120**, 9222 (2004).

21. M. Dzugutov and U. Dahlborg, *J. Non-Cryst. Solids*, **131**, 62 (1991).

22. M. Dzugutov, *Phys. Rev.* **A46**, R2984 (1992).

23. M. Dzugutov, *Phys. Rev. Lett.* **70**, 2924 (1993).

24. J. Roth and A. Denton, *Phys. Rev.* **E61**, 6845 (2000).

25. J. Roth, *Eur. Phys. J.* **B14**, 449 (2000).

26. M. Dzugutov, S. I. Simdyankin, and F. H. Zetterling, *Phys. Rev. Lett.* **89**, 195701 (2002).

27. S. I. Simdyankin, S. N. Taraskin, M. Dzugutov, and S. R. Elliott, *Phys. Rev.* **B62**, 3223 (2000).

28. S. I. Simdyankin, S. N. Taraskin, M. Elenius, S. R. Elliott, and M. Dzugutov, *Phys.Rev.* **B65**, 104302 (2002).

29. T. Mattila, R. M. Nieminen, and M. Dzugutov, *Phys. Rev.* **53**, 192 (1996).

30. B. Sadigh, M. Dzugutov, and S. R. Elliott, *Phys. Rev.* **B59**, 1 (1999).

31. S. I. Simdyankin, M. Dzugutov, S. N. Taraskin, and S. R. Elliott, *Phys. Rev.* **B63**, 184301 (2001).

32. J. P. K. Doye, D. J. Wales, and S. I. Simdyankin, *Faraday Discuss.* **118**, 159 (2001).

33. J. P. K. Doye and D. J. Wales, *Phys. Rev. Lett.* **86**, 5719 (2001).

34. J. Roth and F. Gähler, *Eur. Phys. J.* **B6**, 425 (1998).

35. J. Roth, *Phys. Rev.* **B72**, 014125 (2005).

36. J. Roth, *Phys. Rev.* **B72**, 014126, (2005).

37. J. Hafner, *From Hamiltonians to Phase Diagrams* (Springer-Verlag, Berlin, 1987).

38. M. Engel and H.-R. Trebin, Accepted by *Phys. Rev. Lett.* (2007).

39. J. Dondhue, *The Structures of the Elements* (John Wiley and Sons, New York, 1974).

40. M. I. McMahon and R. J. Nelmes, *Chem. Soc. Rev.* **35**, 943 (2006).

41. M. W. Finnis and J. E. Sinclair, *Phil. Mag.* **A50**, 45 (1984).

42. M. S. Daw and M. I. Baskes, *Rev. Lett.* **50**, 1285 (1983).

43. M. S. Daw and M. I. Baskes, *Phys. Rev.* **B29**, 6443 (1984).

44. F. Ercolessi, M. Parrinello, and E. Tosatti, *Phil. Mag.* **A58**, 213 (1988).

45. M. S. Daw, S. M. Foiles, and M. I. Baskes, *Mater. Sci. Rep.* **9**, 251 (1993).

46. J. Tersoff, *Phys. Rev.* **B39**, 5566 (1989).

47. D. G. Pettifor, in *Many-Atom Interactions in Solids*, Eds. R. M. Nieminen, M. J. Puska,and M. J. Manninen, Vol. 48, Springer Proceedings in Physics, p. 64. Springer-Verlag, (1990).

48. M. I. Baskes, Phys. *Rev. Lett.* **59**, 2666 (1987).

49. M. I. Baskes, *Phys. Rev.* **B46**, 2727 (1992).

50. M. I. Baskes, *Mater. Chem. Phys.* **50**, 152 (1997).

51. Y. Mishin, M. J. Mehl, and D. A. Papaconstantopoulos, *Acta Mater.* **53**, 4029 (2005).

52. J. A. Moriarty and M. Widom, *Phys.Rev.* **B56**, 7905–7917, (1997).

53. M. Widom, I. Al-Lehyani, and J. A. Moriarty, *Phys. Rev.* **B62**, 3648 (2000).

54. I. Al-Lehyani, M. Widom, Y. Wang, N. Moghadam, G. M. Stocks, and J. A. Moriarty, *Phys. Rev.* **B64**, 075109 (2001).

55. M. Mihalkovič, I. Al-Lehyani, E. Cockayne, C. L. Henley, N. Moghadam, J. A. Moriarty, Y. Wang, and M. Widom, *Phys. Rev.* **B65**, 104205 (2002).

56. S. Hocker and F. Gähler, *Phys. Rev. Lett.* **93**, 075901 (2004).

57. F. Ercolessi and J. B. Adams, *Europhys. Lett.* **26**, 583 (1994).

58. P. Brommer and F. Gähler. *Mater. Sci. Eng.* **15**, 295 (2007). http://www.itap.physik.uni-stuttgart.de/~imd/potfit/.

59. M. J. D. Powell, *Comp.* J. **7**, 303 (1965).

60. S. Kirkpatrick, C. D. Gelatt, and M. P. Vecci, *Science*, **220**, 671 (1983).

61. A. Corana, M. Marchesi, C. Martini, and S. Ridella, *ACM Trans. Math. Soft.* **13**, 262 (1987).

62. W. Gropp, E. Lusk, and A. Skjellum, *Using MPI* - 2nd Edition. (The MIT Press, 1999). ISBN 0–262–57132–3.

63. G. Kresse and J. Hafner, *Phys. Rev.* **B47**, 558 (1993).

64. G. Kresse and J. Furthmüller, *Phys. Rev.* **B54**, 11169 (1996).

65. P. E. Blöchl, *Phys. Rev.* **B50**, 17953 (1994).

66. G. Kresse and D. Joubert, *Phys. Rev.* **B59**, 1758 (1999).

67. P. Brommer and F. Gähler, *Phil. Mag.* **86**, 753 (2006).

68. M. Mihalkovĭc, I. Al-Lehyani, E. Cockayne, C. L. Henley, N. Moghadam, J. A. Moriarty, Y. Wang, and M. Widom, *Phys. Rev.* **B65**, 104205 (2002).

69. E. Cockayne and M. Widom, *Phys. Rev. Lett.* **81**, 598 (1998).

70. S. Hocker, F. G¨ahler, and P. Brommer, *Phil. Mag.* **86**, 1051 (2006).

71. T. Róg, K. Murzyn, K. Hinsen, and G. R. Kneller, *J. Comput. Chem.* **24**, 657 (2003). http://dirac.cnrs-orleans.fr/nMOLDYN/index.html.

72. A. A. Griffith, *Philos. Trans.R. Soc. London, Ser.* **A221**, 163 (1921).

73. R. Thomson, C. Hsieh, and V. Rana, *J. Appl. Phys.* **42**, 3154 (1971).

74. F. Rösch, C. Rudhart, J. Roth, H.-R. Trebin, and P. Gumbsch, *Phys. Rev.*, **B72**, 014128 (2005).

75. F. Rösch, H.-R. Trebin, and P. Gumbsch, *Phil. Mag.* **86**, 1015 (2006).

76. F. Rösch, H.-R. Trebin, and P. Gumbsch, *Int. J. Fracture.* **139**, 517 (2006).

77. J. B. Friauf, *Phys. Rev.* **29**, 34 (1927).

78. J. B. Friauf, *J. Am. Chem. Soc.* **49**, 3107 (1927).

79. F. Laves and H.Witte, *Metallwirtsch.* **14**, 645 (1935).

80. F. Laves and H. Witte, *Metallwirtsch.* **15**, 840 (1936).

81. J. H. Wernick, in Intermetallic Compounds. Wiley Series on the Science and Technology of Materials.Ed. J. H. Westbrook, p. 197. (John Wiley & Sons, Incorporated, (1967)).

82. J. D. Livingston, *Phys. Stat. Sol.* **A131**, 415 (1992).

83. P. M. Hazzledine and P. Pirouz, *Scr. Mat. Metall.* **28**, 1277 (1993).

84. P. Gumbsch, S. J. Zhou, and B. L. Holian, *Phys. Rev.* **B55**, 3445 (1997).

85. J. Q. Guo, E. Abe, and A. P. Tsai, *Phys. Rev.* **B62**, R14605 (2000).

86. A. P. Tsai, J. Q. Guo, E. Abe, H. Takakura, and T. J. Sato, *Nature*, **408**, 537 (2000).

87. M. Widom and M. Mihalkovĭc, *Mat. Res. Soc. Symp. Proc.* **805**, LL1.10.1 (2004).

88. M. Mihalkovĭc and M.Widom, *Phil. Mag.* **86**, 519 (2006).

89. C. P. Gómez and S. Lidin, *Phys. Rev.* **B68**, 024203 (2003).

90. R. Tamura, K. Edagawa, C. Aoki, S. Takeuchi, and K. Suzuki, *Phys. Rev.* **B68**, 174105 (2003).

91. R. Tamura, K. Edagawa, Y. Murao, S. Takeuchi, K. Suzuki, M. Ichihara, M. Isobe, and Y. Ueda, , *J. Non-Cryst. Solids.* **334&335**, 173 (2004).

92. K. Nozawa and Y. Ishii, *Phil. Mag.* **86**, 615 (2006).

93. P. Brommer, F. Gähler, and M. Mihalkovĭc, Accepted by *Phil. Mag.* (2007).

CHAPTER 9

SCIENCE AND TECHNOLOGY OF HYDROGEN

Andreas Züttel and Louis Schlapbach

Empa - Materials Sciences and Technology
8600, Dübendorf, Switzerland
E-mail: andreas.zeuettel@empa.ch

Hydrogen exhibits the highest heating value per weight of all chemical fuels. It is regenerative and environment friendly. However hydrogen is not the major fuel of toady's energy consumption because: (1) it is just an energy carrier and has to be produced (on earth it is present almost only in the form of water), hence we have to pay for this energy, a difficult economic task, and (2) it is a gas at room temperature (critical temperature = 33K). So, hydrogen storage and transportation or distribution, are closely linked. For mobile and often also for stationary applications, the volumetric and gravimetric densities of hydrogen are crucial. Essentially six ways allow to store hydrogen: in high pressure gas cylinders, as a liquid in cryogenic tanks, adsorbed on materials with a large specific surface area, absorbed on host metal interstitial sites, chemically bond in covalent and ionic compounds or from oxidation of reactive metals (Li, Na, Mg, Al, Zn, etc) with water that produce the corresponding hydroxide and liberate gaseous hydrogen. This paper reviews the hydrogen storage methods known so far and focuses on the basic limitations of each method and the current state of research. A summary of the history of hydrogen is added.

1. Introduction

When was our today's knowledge on hydrogen and its chemical, physical and biological properties and their applications developed? Artistotle classified the materials world into his "elements" earth, water, air and fire. The term "gas" was introduced quite late. For a long time studies of "air" were strongly related with medical aspects of respiration and the "elasticity of air". Results of quantitative measurements of volume and

331

pressure and later the use of more accurate balances forced critical analytical thoughts about air and air as a single element and contributed to the development of ideas in direction of the ideal gas law and of chemical reactions.

Henry Cavendish (1731-1810) is considered as the discoverer of hydrogen; he clearly distinguished (1766) different qualities of air, air which burns and air which extincts flames. However, he was convinced that air which escaped upon the solution of metals in acids was contained in the metal (old phlogiston theory of Stahl) and liberated from the metal.

The great discoverer of the nature of hydrogen and founder of modern chemistry was Antoine Lavoisier (1743-1794). He showed (1783) that water is not an element, but is composed of two elements which he named hydrogen and oxygen and that they can be recombined to form water. Lavoisier made several other discoveries, e.g. that diamond is made of carbon, and developed e.g. gunpowder manufacturing and testing processes and assisted a failed gunfire which ended with the death of the testing person. Parallel to his scientific work Lavoisier was tax-collector for the royalistic "old regime" in France. During the French Revolution many of the tax-collectors were condemned to death. Lavoisier mounted the scaffold "comme un sage" and was executed by the guillotine. His wife had assisted him in the laboratory work; her efforts to liberate the condemned Lavoisier are judged not to have been very strong and certainly not successful; she got married to a soup manufacturer shortly after Lavoisier's execution.

Hydrogen technology started with hydrogen production by the solution of rather large quantities of Iron in sulfuric acids and its use for balloon flights up to the "incredible height of 1524 Klafter" (around 3000m) by Jacques Charlier (1783). In 1901 Zeppelin built the 1st airship. Over more than 30 years Zeppelins transported passengers more comfortably than today's economy class aircraft's over Europe, over the Atlantic and finally around the world. The Zeppelin area ended very abruptly when the highly inflammable skin of the airship "Hindenburgh" caught fire during the landing operation in Lakehurst (USA) on May 6, 1939; 13 out of the 36 passenger, 22 crew members and one person from the ground lost their life.

Today, hydrogen is a technologically and economically important chemical; its potential as a future energy carrier is high; compact storage of hydrogen remains a challenge. Todays most popular hybrid car Toyota Prius of the first and second generation run with electric power from metal hydride batteries, i.e. with electric current produced over hydrogen which is stored in its atomic form on interstitial sites of a host metal lattice. Hydrogen storage and transportation (or distribution) are closely linked together. Hydrogen can be distributed continuously in pipelines or batch wise by ships, trucks, railway or airplanes. All batch transportation requires a storage system but also pipelines can be used as pressure storage system.

The world energy consumption increased from $5 \cdot 10^{12}$ kWh/year in 1860 to $1.4 \cdot 10^{14}$ kWh/year today. More than $1.2 \cdot 10^{14}$ kWh/year (80%) are based on fossil fuels (coal, oil and gas). The population of human beings increased during the last century by a factor of 6 but the energy consumption by a factor of 80.[2] The worldwide average continuos power consumption today is 2 kW/person. In the USA the power consumption is in average, 10 kW/person and in Europe about 5 kW/person and two billion of people on earth do not consume any fossil fuels. (Fig. 1)

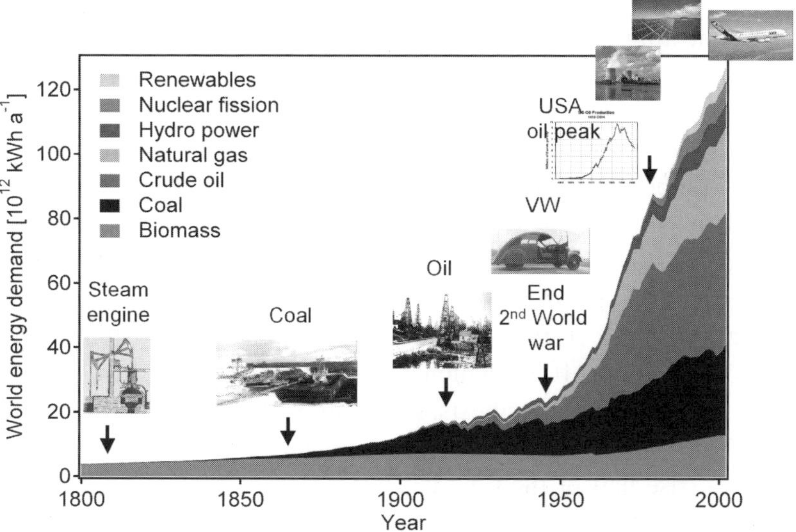

Fig. 1. World energy consumption over the last 200 years.[1]

Considering the historical development of the energy carriers towards hydrogen rich fuels and the necessity to avoid the carbon dioxide emission, hydrogen is the energy carrier of the future:

$$C \text{ (coal)} \rightarrow -CH_2- \text{ (oil)} \rightarrow CH_4 \text{ (natural gas)} \rightarrow H_2 \text{ (hydrogen)}$$

The series also shows a development from a solid to a liquid and finally gas state energy carrier. It has been estimated that hydrogen makes up more than 90% of all the atoms or 75% of the mass of the universe.[3] Hydrogen is the most abundant element on earth, less than 1% is present as molecular hydrogen gas H_2, the overwhelming part is chemically bound as H_2O in water and some is bound to liquid or gaseous hydrocarbons. The renewable production of hydrogen by means of electrolysis consumes electricity, which is physically work. The chemical energy per mass of hydrogen (39.4 kWh·kg^{-1}) is three times greater than that of other chemical fuels, e.g. liquid hydrocarbons (13.1 kWh·kg^{-1}).[3] In other words the energy content of 0.33 kg of hydrogen corresponds to the energy content of 1 kg of oil. The energy content of a fuel is usually called the *heating value*. The difference between the upper and the lower heating value is the enthalpy of vaporization of water according to the state of the water as a result of the combustion. For hydrogen the lower heating value is 33.3 kWh·kg^{-1}, and the upper heating value is 39.4 kWh·kg^{-1}.

There is a technical and an economic challenge to overcome before the hydrogen energy economy becomes reality, the technical challenge is the real time production, the safe and comfortable storage and the efficient combustion of hydrogen. In order to satisfy the world demand on fossil fuels, more than $5 \cdot 10^{12}$ kg hydrogen have to be produced per year. This is roughly 100 times the today's hydrogen production. The solar constant is 1.362 kW·m^{-2} and approximately 50% of the solar radiation reach the surface of the earth. Photovoltaic systems have an efficiency of about 10%, which is the same for the best biological systems based on photosynthesis. Corn is one of the most efficient plants with a photo-conversion efficiency of about 10%.[4] Photovoltaic cells are in the best case half of the time irradiated. Under these conditions about 541000km^2 (90m^2/person) covered with photovoltaic cells are necessary to produce the word energy consumption of today. This surface area corresponds to a square with a side length of 740 km (Fig. 2).

Fig. 2. Lower picture: Earth view with a square in the Sahara corresponding to the surface area necessary to be covered with photovoltaic cell in order to produce the world energy consumption of $1.6 \cdot 10^{14}$ kWh/year. On the upper picture, earth view during the night showing the illuminated regions were energy is consumed.

The economic challenge is the cost of the hydrogen production. World economy today is based on free energy naturally stored sun energy in fossil fuels over millions of years. The price we are used to pay for fossil fuels is the mining and treatment cost only. In order to adapt the world to a synthetic fuel like hydrogen, the world economy has to be convinced to pay for the energy content of the fuel too. The sun's energy is for free but the conversion of the sun's energy into an energy carrier e.g. biomass or electricity requires an investment. Renewable energy requires in general an investment of approximately 5 years the cost of the energy production. This means an investment of approximately 10^{13} to 10^{14} € (1€ ≈ 1 US$) or in words 10 to 100 trillion €, i.e. the same amount as is paid worldwide for renewable energy within 5 years. The investment corresponds to approximately 20% of the yearly income of a person in the highly industrialized countries.

The today worldwide consumed hydrogen as a chemical raw material (about $5 \cdot 10^{10}$ kg/year) is to a large extent produced using fossil fuels by means of the reaction $-CH_2- + 2H_2O \rightarrow 3H_2 + CO_2$ at elevated temperature (>850°C). This reaction consumes energy is 8.9 kWh·kg^{-1} hydrogen. However, the hydrogen production from fossil fuels is not renewable and produces at least the same amount of carbon dioxide as the direct combustion of the fossil fuel.

Hydrogen is a renewable fuel only if hydrogen is produced directly from solar light or indirectly via electricity from a renewable source e.g. wind power or hydro power. The direct thermal dissociation of H_2O requires temperatures higher than 2500°C.[5] Furthermore, the thermal dissociation not only produces H_2 but also atomic hydrogen and hydroxyl ions. This production method is therefore a subject of research activities and various problems e.g. the separation of hydrogen from oxygen, have to be solved before the thermal dissociation becomes an applicable method for hydrogen production. Electricity from a renewable energy source, e.g. wind power, photovoltaics, hydropower and geothermal can be used for the electrolysis of water. Electrolysis at ambient temperature and ambient pressure requires a minimum voltage of 1.481V and therefore a minimum energy of 39.7 kWh·kg^{-1} hydrogen. Electrolyser systems today consume approximately 47 kWh·kg^{-1} hydrogen, i.e. the efficiency is approximately 85%.[6]

Figure 3 shows the principle of the hydrogen cycle. The synthetic fuel hydrogen is produced with the dissociation of water applying solar energy. The hydrogen is then stored and transported and finally combusted with air to form water and delivers the stored energy in form of work and heat.

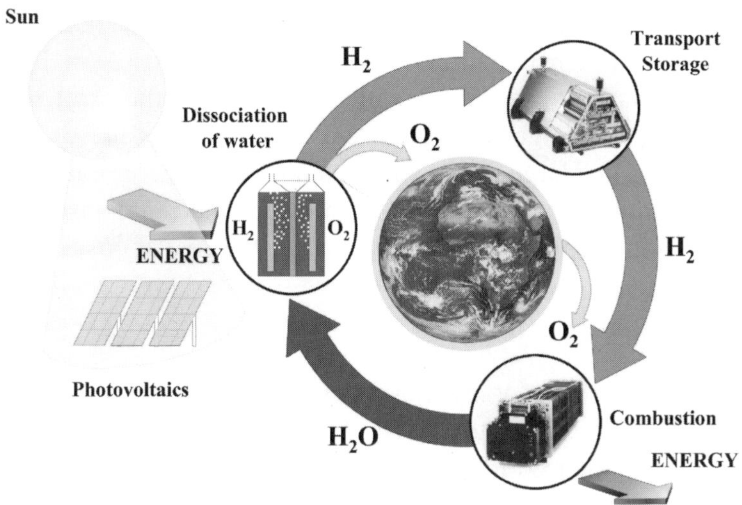

Fig. 3. The hydrogen cycle: the energy from the sunlight is converted into electricity by means of photovoltaic cells. The electricity is used to dissociate water into hydrogen and oxygen. The latter is released in the atmosphere and hydrogen is stored, transported and distributed. Finally, hydrogen together with the oxygen is combusted and the energy is released as work and heat leaving water or steam into the atmosphere. Therewith the hydrogen cycle is closed.

Hydrogen can be transported in pipelines similar to natural gas. There are networks for hydrogen already operating today, a 1500 km network in Europe and a 720 km network in the USA. The oldest hydrogen pipe network is in the Ruhr area in Germany now operated for more than 50 years. The tubes with a typical diameter of 25-30 cm are built using conventional pipe steel and operate at a pressure of 10 to 20 bar. The volumetric energy density of hydrogen gas is 36% of the volumetric energy density of natural gas at the same pressure. In order to transport the same amount of energy the hydrogen flux has to be 2.8 times larger as compared to natural gas. Yet, the viscosity of hydrogen ($8.92 \cdot 10^{-6}$ Pa s) is significantly smaller than that of natural Gas ($11.2 \cdot 10^{-6}$ Pa s). The minimum power P required to pump a gas through a pipe is given by

$$P = 8 \pi l v^2 \eta \qquad (1)$$

where l is the length of the pipe, v the velocity and η the dynamic viscosity of the gas. The transmission power per energy unit is therefore

2.2 times larger for hydrogen as compared to natural gas. The total energy loss for transportation of hydrogen is about 4% of the energy content. Because of the great length, and therefore the great volume of piping systems, a slight change in the operating pressure of a pipeline system results in a large change of the amount of hydrogen gas contained within the piping network. Therefore, the pipeline can be used to handle fluctuations in supply and demand, avoiding the cost of onsite storage.

2. Hydrogen Storage

The ordinary isotope of hydrogen, H. is known Protium and has an atomic weight of 1 (1 proton and 1 electron). The preparation of a stable isotope was announced in 1932, deuterium (D) with an atomic weight of 2 (1 proton and 1 neutron, 1 electron). Two years later an unstable isotope, tritium (T), with an atomic weight of 3 (1 proton and 2 neutron, 1 electron) was discovered. Tritium has a half-life of about 12.5 years.[6] One atom of deuterium is found in about 6000 ordinary hydrogen atoms (0.017%). Tritium atoms are also present 10^{-18}%.[7] All the isotopes of hydrogen react together and form, due to the single electron in the atom, covalent molecules like H_2, D_2 and T_2, respectively. Hydrogen has a very ambivalent behavior to other elements, it occurs as anion (H^-) or cation (H^+) in ionic compounds, it participates with its electron to form covalent bonds e.g. with carbon, and it can even behave like a metal and form alloys at ambient temperature.

The hydrogen molecule H_2 can be found in various forms depending on the temperature and the pressure which are shown in the phase diagram[8] (Fig. 4). At low temperature hydrogen is a solid with a density of 70.6 kg·m^{-3} at -262°C and hydrogen is a gas at higher temperatures with a density of 0.09 kg·m^{-3} at 0°C and a pressure of 1 bar. A small zone starting at the triple point and ending at the critical point exhibits the liquid hydrogen with a density of 70.8 kg·m^{-3} at -253°C. At ambient temperature (298.15K) hydrogen is a gas and can be described by the Van der Waals equation:

$$p(V) = \frac{n \cdot R \cdot T}{V - n \cdot b} - a \cdot \frac{n^2}{V^2} \qquad (2)$$

where p is the gas pressure, V the volume, T the absolute temperature, n the number of mols, R the gas constant (R = 8.314 J·K⁻¹·mol⁻¹), a is the dipole interaction or repulsion constant (a = 2.476·10⁻² m⁶·Pa·mol⁻²) and b is the volume occupied by the hydrogen molecules themselves (b = 2.661·10⁻⁵ m³·mol⁻¹).[3] The strong repulsion interaction between hydrogen molecules is responsible for the low critical temperature (T_c = 33K) of hydrogen gas.

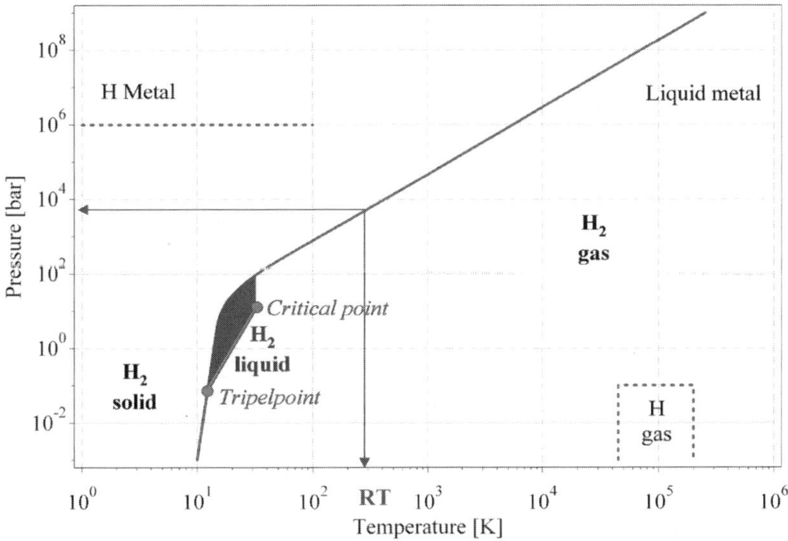

Fig. 4. Primitive phase diagram for hydrogen.[8] Liquid hydrogen only exists between the solidus line and the line from the triple point at 21.2K and the critical point at 32K.

Hydrogen storage implies basically the reduction of the enormous volume of the hydrogen gas. 1 kg of hydrogen at ambient temperature and atmospheric pressure takes a volume of 11 m³. In order to increase the hydrogen density in a storage system either work can be applied to compress hydrogen or the temperature has to be decreased below the critical temperature or finally the repulsion has to be reduced by the interaction of hydrogen with another material. The second important criterion for a hydrogen storage system is the reversibility of the hydrogen uptake and release. This criteria excludes all covalent hydrogen carbon compounds as hydrogen storage materials because the hydrogen is only released from carbon hydrogen compounds if they are heated to temperatures above 800°C or the carbon is oxidized. There are basically six methods in order to store hydrogen reversibly with a high volumetric and gravimetric density (Table 1).

Table 1. The six basic hydrogen storage methods and phenomena. The gravimetric density ρ_m, the volumetric density ρ_v, the working temperature T and pressure p are listed. RT stands for room temperature (25°C).

Storage method	ρ_m [mass%]	ρ_v [kg H$_2$ m^{-3}]	T [°C]	P [bar]	Phenomena and remarks
high pressure gas cylinders	13	<40	RT	800	compressed gas (molecular H$_2$) in light weight composit cylinder (tensile strength of the material is 2000 MPa).
liquid hydrogen in cryogenic tanks	size dep.	70.8	-252	1	liquid hydrogen (molecular H$_2$), continuous loss of a few % per day of hydrogen at RT.
adsorbed hydrogen	≈2	20	-80	100	physisorption (molecular H$_2$) on materials e.g. carbon with a very large specific surface area, fully reversible.
absorbed on interstitial sites in a host metal	≈2	150	RT	1	hydrogen (atomic H) intercalation in host metals, metallic hydrides working at RT are fully reversible.
complex compounds	<18	150	>100	1	complex compounds ([AlH$_4$]$^-$ or [BH$_4$]$^-$), desorption at elevated temperature, adsorption at high pressures.
metals and complexes together with water	<40	>150	RT	1	chemical oxidation of metals with water and liberation of hydrogen, not directly reversible?

The following chapters focus on these methods and balance there advantages and disadvantages.

2.1. *High pressure gas cylinders*

The most common storage system are high pressure gas cylinders with a maximum pressure of 20 MPa. New light weight composite cylinders are developed which support pressure up to 80 MPa and therefore, the hydrogen can reach a volumetric density of 36 kg·m^{-3}, approximately half as much as in its liquid form at the normal boiling point. The gravimetric hydrogen density decreases with increasing pressure due to the increasing thickness of the walls of the pressure cylinder. The wall thickness of a cylinder capped with two hemispheres is given by the following equation:

$$\frac{d_w}{d_o} = \frac{\Delta p}{2 \cdot \sigma_v + \Delta p} \tag{3}$$

where d_w is the wall thickness, d_o the outer diameter of the cylinder, Δp the overpressure and σ_v the tensile strength of the material. The tensile strength of materials varies from 50 MPa for aluminum to more than 1100 MPa for high quality steel. Future, developments of new composite materials have a potential to increase the tensile strength above that of steel with a materials density which is less than half of the density of steel. Figure 4 shows the volumetric density of hydrogen inside the cylinder and the ratio of the wall thickness to the outer diameter of the pressure cylinder for stainless steel with a tensile strength of 460 MPa. The volumetric density of hydrogen increases with pressure and reaches a maximum, depending on the tensile strength of the material, above 100 MPa. However, the gravimetric density decreases with increasing pressure and the maximum gravimetric density is found for zero overpressure! Therefore, the increase in volumetric storage density is sacrificed by the reduction of the gravimetric density in pressurized gas systems.

The safety of pressurized cylinders is an issue of concern especially in highly populated regions. However, future pressure vessels are envisaged to consist of three layers: an inner polymer liner, over wrapped with a carbon-fiber composite (which is the stress-bearing component) and an outer layer of an aramid-material capable to withstand mechanical and corrosion damage. The target that the global industry has set for itself is a 70 MPa cylinder with a mass of 110 kg resulting in a gravimetric storage density of 6mass% and a volumetric storage density[9] of 30 kg·m^{-3}.

Hydrogen can be compressed using standard piston-type mechanical compressors. Slight modifications of the seals in order to compensate for the higher diffusivity of hydrogen are sometimes necessary. The theoretical work necessary for the isothermal compression of hydrogen is given by the following equation:

$$\Delta G = R \cdot T \cdot \ln(\frac{p}{p_0}) \tag{4}$$

where R stands for the gas constant (R = 8.314 J·mol^{-1}·K^{-1}), T for the absolute temperature and p and p_0 for the end pressure and the starting pressure, respectively. The error of the work calculated with equation 3 in the pressure range of 0.1 to 100 MPa is less than 6%. The isothermal compression of hydrogen from 0.1 MPa to 80 MPa consumes therefore 2.21 kWh/kg. In a real process the work consumption for compression is significantly higher because the compression is not isothermal. Metal hydrides can be used to compress hydrogen from a heat source only. Compression ratios of grater than 20:1 are possible with final pressures[10] of more than 10 MPa.

The relatively low hydrogen density together with the very high gas pressures, diffusion problems and the cyclic stability in the systems are important drawbacks of the technically simple and on the laboratory scale well established high pressure storage method.

2.2. *Liquid hydrogen*

Liquid hydrogen is stored in cryogenic tanks at 21.2K at ambient pressure. Due to the low critical temperature of hydrogen (33K) liquid hydrogen can only be stored in open systems, because there is no liquid phase existent above the critical temperature. The pressure in a closed storage system at room temperature could increase to about 10^4 bar. The volumetric density of liquid hydrogen is 70.8 kg·m^{-3} and slightly higher than that of solid hydrogen (70.6 kg·m^{-3}). The challenges of the liquid hydrogen storage are the energy efficient liquefaction process and the thermal insulation of the cryogenic storage vessel in order to reduce the boil-off of hydrogen.

Normal hydrogen at room temperature contains 25% of the para form and 75% of the ortho form, according to the nuclear spin. The ortho form cannot be prepared in the pure state. Since the two forms differ in energy, the physical properties also differ. The melting and boiling points of para hydrogen are about 0.1K lower than those of normal hydrogen. At zero Kelvin, all the molecules must be in a rotational ground state i.e. in the para form.

When hydrogen is cooled from room temperature (RT) to the normal boiling point (nbp = 21.2K) the ortho hydrogen converts from an equilibrium concentration of 75% at RT to 50% at 77K and 0.2% at nbp. The self conversion rate is an activated process and very slow, the half-life time of the conversion is greater than one year at 77K. The conversion reaction from ortho- to para-hydrogen is exothermic and the heat of conversion is also temperature dependent. At 300K the heat of conversion is 270 $kJ \cdot kg^{-1}$ and increases as the temperature decreases, where it reaches 519 $kJ \cdot kg^{-1}$ at 77K. At lower temperatures than 77K the enthalpy of conversion is 523 $kJ \cdot kg^{-1}$ and almost constant. The enthalpy of conversion is greater than the latent heat of vaporization (H_V = 451.9 $kJ \cdot kg^{-1}$) of normal and para hydrogen at the nbp. If the unconverted normal hydrogen is placed in a storage vessel, the enthalpy of conversion will be released in the vessel, which leads to the evaporation of the liquid hydrogen. The transformation from ortho- to para-hydrogen can be catalyzed by a number of surface active and paramagnetic species, e.g. normal hydrogen can be adsorbed on charcoal cooled with liquid hydrogen and desorbed in the equilibrium mixture. The conversion may take only a few minutes if a highly active form of charcoal is used. Other suitable ortho-para catalysts are metal such as tungsten or nickel, any paramagnetic oxides like chromium or gadolinium oxides. The nuclear spin is reversed without breaking the H-H bond.

The simplest liquefaction cycle is the Joule-Thompson cycle (Linde cycle). The gas is first compressed, then cooled in a heat exchanger, before it passes through a throttle valve where it undergoes an isenthalpic Joule-Thomson expansion, producing some liquid. The cooled gas is separated from the liquid and returned to the compressor via the heat exchanger.[11] The Joule-Thompson cycle works for gases, such as nitrogen, with a inversion temperature above room temperature. Hydrogen, however, warms upon expansion at room temperature. In order for hydrogen to cool upon expansion, its temperature must be below its inversion temperature of 202K. Therefore, hydrogen is usually pre cooled using liquid nitrogen (78K) before the first expansion step occurs. The free enthalpy change[12] between gaseous hydrogen at 300K and liquid hydrogen at 20K is 11640 kJ·kg^{-1}. The theoretical energy (work) needed to liquefy hydrogen from RT is W_{th} = 3.23 kWh·kg^{-1}, the technical work[13] is about 15.2 kWh·kg^{-1} roughly one third of the higher heating value of the hydrogen combustion.

The boil-off rate of hydrogen from a liquid hydrogen storage vessel due to heat leak is a function of the size, the shape and the thermal insulation of the vessel. Theoretically the best shape is a sphere since it has the least surface to volume ratio and stress and strain are distributed uniformly. However, large size, spherical containers are expensive because of their manufacturing difficulty. Since boil off losses due to heat leak is proportional to the surface to volume ratio, the evaporation rate diminishes drastically as the storage tank size increases. For double-walled vacuum-insulated spherical dewars, boil-off losses are typically 0.4% per day for tanks having a storage volume of 50 m^3, 0.2% for 100m^3 tanks, and 0.06% for 20000m^3 tanks.[14] The boil off hydrogen can be oxidized with air on a catalyst like Pd (cold combustion) or used in a fuel cell to produce electricity.

The relatively large amount of energy necessary for the liquefaction and the continuous boil-off of hydrogen limit the possible applications for liquid hydrogen storage systems. Applications where the cost of hydrogen is not an important issue and the hydrogen is consumed in a rather short time, e.g. air and space, are possible.

2.3. *Physisorption of hydrogen*

The adsorption of a gas on a surface is a consequence of the field force at the surface of the solid, called the adsorbent, which attracts the molecules of the gas or vapor, called adsorbate. The origin of the physisorption of gas molecules on the surface of a solid are resonant fluctuations of the charge distributions and are therefore called dispersive interactions or Van der Waals interactions. In the physisorption process a gas molecule interacts with several atoms at the surface of the solid. The interaction is composed of an attractive term which diminishes with the distance between the molecule and the surface to the power of -6 and a repulsive term which diminishes with the distance to the power of −12. Therefore, the potential energy of molecule shows a minimum at a distance of approximately one molecular radius of the adsorbate and the energy minimum[15] is of the order of 0.01 to 0.1 eV (1 to 10 kJ·mol^{-1}). Due to the weak interaction significant physisorption is only observed at low temperatures (< 273K). Once a monolayer of adsorbate molecules is formed the gaseous molecule interact with a surface of the liquid or solid adsorbate. Therefore, the binding energy of the second layer of adsorbate molecules is similar to the latent heat of sublimation or vaporization of the adsorbate. Consequently, the adsorption at a temperature equal or greater to the boiling point of the adsorbate at a given pressure leads to the adsorption of one single monolayer.[16] In order to estimate the quantity of adsorbate in the monolayer the density of the liquid adsorbate and the volume of the molecule can be used. If a liquid is assumed to consist of a closed packed fcc structure, the minimum surface area S_{ml} for one mol of adsorbate in a monolayer on a substrate can be calculated from the density of the liqud ρ_{liq} and the molecular mass of the adsorbate $M_{ads.}$:

$$S_{ml} = \frac{\sqrt{3}}{2} \cdot \left(\sqrt{2 \cdot N_A} \cdot \frac{M_{ads}}{\rho_{liq}} \right)^{\frac{2}{3}} \tag{5}$$

N_A stands for the Avogadro constant ($N_A = 6.022 \cdot 10^{23}$ mol^{-1}). The monolayer surface area for hydrogen is $S_{ml}(H_2) = 85917$ m^2·mol^{-1}. The amount of adsorbate m_{ads} on a substrate material with a specific surface area S_{spec} is then given by $m_{ads} = M_{ads} \cdot S_{spec}/S_{ml}$. In the case of carbon as the substrate and hydrogen as the adsorbate, the maximum specific surface area of carbon is $S_{spec} = 1315$ m^2·g^{-1} (single side graphene sheet) and the maximum amount of adsorbed hydrogen is $m_{ads} = 3.0$ mass%. From this theoretical approximation we may conclude that the amount of adsorbed hydrogen is proportional to the specific surface area of the adsorbent with $m_{ads}/S_{spec} = 2.27 \cdot 10^{-3}$ mass%·m^{-2} g and can only be observed at very low temperatures.

Much work on reversible hydrogen sorption by carbon nanostructures was stimulated by the findings published in an article from Dillon *et al.*[17] This paper describes the results of a brief hydrogen desorption experiment. The authors estimated the hydrogen storage capacity of carbon nanotubes at that time to be 5 to 10 mass%. The investigation was carried out on a carbon sample containing an estimated (TEM micrographs) amount of 0.1 to 0.2 mass% of single wall carbon nanotubes. The amount of hydrogen desorbed in the high-temperature peak, which is roughly 5 to 10 times smaller than the low temperature physisorption peak, was 0.01 mass%. The authors concluded "Thus the gravimetric storage density per SWNT ranges from 5 to 10 mass%". Three years later in a report to the DOE[18] this peak has moved significantly by 300 K up to 600 K. Apparently the reported results are inconsistent. Hirscher *et al.*[19] clarified the situation and showed that the desorption of hydrogen originates from Ti-alloy particle, introduced during the ultrasonic treatment, in the sample rather than from the carbon nanotubes.

The main difference between carbon nanotubes and high surface area graphite is the curvature of the graphene sheets and the cavity inside the tube. In microporous solids with capillaries which have a width not exceeding a few molecular diameters, the potential fields from opposite walls will overlap so that the attractive force acting on adsorbate molecules will be increased as compared with that on a flat carbon surface.[20] This phenomenon is the main motivation for the investigation of the hydrogen interaction with carbon nanotubes.

The investigation of the absorption of the hydrogen inside the tubes has shown that it is energetically more favorable for the hydrogen atoms to recombine and form molecules, which are then physisorbed inside the nanotube. Ma *et al.*[21] performed a molecular dynamics simulation for H implantation. The hydrogen atoms (20 eV) were implanted through the side walls of a single wall carbon nanotube (5,5) consisting of 150 atoms and having a diameter of 0.683 nm. They found that the hydrogen atoms recombine to molecules inside the tube and arrange themselves to a concentric tube. The hydrogen pressure inside the SWNT increases as the number of injected atoms increases and reaches 35 GPa for 90 atoms

(5mass%). This simulation does not exhibit a condensation of hydrogen inside the nanotube. The critical temperature of hydrogen (H_2) is 33.25K.[22] Therefore, at temperatures above 33.25 K and at all pressures hydrogen does not exist as a liquid phase, hydrogen is either a gas or a solid. The density of liquid and solid hydrogen at the melting point (T_m = 14.1 K) is 70.8 kg m^{-3} and 70.6 kg m^{-3}, respectively. Measurement of the latent heat of condensation of nitrogen on carbon black[23] showed, that the heat for the adsorption of one monolayer is between 11 to 12kJ mol^{-1} (0.11 to 0.12 eV) and drops for subsequent layers to the latent heat of condensation for nitrogen which is 5.56 kJ mol^{-1} (0.058eV). If we assume, that hydrogen behaves similar to nitrogen, hydrogen would only form one monolayer of liquid at the surface of carbon at temperatures above the boiling point. Geometrical considerations of the nanotubes lead to the specific surface area and therefore, to the maximum amount of condensed hydrogen in a surface monolayer. Figure 5 shows the maximum amount of hydrogen in mass% for the physisorption of hydrogen on carbon nanotubes.[24] The maximum amount of adsorbed hydrogen is 3.0 mass% for single wall carbon nanotubes (SWNT) with a specific surface area of 1315 m^2 g^{-1} at a temperature of 77K.

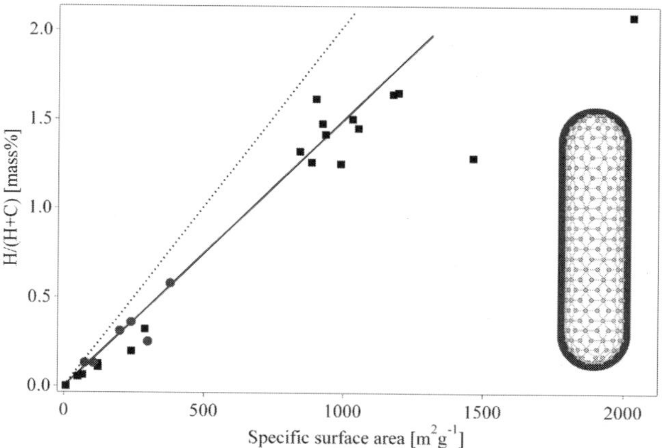

Fig. 5. Reversible amount of hydrogen (electrochemical measurement at 298K) versus the B.E.T. surface area (round markers) of a few carbon nanotube samples including two measurements on high surface area graphite (HSAG) samples together with the fitted line. Hydrogen gas adsorption measurements at 77K from Nijkamp *et al.*[25] (square markers) are included. The dotted line represents the calculated amount of hydrogen in a monolayer at the surface of the substrate.

From adsorption–desorption experiments it is evident that reversible physisorption exclusively takes place with all samples. The amount of adsorbed hydrogen correlates well with the specific surface area of the sample[25] (Fig. 5). A few papers on electrochemical measurements at room temperature of the hydrogen uptake and release have been published.[26,27,28,29] The electrochemical hydrogen absorption is reversible. The maximum discharge capacity measured at 298K is 2 mass% with a very small discharge current (discharge process for 1000h). The few round markers together with the fitted line in Fig. 3 are electrochemical results. It is remarkable, that the measurements of the hydrogen uptake in the gas phase at 77K exhibit the same quantities as the electrochemical measurements at room temperature 298K. In the electrochemical charge process hydrogen atoms are left back at the surface of the electrode when the electron transfer from the conductor to the water molecules takes place. The hydrogen atoms recombine to hydrogen molecules. This process goes on until the surface is completely covered with a monolayer of physisorbed H_2 molecules. Any further hydrogen does not interact with the attractive Van der Waals forces of the surface. The hydrogen molecules become very mobile and form gas bubbles, which are released from the electrode surface. The formation of a stable monolayer of hydrogen at the electrode surface at room temperature is only possible if either the hydrogen atoms or the hydrogen molecules are immobile, i.e. their surface diffusion has to be cinematically hindered by a large energy barrier probably due to the adsorbed electrolyte (H_2O) molecules in the second layer. Another possible reaction path was first reported[30] as a result of density-functional calculation. The result of the calculation is that hydrogen atoms tend to chemisorb at the exterior surface of a nanotube. The atoms can then flip in and finally at high coverage recombine to hydrogen molecules forming a concentric cylinder in the cavity of the nanotube. If the binding energy of the chemisorbed hydrogen is relatively low as compared to the energy in hydrocarbons, the absorbed amount of hydrogen is proportional to the surface area of the carbon sample and could also desorb at rather positive electrochemical potential.

In conclusion: the reversible hydrogen sorption process is based on the physisorption. The amount of adsorbed hydrogen is proportional to the BET surface area of the nanostructured carbon sample. The amount of hydrogen adsorbed from the gas phase at 77K and electrochemically at room temperature is $1.5 \cdot 10^{-3}$ mass%·m^{-2} g. Together with the maximum specific surface area of carbon (1315 m^2 g^{-1}) the maximum absorption capacity of carbon nanostructures is 2 mass%. The experimental results are in good agreement with the theoretical estimations if we take into account that the measurements were carried out at a temperature of 77K which is still far above the critical temperature of hydrogen of 32K and therefore the monolayer of hydrogen is not complete at 77K. No evidence for an influence of the geometric structure of the nanostructured carbon on the amount of hydrogen absorbed was found. It's quite obvious, that the curvature of nanotubes may only influence the adsorption energy but not the amount of adsorbed hydrogen. Furthermore, all attempts to open the nanotubes and absorb hydrogen inside the tubes did not show an increased absorption of hydrogen molecules.

The big advantage of the physisorption for hydrogen storage are the low operating pressure, the relatively low cost of the materials involved and the simple design of the storage system. The rather small amount of hydrogen adsorbed on carbon together with the low temperatures necessary are significant drawbacks of the hydrogen storage based on physisorption.

2.4. *Metalhydrides*

Metals, intermetallic compounds and alloys in general react with hydrogen and form mainly solid metal-hydrogen compounds. Hydrides exist as ionic, polymeric covalent, volatile covalent and metallic hydrides (Fig. 6) according to the Allred-Rochow electronegativity.[31] The demarcation between the various types of hydrides is not sharp, they merge into each other according to the electronegativities of the elements concerned. This chapter focuses on metallic hydrides, i.e. metals and intermetallic compounds, which form together with hydrogen metallic hydrides. Hydrogen reacts at elevated temperature with many transition metals and their alloys to form hydrides. The electropositive elements are the most reactive, i.e., scandium, yttrium, the lanthanides, the actinides, and the members of the titanium and vanadium groups. The binary hydrides of the transition metals are predominantly metallic in character and are usually referred to as metallic hydrides. They are good conductors of electricity, possess a metallic or graphite-like appearance, and can often be wetted by mercury.

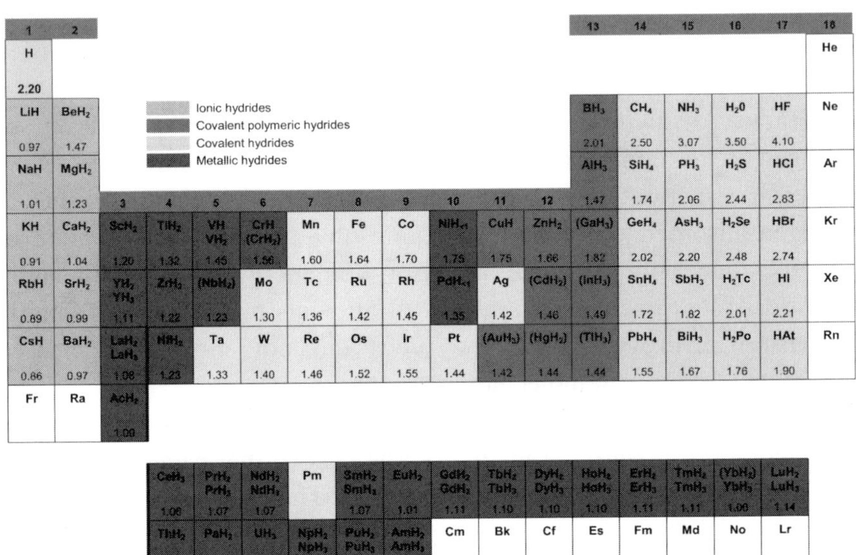

Fig. 6. Table of the binary hydrides and the Allred-Rochow electronegativity.[31] Most elements react with hydrogen to form ionic, covalent or metallic binary hydrides.

Many of these compounds (MH_n) show large deviations from ideal stoichiometry (n = 1, 2, 3) and can exist as multi-phase systems. The lattice structure is that of a typical metal with atoms of hydrogen on the interstitial sites; for this reason they are also called interstitial hydrides. This type of structure has the limiting compositions MH, MH_2, and MH_3; the hydrogen atoms fit into octahedral or tetrahedral holes in the metal lattice, or a combination of the two types. The hydrogen carries a partial negative charge, depending on the metal an exception is, for example[32], $PdH_{0.7}$. Only a small number of the transition metals are without known stable hydrides. A considerable "hydride gap" exists in the periodic table, beginning at group 6 (Cr) up to group 11 (Cu), in which the only hydrides are palladium hydride ($PdH_{0.7}$), the very unstable nickel hydride

($NiH_{<1}$), and the poorly defined hydrides of chromium (CrH, CrH_2) and copper (CuH). In palladium hydride, the hydrogen has high mobility and probably a very low charge density. In the finely divided state, platinum and ruthenium are able to adsorb considerable quantities of hydrogen, which thereby becomes activated. These two elements, together with palladium and nickel, are extremely good hydrogenation catalysts, although they do not form hydrides.[33] Especially interesting are the metallic hydrides of intermetallic compounds, in the simplest case the ternary system AB_xH_n, because the variation of the elements allows to tailor the properties of the hydrides. The A element is usually a rear earth or an alkaline earth metal and tends to form a stable hydride. The B element is often a transition metal and forms only unstable hydrides. Some well defined ratios of B to A in the intermetallic compound $x =$ 0.5, 1, 2, 5 have been found to form hydrides with a hydrogen to metal ratio of up to two.

The reaction of hydrogen gas with a metal is called the absorption process and can be described in terms of a simplified one-dimensional potential energy curve (one-dimensional Lennard-Jones potential[34], Fig. 7). Far from the metal surface the potential of a hydrogen molecule and of 2 hydrogen atoms are separated by the dissociation energy ($H_2 \rightarrow 2H$, $E_D = 435.99$ kJ·mol^{-1}).

The first attractive interaction of the hydrogen molecule approaching the metal surface is the Van der Waals force leading to the physisorbed state ($E_{Phys} \approx 10$ kJ·mol^{-1}) approximately one hydrogen molecule radius (≈ 0.2 nm) from the metal surface. Closer to the surface the hydrogen has to overcome an activation barrier for dissociation and formation of the hydrogen metal bond. The height of the activation barrier depends on the surface elements involved. Hydrogen atoms sharing their electron with the metal atoms at the surface are then in the chemisorbed state ($E_{Chem} \approx 50$ kJ·mol$^{-1}H_2$). The chemisorbed hydrogen atoms may have a high surface mobility, interact with each other and form surface phases at sufficiently high coverage. In the next step the chemisorbed hydrogen atom can jump in the subsurface layer and finally diffuse on the interstitial sites through the host metal lattice. The hydrogen atoms contribute with their electron to the band structure of the metal.

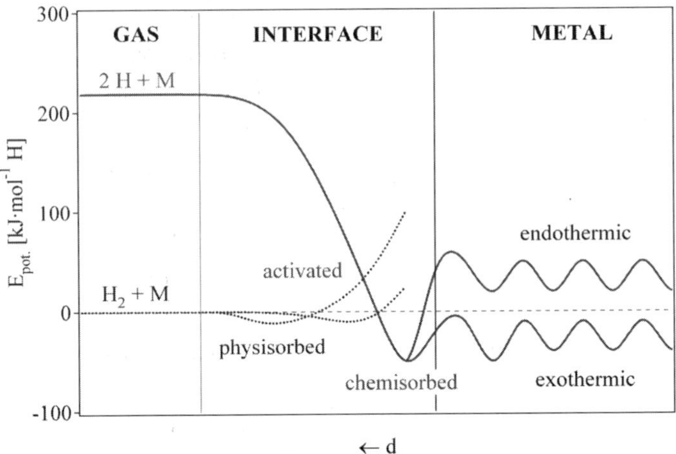

Fig. 7. One-dimensional potential energy curve (one-dimensional Lennard-Jones potential for a hydrogen metal system. Far from the metal surface the potential of a hydrogen molecule and of 2 hydrogen atoms are separated by the dissociation energy. The first attractive interaction of the hydrogen molecule approaching the metal surface is the Van der Waals force leading to the physisorbed state. Closer to the surface the hydrogen has to overcome an activation barrier for dissociation and formation of the hydrogen metal bond. Hydrogen atoms sharing their electron with the metal atoms at the surface are in the chemisorbed state. In the next step the chemisorbed hydrogen atom can jump in the subsurface layer and finally diffuse on the interstitial sites through the host metal lattice. The hydrogen atoms contribute with their electron to the band structure of the metal.

The hydrogen is at small hydrogen to metal ratio (H/M < 0.1) exothermically dissolved (solid-solution, α-phase) in the metal (Fig. 8). The metal lattice expands proportional to the hydrogen concentration by approximately 2 to 3Å^3 per hydrogen atom.[35] At greater hydrogen concentrations in the host metal (H/M > 0.1) a strong H-H interaction due to the lattice expansion becomes important and the hydride phase (β-phase) nucleates and grows. The hydrogen concentration in the hydride phase is often found to be H/M = 1. The volume expansion between the coexisting α- and the β-phase corresponds in many cases 10 to 20% of the metal lattice. Therefore, at the phase boundary large stress is built up and often leads to a decrepitation of brittle host metals such as intermetallic compounds. The final hydride is a powder with a typical particle size of 10 to 100 μm.

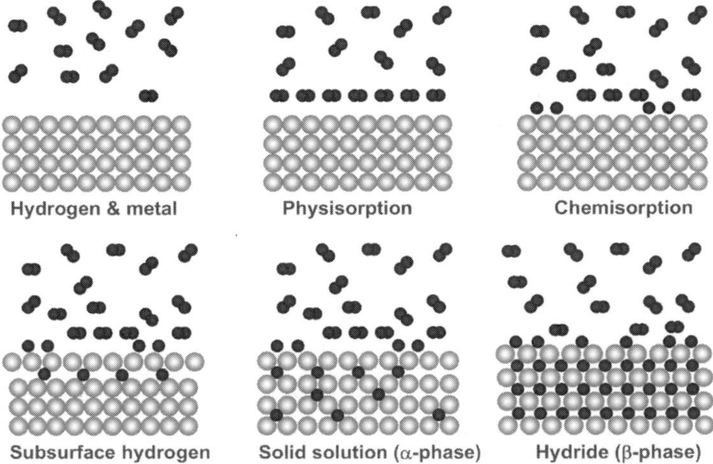

Fig. 8. The six distinct states of hydrogen absorption in metals. On the upper left hand side the metal and hydrogen gas are in two separate phases. Hydrogen molecules can be found in the physisorbed state at low temperature. Hydrogen molecules interact with the electrons at the metal surface, dissociate and bind to the metal atoms at the surface and form the chemisorbed state. Hydrogen can jump in a subsurface layer and diffuse into the metal host lattice as a solid solution (α-phase). Finally, hydrogen-hydrogen interaction (volume expansion) becomes important and the hydride phase (β-phase) is formed.

The thermodynamic aspects of the hydride formation from gaseous hydrogen is described by means of pressure-composition isotherms (Fig. 9). While the solid solution and hydride phase coexists, the isotherms show a flat plateau, the length of which determines the amount of H_2 stored. In the pure β-phase, the H_2 pressure rises steeply with the concentration. The two-phase region ends in a critical point T_C, above which the transition from α- to β-phase is continuous. The equilibrium pressure p_{eq} as a function of temperature is related to the changes ΔH and ΔS of enthalpy and entropy, respectively, by the Van't Hoff equation:

$$\ln\left(\frac{p_{eq}}{p_{eq}^0}\right) = \frac{\Delta H}{R} \cdot \frac{1}{T} - \frac{\Delta S}{R} \tag{6}$$

As the entropy change corresponds mostly to the change from molecular hydrogen gas to dissolved solid hydrogen, it amounts approximately to the standard entropy of hydrogen ($S^0 = 130$ $J \cdot K^{-1} mol^{-1}$) and is therefore, $\Delta S_f \approx -130$ $J \cdot K^{-1} mol^{-1} H_2$ for all metal-hydrogen systems. The enthalpy term characterizes the stability of the metal-hydrogen bond. To reach an equilibrium pressure of 1 bar at 300K, ΔH should make 19.6 kJ mol^{-1}H.

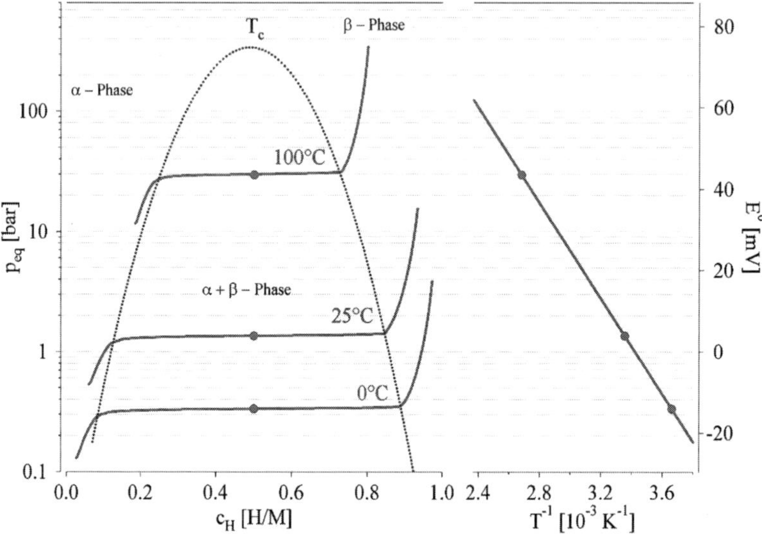

Fig. 9. Pressure composition isothermes for the hydrogen absorption in a typical intermetallic compound on the left hand side. The solid solution (α-phase), the hydride phase (β-phase) and the region of the coexistence of the two phases. The coexistange region is characterized by the flat plateau and ends at the critical temperature T_c. The construction of the Van't Hoff plot is shown on the right hand side. The slope of the line is equal to the enthalpy of formation divided by the gas constant and the interception is equal to the entropy of formation divided by the gas constant.

The stability of metal hydrides is usually presented in the form of Van't Hoff plots according to Eq. 6 (Fig. 10). The most stable binary hydrides have enthalpies of formation of $\Delta H_f = -226$ kJ·mol^{-1}H$_2$ e.g. HoH$_2$. The least stable hydrides are FeH$_{0.5}$, NiH$_{0.5}$ and MoH$_{0.5}$ with enthalpies of formation of $\Delta H_f = +20$ kJ·mol^{-1}H$_2$, $\Delta H_f = +20$ kJ·mol^{-1}H$_2$, and $\Delta H_f = +92$ kJ·mol^{-1}H$_2$, respectively.[36]

Metal hydrides have due to the phase transition upon hydrogen absorption the very appreciated property of absorbing large amounts of hydrogen at a constant pressure, i.e. the pressure does not increase with the amount of hydrogen absorbed as long as the phase transition takes place. The characteristics of the hydrogen absorption and desorption can be tailored by partial substitution of the constituent elements in the host lattice. Some metal hydrides absorb and desorb hydrogen at ambient temperature and close to atmospheric pressure. Several families of intermetallic compounds listed in Table 2 are interesting for hydrogen storage. They all consist of an element with a high affinity to hydrogen, the A-element, and an element with a low affinity to hydrogen, the B-element. The latter is often at least partially nickel, since nickel is an excellent catalyst for the hydrogen dissociation.

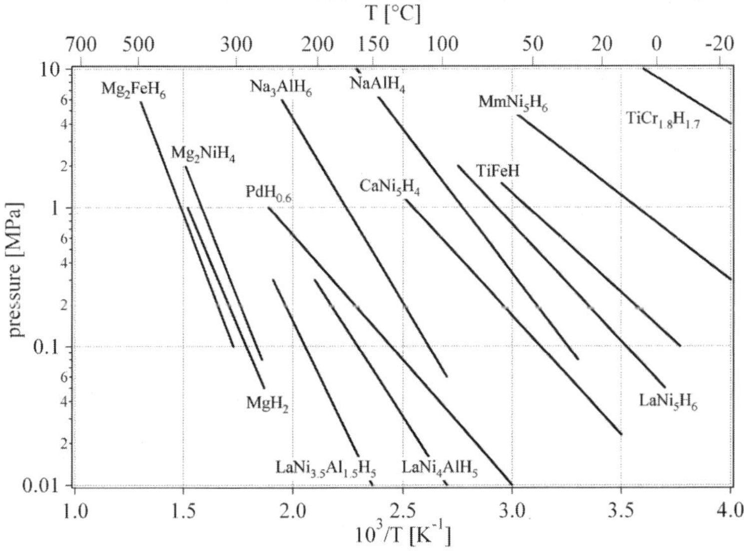

Fig. 10. Van't Hoff plots of some selected hydrides. The stabilisation of the hydride of $LaNi_5$ by the partial substitution of nickel with aluminum in $LaNi_5$ is shown as well as the substitution of lanthanum with mischmetal (e.g. 51% La, 33% Ce, 12% Nd, 4% Pr).

One of the most interesting features of the metallic hydrides is the extremely high volumetric density of the hydrogen atoms present in the host lattice. The highest volumetric hydrogen density known today is 150 $kg·m^{-3}$ found in Mg_2FeH_6 and $Al(BH_4)_3$. Both hydrides are complex hydrides and will be discussed in the next chapter. Metallic hydrides reach a volumetric hydrogen density of 115 $kg·m^{-3}$ e.g. $LaNi_5$. Most metallic hydrides absorb hydrogen up to a hydrogen to metal ratio of H/M = 2. Greater ratios up to H/M = 4.5 e.g. $BaReH_9$, have been found[37], however all hydrides with a hydrogen to metal ratio of more than 2 are ionic or covalent compounds and belong to the complex hydrides.

Table 2. The most important families of hydride forming intermetallic compounds including the prototype and the structure. A is an element with a high affinity to hydrogen, and B is an element with a low affinity to hydrogen.

Intermetallic compound	Prototype	Hydrides	Structure
AB_5	$LaNi_5$	$LaNiH_6$	Haucke phases, hexagonal
AB_2	ZrV_2, $ZrMn_2$, $TiMn_2$	$ZrV_2H_{5.5}$	Laves phase, hexagonal or cubic
AB_3	$CeNi_3$, YFe_3	$CeNi_3H_4$	hexagonal, $PuNi_3$-typ
A_2B_7	Y_2Ni_7, Th_2Fe_7	$Y_2Ni_7H_3$	hexagonal, Ce_2Ni_7-typ
A_6B_{23}	Y_6Fe_{23}	$Ho_6Fe_{23}H_{12}$	cubic, Th_6Mn_{23}-typ
AB	$TiFe$, $ZrNi$	$TiFeH_2$	cubic, CsCl- or CrB-typ
A_2B	Mg_2Ni, Ti_2Ni	Mg_2NiH_4	cubic, $MoSi_2$- or Ti_2Ni-typ

Metal hydrides are very effective to store large amounts of hydrogen in a safe and compact way. All the reversible hydrides working around ambient temperature and atmospheric pressure consist of transition metals and therefore the gravimetric hydrogen density is limited to less than 3 mass%. To explore the properties of the light weight metal hydrides is still a challenge.

2.5. *Complex hydrides*

The groups one, two and three light metals, e.g. Li, Mg, B, Al, build a large variety of metal-hydrogen complexes. They are of especially interest because of their light weight and the number of hydrogen atoms per metal atom which is in many cases 2. The main difference of the complex hydrides to the above described metallic hydrides is the transition to an ionic or covalent compound of the metals upon hydrogen absorption. Hydrogen in the complex hydrides is often located in the corners of a tetrahedra with boron or aluminum in the center. The negative charge of the anion, $[BH_4]^-$ and $[AlH_4]^-$ is compensated by a cation e.g. Li or Na. Hydride complexes of borane, the tetrahydroborates $M(BH_4)$, and of alane, the tetrahydroaluminates $M(AlH_4)$, are interesting storage materials, they are known to be stable and decompose only at elevated temperatures and often above the melting point of the complex.

Bogdanović and Swickardi[38] presented in 1996 for the first time adsorption and desorption pressure-concentration isotherms of catalyzed $NaAlH_4$ at temperature of 180°C and 210°C. The isotherms exhibit an absence of hysteresis and a nearly horizontal pressure plateau. Furthermore, the catalyzed system reversibly absorbed and desorbed hydrogen up to 4.2 mass% and the mechanism of the two step reaction was described. A more detailed study of the $NaAlH_4$ with an improved catalyst was published in 2000 by Bogdanović *et al.*[39] A desorption hydrogen pressure of 2 bar at 60°C was found and the enthalpy for the dissociation reaction was determined to be 37 kJ·mol^{-1} and 47 kJ·mol^{-1} for the first dissociation step of Ti-doped $NaAlH_4$:

$$3\ NaAlH_4 \rightarrow Na_3AlH_6 + 2\ Al + 3\ H_2\ (3.7\ wt\%\ H)$$

and the second

$$Na_3AlH_6 \rightarrow 3NaH + Al + 3/2\ H_2\ (3.0\ wt\%\ H),$$

respectively. Therefore, the equilibrium hydrogen pressure at room temperature is approximately 1 bar. Furthermore, the reaction is reversible, a complete conversion to product was achieved at 270°C under 175 bar hydrogen pressure in 2-3 hours.[40]

The first report of a pure alkali metal tetrahydroboride appeared in 1940 by Schlesinger and Brown[41] who synthesized the lithiumtetrahydroboride (lithiumborohydride) ($LiBH_4$) by the reaction of ethyllithium with diborane (B_2H_6). The direct reaction of the corresponding metal with diborane in etheral solvents under suitable conditions produces high yields of the tetrahydroborides[42]

$$2\ MH + B_2H_6 \rightarrow 2\ MBH_4$$

where M = Li, Na, K etc.

Direct synthesis from the metal, boron and hydrogen at 550 - 700°C and 30 - 150 bar H_2 has been reported to yield the lithium salt, and it has been claimed that such a method is generally applicable to groups IA and IIA metals.[43] The reaction involving either the metal or the metal hydride, or the metal together with triethylborane in an inert hydrocarbon has formed the basis of a patent:

$$M + B + 2H_2 \rightarrow MBH_4,$$

where M = Li, Na, K etc.

The stability of metal tetrahydroborides has been discussed in relation to their percentage ionic character, and those compounds with less ionic

                    Andreas Züttel and Louis Schlapbach

character than diborane are expected to be highly unstable.[44] Steric effects have also been suggested to be important in some compounds.[45, 46] The special feature exhibited by the covalent metal hydroborides is that the hydroboride group is bonded to the metal atom by bridging hydrogen atoms similar to the bonding in diborane, which may be regarded as the simplest of the so called "electron-deficient" molecules. Such molecules possess fewer electrons than those apparently required to fill all the bonding orbital's, based on the criterion that a normal bonding orbital involving two atoms contains two electrons. The molecular orbital bonding scheme for diborane has been discussed extensively.[46]

The compound with the highest gravimetric hydrogen density at room temperature known today is $LiBH_4$ (18 mass%). Therefore, this complex hydride could be the ideal hydrogen storage material for mobile applications. $LiBH_4$ desorbs three of the for hydrogen in the compound upon melting at 280°C and decomposes into LiH and boron. The desorption process can be catalyzed by adding SiO_2 and significant thermal desorption was observed starting at 100°C.[47] Recently it has been shown, that the hydrogen desorption reaction is reversible and the end products lithimhydride and boron absorb hydrogen[48] at 690°C and 200 bar to form $LiBH_4$. The scientific understanding of the mechanism of the thermal hydrogen desorption from $LiBH_4$ and the absorption remains a challenge and more research work has to be carried out. Very little is known today about $Al(BH_4)_3$ a complex hydride[49] with a very high gravimetric hydrogen density of 17 mass% and the highest known volumetric hydrogen density of 150 $kg \cdot m^{-3}$. Furthermore, $Al(BH_4)_3$ has a melting point of −65°C and is liquid at room temperature. Beside of the the covalent hydrocarbons, this is the only hydride liquid at room temperature.

2.6. *Chemical reaction with water*

Hydrogen can be generated from metals and chemical compounds reacting with water. The common experiment -shown in many chemistry classes – where a piece of sodium floating on water produces hydrogen, demonstrates such a process. The sodium is transformed to sodium hydroxide in this reaction. The reaction is not directly reversible but the sodium hydroxide could later be removed and reduced in a solar furnace back to metallic sodium. Two sodium atoms react with two water molecules and produces one hydrogen molecule. The hydrogen molecule produces again a water molecule in the combustion, which can be recycled to generate more hydrogen gas. However, the second water molecule necessary for the oxidation of the two sodium atoms has to be added. Therefore, sodium has a gravimetric hydrogen density of 3 mass%.

The same process carried out with lithium leads to a gravimetric hydrogen density of 6.3 mass%. The major challenge with this storage method is the reversibility and the control of the thermal reduction process in order to produce the metal in a solar furnace. The process has been successfully demonstrated with zinc.[50] The 1st, endothermic step is the thermal dissociation of $ZnO(s)$ into $Zn(g)$ and O_2 at 2300 K using concentrated solar energy as the source of process heat. The second, non-solar, exothermic step is the hydrolysis of $Zn(l)$ at 700 K to form H_2 and $ZnO(s)$; the latter separates naturally and is recycled to the first step.

Hydrogen and oxygen are derived in different steps, thereby eliminating the need for high-temperature gas separation. A 2nd law analysis performed on the closed cyclic process indicates a maximum energy conversion effciency of 29% (ratio of ΔG ($H_2 + 0.5\ O_2 \rightarrow H_2O$) for the H_2 produced to the solar power input), when using a solar cavity-receiver operated at 2300K and subjected to a solar flux concentration ratio of 5000. The major sources of irreversibility are associated with the re-radiation losses from the solar reactor and the quenching of $Zn(g)$ and O_2 to avoid their recombination. An economic assessment for a large-scale chemical plant, having a solar thermal power input into the solar reactor of 90 MW and a hydrogen production output from the hydrolyser of 61 GWh/year, indicates that the cost of solar hydrogen ranges between 0.13 and 0.15\$/kWh (based on its low heating value and a heliostat feld cost at 100-150\$/m^2) and, thus, might be competitive vis-à-vis other renewables-based routes such as electrolysis of water using solar-generated electricity. The economic feasibility of the proposed solar process is strongly dependent on the development of an effective Zn/O_2 separation technique (either by quench or by in situ electrolytic separation) that eliminates the need for an inert gas.

3. Conclusion

The hydrogen revolution following the industrial age has just started. Hydrogen production, storage and conversion has reached a technological level although plenty of improvements and new discoveries are still possible. The hydrogen storage is often considered as the bottleneck of the renewable energy economy based on the synthetic fuel hydrogen. Six different hydrogen storage methods have been described. Among the well established high pressure cylinders for laboratory applications and the liquid hydrogen for air and space applications, metal hydrides and complex hydrides offer a very safe and efficient way to store hydrogen (Fig. 11).

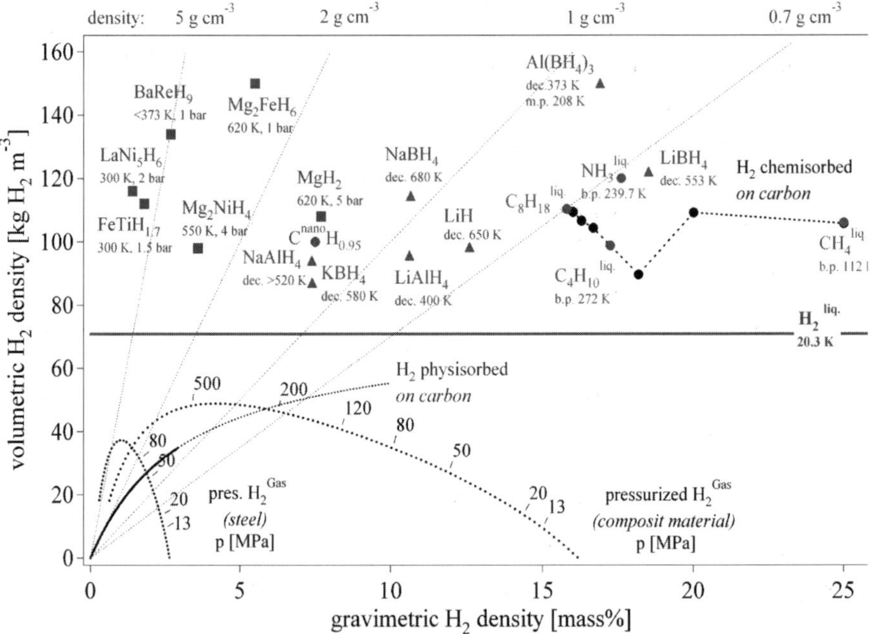

Fig. 11. Volumetric and gravimetric hydrogen density of some selected hydrides. Mg$_2$FeH$_6$ shows the highest known volumetric hydrogen density of 150 kg·m^{-3}, which is more than the double of liquid hydrogen. BaReH$_9$ has the largest H/M ratio of 4.5, i.e. 4.5 hydrogen atoms per metal atom. LiBH$_4$ exhibits the highest gravimetric hydrogen density of 18 mass%. Pressurized gas storage is shown for steel (tensile strength σ_v = 460 MPa, density 6500 kg·m^{-3}) and a hypothetical composite material (σ_v = 1500 MPa, density 3000 kg·m^{-3}).

Further scientific research and technical developments will lead to new materials with a higher volumetric and gravimetric hydrogen density. The best materials today show a volumetric storage density of 150 kg·m^{-3} which can be improved by approximately 50% according to theoretical estimations.

"Our ignorance is not so vast as our failure to use what we know."
(after M. King Hubbert, (October 5th, 1903 -- October 11th, 1989)
"Nuclear energy and the fossil fuels", Drilling and Production Practice (1956))

Acknowledgment

This work was supported by the Swiss federal office of energy (Bundesamt für Energie, BfE) in contract with the International Energy Agency (IEA), the Swiss Federal Office of Education and Science (BBW), the European Commission and the Science Faculty of the University of Fribourg in Switzerland.

References

1. Jean-Marie Martin-Amouroux, IEPE, Grenoble, France.
2. Bulletin, *Magazin der Eidgenössischen Technischen Hochschule Zürich* 276 (2000).
3. Handbook of Chemistry and Physics, Ed. R. C. Weast, 57th (CRC Press) (1976).
4. N. A. Campbell, *Biology*, 4th Edition, (The Benjamin/Cummings Publishing Company Inc., Menlo Park, California USA) (1996), ISBN: 0-8053-1940-9, p. 182.
5. A. Steinfeld, R. Palumbo, *Solar Thermochemical Process Technology*, Encyclopedia of Physical Science and Technology, 15Ed. R. A. Meyers (Academic Press) (2001), 237, ISBN 0-12-227410-5.
6. Product specification (2001) from GTec SA, rte de Clos-Donroux 1, CH-1870 Monthey, Switzerland.
7. M. D. Baltimore, Williams & Wilkins, *Health Physics & Radiological Health Handbook*, 3rd edition, 6 (1998).
8. W. B. Leung, N. H. March and H. Motz, *Physics Letters* **56A** 425 (1976).
9. R. S. Irani, *MRS Bulletin*, 680, September (2002).
10. E. L. Huston, *A Liquid and Solid Storage of Hydrogen*, Proceedings of the 5th World Hydrogen Energy Conference. Vol. 3; July 15-20, 1984, Toronto, Canada.
11. T. M. Flynn, *A Liquification of Gases,* McGraw-Hill Encyclopedia of Science &Technology. 7th edition. Vol. 10. p. 106, (New York: McGraw-Hill); (1992).
12. Gary, Chen and Samim, Anghaie based on NASA/NIST databases, http://www.inspi.ufl.edu/data/h_prop_package.html

13. M. von Ardenne, G. Musiol and , S. Reball, *Effekte der Physik*, (Verlag Harry Deutsch) p. 712 (1990).

14. A. Abe, M. Nakamura, I. Sato, H. Uetani and T. Fujitani, *International Journal of Hydrogen Energy*, **23**, 115 (1998).

15. F. London, *Z. Physik.* **63** 245 (1930) and *Z. Physik. Chem.* **11** 222 (1930).

16. S. Brunauer, P. H. Emmett and E. Teller, *J. Amer. Chem. Soc.* **60** 309 (1938).

17. A. C. Dillon, K. M. Jones, T. A. Bekkedahl, C. H. Kiang, D. S. Bethune, and M. J. Heben, *Nature* **386** 377 (1997).

18. A. C. Dillon *et al. Proceedings of the 2000 DOE/NREL hydrogen program review*, May 8-10, (2000).

19. M. Hirscher, M. Becher, M. Haluska, U. Dettlaff-Weglikowska, A. Quintel, G. S. Duesberg, Y. M. Choi, P. Downes, M. Hulman, S. Roth, I. Stepanek, P. Bernier, *Applied Physics* **A72** 129 (2001).

20. S. J. Gregg and K. S. W. Sing, *Adsorption, Surface Area and Porosity*, (Academic Press, London and New York) (1967).

21. Y. Ma, Y. Xia, M. Zhao, R. Wang, and L. Mei, *Phys. Rev.* **B63** 115422 (2001).

22. W. B. Leung and N. H. March, and H. Motz, *Physics Letters* **56** 425 (1976).

23. R. A. Beebe, J. Biscoe, W. R. Smith, and C. B. Wendell, *J. Am. Chem. Soc.* **69** 95 (1947).

24. A. Züttel, P. Sudan, Ph. Mauron, T. Kyiobaiashi, Ch. Emmenegger, and L. Schlapbach, *Int. J. of Hydrogen Energy* **27** 203 (2002).

25. M. G. Nijkamp, J. E. M. J. Raaymakers, A. J. van Dillen, and K. P. de Jong, *Appl. Phys.* **A72** 619 (2001).

26. Ch. Nützenadel, A. Züttel, and L. Schlapbach, *Electronic Properties of Novel Materials, Science and Technology of Molecular Nanostructures*, Eds. H. Kuzmany, J. Fink, M. Mehring, and S. Roth (American Institute of Physics, New York), 462 (1999).

27. Ch. Nützenadel, A. Züttel, Ch. Emmenegger, P. Sudan, and L. Schlapbach, *Sciences and Applications of Nanotubes*", Ed. M. F. Thorpe (Kluwer Academic Publishing/ Plenum Press, "Fundamental Materials Research Series") 205 (XXXX).

28. A. Züttel, P. Sudan, Ph. Mauron, Ch. Emmenegger, T. Kiyobayashi, and L. Schlapbach, *Journal of Metastable and Nanocrystalline Materials,* **11** 95(2001).

29. S. M. Lee, K.S. Park, Y. C. Choi, Y. S. Park, J. M. Bok, D. J. Bae, K. S. Nahm, Y. G. Choi, S. Ch. Yu, N. Kim, T. Frauenheim, and Y. H. Lee, *Synthetic Metals* **113** 209 (2000).

30. S. M. Lee, K. H. An, Y. H. Lee, G. Seifert, T. Frauenheim, *J. Korean Phys. Soc.* **38** 686 (2001) and S. M. Lee and Y. H. Lee, *Appl. Phys. Lett.* **76** 2879 (2000).

31. J. E. Huheey, *Inorganic Chemistry* (Harper & Row: New York) (1983).

32. G. R. Pearson, *Chem. Rev.* **85** 41 (1985).

33. W. M. Müller, I. R. Blackledge, G. G. Libowitz, *Metal Hydrides* (Academic Press, New York) (1968).

34. J. E. Lennard-Jones, *Trans. Faraday Soc.* **28** 333 (1932).

35. Y. Fukai, *Z. Phys. Chem.* **164** 165 (1989).

36. R. Griessen and T. Riesterer, *Hydrogen in Intermetallic Compounds I Electronic, Thermodynamic, and Crystallographic Properties, Preparation*, Ed. L. Schlapbach, *Topics in Applied Physics* **63** (Springer Series) 219 (1988).

37. K. Yvon, *Chimia* **52** 613 (1998).

38. B. Bogdanovic, M. Schwickardi, *J. Alloys Comp.* **253-254** 1 (1997).

39. B. Bogdanovic, R. A. Brand, A. Marjanovic, M. Schwickardi, J. Tolle, *J. Alloys Comp*, **302** 36 (2000).

40. N. Dymova, N. G. Eliseeva, S. I. Bakum, Y. M. Dergachev, *Dok. Akad. Nauk USSR* **215** 1369 (1974).

41. H. J. Schlesinger and H. C. Brown, *J. American Chemical Society* **62** 3429(1940).

42. H. J. Schlesinger, H. C. Brown, H. R. Hoekstra, L. R. Rapp, *J. American Chemical Society* **75** 199 (1953).

43. D. Goerrig, *Ger. Pat.* 1, 077, 644 (1958).

44. G. N. Schrauzer, *Naturwissenschaften* **42** 438 (1955).

45. S. J. Lippard and D. A. Ucko, *Inorg. Chem.* **7** 1051 (1968).

46. W. N. Lipscomb, *Boron Hydrides*, (Benjamin, New York) (1963).

47. A. Züttel , P. Wenger , S. Rensch ,P. Sudan, P. Mauron and C. Emmenegger, *Journal of Power Sources* **5194** 1 (2003).

48. P. Sudan and A. Züttel, *Letter to Nature Materials* (2003), submitted.

49. A. Almenningen, G. Gundersen and A. Haaland, *Acta Chemica Scandinavica* **22** 328 (1968).

50. A. Steinfeld, *International Journal of Hydrogen Energy* **27** 611 (2002).

CHAPTER 10

HYDROGEN STORAGE MATERIALS — RECENT DEVELOPMENT AND FUTURE STRATEGY OF JAPAN

Etsuo Akiba

Energy Technology Research Institute
National Institute of Advanced Industrial Science and Technology (AIST)
AIST Central 5, 1-1-1, Higashi, Tsukuba, Ibaraki 305-8565 Japan
E-mail: e.akiba@aist.go.jp

Hydrogen storage materials can absorb hydrogen more dense in volume than other media. Therefore, hydrogen storage materials are expected to be used for on board applications. Research and development (R & D) of both materials and systems have been intensively carried out under Japanese national projects. In this review, the state-of-art of R & D on hydrogen storage materials in Japan will be introduced and the perspective of these materials and systems will be discussed.

1. Introduction

Hydrogen is a secondary energy that can be generated from various primary energies (energy resources). In addition, hydrogen is only a fuel that can be converted from/to electricity. Therefore, hydrogen is expected to be used by mid-21st century as one of the major secondary energies same as electricity.

At present, hydrogen is mainly used for fuel cell vehicles as a fuel. Hydrogen gas at room temperature and 1 bar has 1/3000 of volume energy density of gasoline. Considering that efficiency of fuel cell is more than three times of conventional internal combustion engine, volume of hydrogen in a stored form should be reduced to the level of 1/1000 of hydrogen gas at 1 bar, otherwise the cruising range of fuel cell vehicles may be shorter than conventional vehicles on road at present. Therefore, to reduce volume density of hydrogen is a real key issue to realize the hydrogen economy. In this review, Japanese national

365

projects related to hydrogen storage will be introduced. Because real commercialization of hydrogen energy may be in a few ten years or later, funding to R & D of hydrogen energy technology should be dominated by the government and public organizations.

2. The "Development for Safe Utilization and Infrastructure of Hydrogen" Project

Major funding source for R & D of fuel cell and hydrogen is New Energy Development Organization (NEDO) that is an independent organization for funding on energy and industrial R & D and is closely related to Ministry of Economy, Trade and Industry (METI). METI provides funds to NEDO and NEDO subsidizes to industries, universities and national institutes to carry out R & D projects. NEDO itself does not have its own research facilities and a pure funding agency.

NEDO is carrying out the research and development of elemental technologies necessary for making hydrogen available as a common source of energy as a fuel for fuel cell vehicles. NEDO is also conducting studies on scenarios to realize the hydrogen economy. The major parts of R & D are elemental technologies related to the production, transportation, storage and refueling of hydrogen. The purpose of these activities is to improve their performance, cost efficiency, reliability and durability, as well as to reduce the size of related equipment. Through these efforts, NEDO aims to reduce the cost of hydrogen-related equipment while improving its performance. Of course, NEDO promotes practical use of fuel cells and hydrogen energy to realize the hydrogen economy. The project is a five-year term from the fiscal year of 2003 to 2007. Figure 1 illustrates outline of the "Hydrogen infrastructure development and safe utilization of hydrogen".

Hydrogen storage, especially R & D of hydrogen storage materials is the most significant subject of this project. Figure 2 shows the budget of the "Hydrogen infrastructure development and safe utilization of hydrogen" for the fiscal year of 2007. More than 1/3 of the total budget comes to the hydrogen storage. If that of storage by compressed gas and liquefied hydrogen is added, it reaches to 41% of the total amount. The reason why hydrogen storage is considered to be the most important

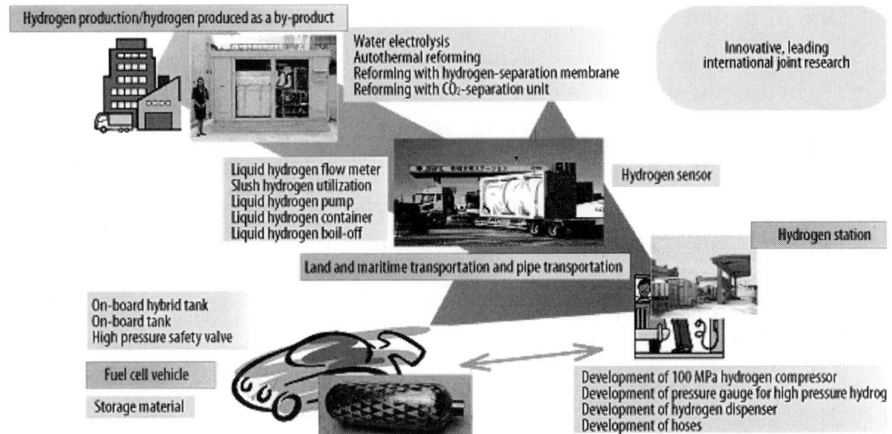

Fig. 1. Outline of Hydrogen infrastructure development and safe utilization of hydrogen. (http://www.nedo.go.jp/english/activities/portal/gaiyou/p03015/p03015.html)

subject by the Japanese government is that on board hydrogen storage is a real challenge and it needs to realize the hydrogen economy because hydrogen storage in at least the same level to the conventional fuel is essential for the hydrogen economy.

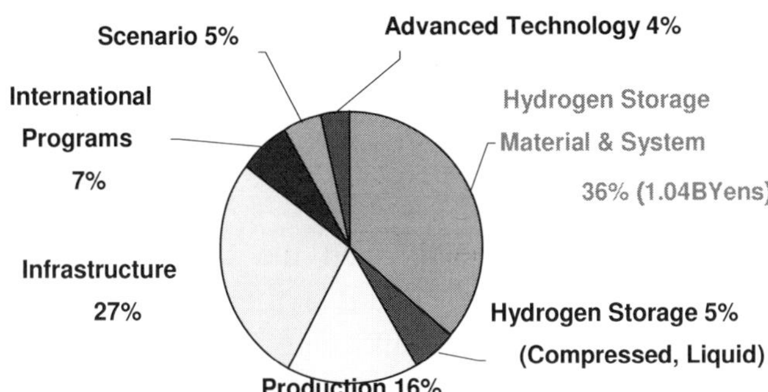

Fig. 2. The budget of "Hydrogen infrastructure development and safe utilization of hydrogen" for the fiscal year of 2007. The total budget is 2.87 billion yen.

In the project, there are two directions of the studies. One is a short-medium term target that will be completed in a few to several years. The other is a long term target to develop hydrogen storage materials can be used for on board application under moderate conditions.

One of the most typical short-medium term targets is the "hybrid tank system" or "high pressure metal hydride tank system". As shown in Fig. 3, hydrogen storage materials are usually fine powder. The materials occupy around 50% of the space of the container. If one puts hydrogen of 35 MPa or so to the space that is not filled by the materials, hydrogen storage capacities of the container should be between a 35 MPa compressed gas container and that filled by the hydrogen storage materials under lower hydrogen pressure less than a few MPa. Under Japanese regulation, if one handles pressure above 1 MPa, permission should be given from local government. Usually, it needs considerable costs, time and paperwork. Therefore, metal hydride hydrogen tanks have been designed for operation under 1 MPa in Japan for a long time.

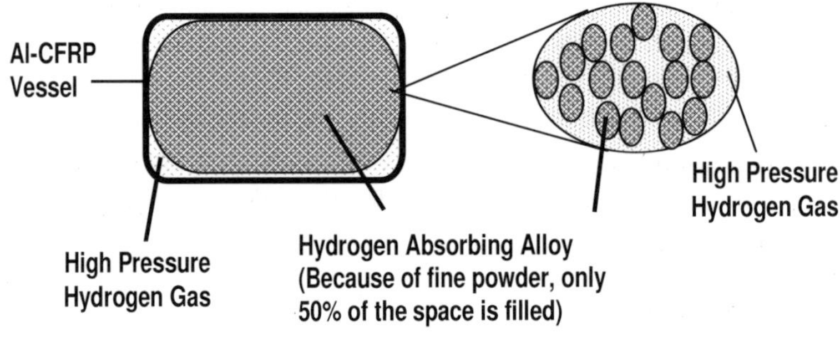

Fig. 3. Schematic drawing of the hybrid tank system combined with high pressure hydrogen gas and hydrogen storage materials.

In 2005, Toyota Motor Co. published the concept of the "high pressure metal hydride tank system"[1] and it is often said the "hybrid tank system". They reported that the tank that is combined hydrogen absorbing alloy with hydrogen capacity of 1.9 mass% and 35 MPa compressed gas reached hydrogen volume capacity of 41 g/L and weight capacity of 1.7 mass%. It was also shown that refueling time was less than 5 minutes. It should be noted that this systems does not need cooking liquid supply outside of vehicle because radiator on board is enough to remove heat generated when the alloy absorbs hydrogen.

Recently, Toyota reported improved hybrid tank using the alloy with higher capacity than the first report. They used the alloy with 2.5 mass%

that has a BCC (Body Centered Cubic) structure.[2] The volume and weight capacity of the tank are 50 g/L and 2.2 mass%, respectively. If one compared to the DOE target of the volume density in 2010 (45 g/L), this tank has already exceeded it.

Under the "Hydrogen infrastructure development and safe utilization of hydrogen", development of the hybrid tank has been carried out. A container for the hybrid tank is under development and hydrogen absorbing alloys optimized for this purpose are also developed by a few companies.

As for long-term targets, various hydrogen storage materials have been investigated under the "Hydrogen infrastructure development and safe utilization of hydrogen". The details of the materials development will be shown later.

3. Fundamental Research Project on Advanced Hydrogen Science

To construct a hydrogen energy society, it is necessary to establish technology that enables the transportation and storage of large volumes of hydrogen in a compact form. Handling capabilities of pressurized and liquefied hydrogen are essential. Unfortunately, there has not been a sufficient accumulation of in-depth data or scientific knowledge regarding the embrittlement of metals, caused by their absorption of hydrogen under high-pressure or liquefied hydrogen atmosphere. Hydrogen tribology is also essential for the manufacture and use of storage/transportation systems, parts, and other materials. To provide industries with guidance and information on materials, device engineering, deterioration evaluation methods, etc. based on fundamental knowledge of embrittlement and fatigue of materials, this project was established. The project is a seven-year term from the fiscal year of 2006 to 2012.

In order to conduct this project, the Research Center for Hydrogen Industrial Use and Storage ("*HYDROGENIUS*") was established in National Institute of Advanced Industrial Science and Technology (AIST) in July 2006 at the campus of Kyushu University.[3]

4.   Other Hydrogen Storage Related Projects

4.1.   *Establishment of Codes & Standards for hydrogen economy society*

In order to realize the hydrogen economy, it is essential to develop infrastructure as well as new technology. The review and establishment of laws, Codes & Standards, and regulations is particularly important for developing infrastructure of hydrogen systems and for promoting the widespread use of fuel cells. This project aims to develop methods to examine and evaluate the new technologies and products that are necessary to obtain data to establish domestic and international technology standards.

The project is a five-year term from the fiscal year of 2005 to 2009.

4.2.   *Advanced fundamental research on hydrogen storage materials (HYDRO ✶STAR)*[4]

Hydrogen storage materials realize compact and energy effective hydrogen storage on board. However, it needs breakthroughs to reach sufficient performance for applications. One of the most possible ways to make breakthroughs is fundamental research of the materials. A proposal for fundamental research of hydrogen storage materials has been made and it has started from June 2007 as a five year plan under the name of Advanced Fundamental Research on Hydrogen Storage Materials (HYDRO✶STAR). This project contains five major activities; research of metallic materials, research of non-metallic materials, studies of physical properties of metal hydrides using synchrotron radiation, computational science for hydrides and neutron total scattering studies of hydrides. Under this project, the third generation synchrotron source Spring-8, Hyogo, Japan and J-PARC, one of most powerful spallation neutron sources, Ibaraki, Japan are used for detailed characterization of hydride materials under various conditions including hydrogen atmosphere up to 40 GPa.

5. Recent Progress in Material Development under the "Development for Safe Utilization and Infrastructure of Hydrogen" Project

5.1. *Metallic materials*

Mg-based alloys. Mg based alloys are one of the promising candidates for on board application because Mg absorbs hydrogen up to 7.6 mass%. The author and co-workers proposed a concept of Mg based alloys with BCC structure. BCC lattice has larger numbers of hydrogen site in the lattice than the closest packed FCC and HCP ones. Using the ball milling method, Mg-Ti and Mg-Co BCC alloys were successfully synthesized. Among them, $Mg_{60}Co_{40}$ alloys absorbed hydrogen of 2.7 mass% at 373 K.[5] Figure 4 shows the pressure-composition isotherms of $Mg_{60}Co_{40}$ BCC alloy.

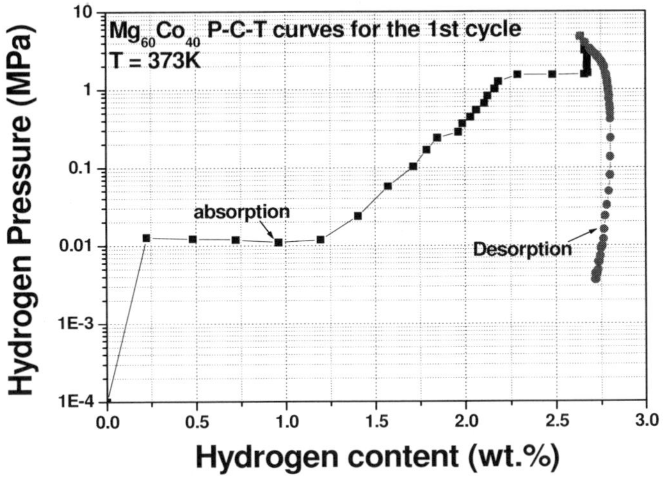

Fig. 4. Pressure-composition isotherms of $Mg_{60}Co_{40}$ alloy.[5]

Mg has a significantly high vapor pressure even at the lower temperature than its melting point. Therefore, preparation of Mg based alloy is one of the major subjects of the project. There has been reported that ball milling is a convenient and suitable technique to synthesize Mg based alloy at ambient temperature. However, because ball milled powder is sensitive to air, milling should be carried out under inert atmosphere and handle the milled powder also under inert conditions. In addition, the amount of product by one operation is usually less than 1 g. Therefore, a method that realizes to synthesize under air in a large amount has been expected.

If the thin layers are stacked in certain numbers and the stack is cold rolled, it becomes a thin layer again. By repeating this process, atomic mixing may occur in the cold rolled stacks. The process is shown in Fig. 5. The amount of constituent metals can be controlled by the thickness of each element. Kikuchi developed this method and named as "Super-Lamination".

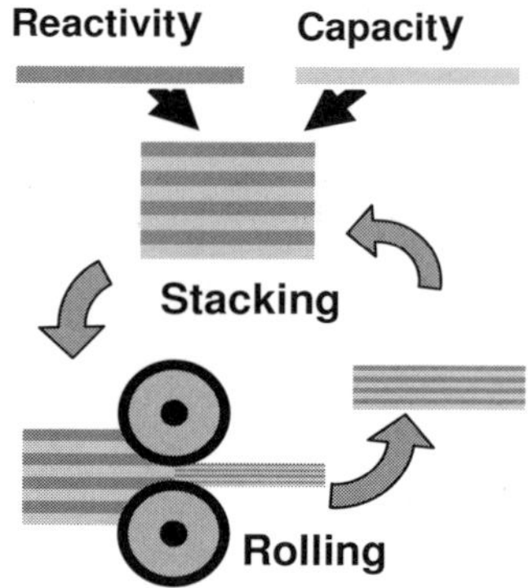

Fig. 5. Schematic drawing of cold rolling to synthesize Mg-based alloys.

However, this method could not synthesize an alloy only by super lamination. It needs further heat treatment to synthesize alloys. Kikuchi *et al.* synthesized Mg-Pd, Mg-Al, Mg-Cu and other Mg-based alloys by this method followed by mild heat treatment. Because morphology made by this method is fine and homogeneous, they found reversible hydrogenation/dehydrogenation in Mg-Al and Mg-Cu alloys that occur disproportionation with hydrogenation. This means recombination of atoms readily occur in the alloy synthesized by this method.

5.2. *Inorganic materials*

Chen *et al.* reported that Li nitride absorbed hydrogen around 6% each in hydrogenation of nitride to imide and imide to amide.[6] However, reported reaction temperature was above 530K. Enthalpy change of hydride formation (ΔH) was reported to be $-$ 66 kJ/molH$_2$. To reduce reaction temperature or increase ΔH, Luo,[7] Xiong *et al.*,[8] Leng *et al.*,[9] and Nakamori *et al.*[10] independently developed the Li-Mg-N-H system. Luo[7] reported that ΔH is -34 kJ/molH$_2$. The reactions reported by them are shown below. Leng *et al.* and Nakamori *et al.* have carried out investigation of these systems under the "Hydrogen infrastructure development and safe utilization of hydrogen".

Luo *et al.*[7] and Xiong *et al.*[8]

$$Mg(NH_2)_2 + 2LiH \rightarrow Li_2Mg(NH)_2 + 2H_2 \tag{1}$$

(originally Luo *et al.*[7]

$$2LiNH_2 + MgH_2 \rightarrow Li_2Mg(NH)_2 + 2H_2)$$

Leng *et al.*[9]

$$3Mg(NH_2)_2 + 8LiH \rightarrow Mg_3N_2 + 4Li_2NH + 8H_2 \tag{2}$$

Nakamori *et al.*[10]

$$3Mg(NH_2)_2 + 12LiH \rightarrow Mg_3N_2 + 4Li_3N + 12H_2 \tag{3}$$

Figure 6 shows the difference between lithium amide/hydride system and Li-Mg-N-H system. By combination of magnesium amide and lithium hydride, dehydrogenation temperature was significantly reduced but the hydrogen capacity was kept around 7 mass%.

According to Leng *et al.*, 7 mass% of hydrogen was desorbed and absorbed at 393~473K.[9] Nakamori *et al.* also reported that 9 mass% of hydrogen was absorbed at 523K.[10] These proposed reactions differ in the ratio of magnesium amide and lithium hydride. Recently, the reaction mechanism is found to be the same in above chemical equations.[11] Among inorganic hydrides, this Li-Mg-N-H system has a potential to charge by pressurized gas at room temperature and recharge hydrogen on board using the exhaust heat from polymer electrolyte fuel cell considering the value of ΔH.[7] Thermal conductivities, cycle life, safety issues and mass production method have been investigated under the Project. It has been reported that during reactions ammonia is produced as a by-product. Ammonia degrades Pt catalyst of polymer electrolyte fuel cell if the concentration in hydrogen is above sub-zero ppm. Therefore, to prevent formation of ammonia or to absorb ammonia by some chemicals should be investigated for on board applications.

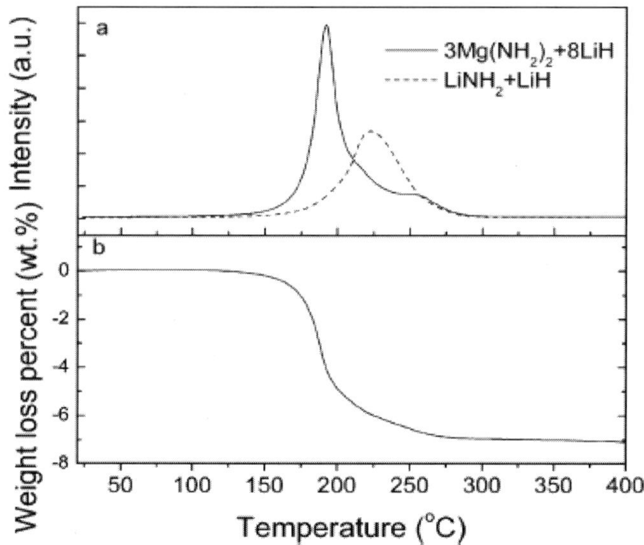

Fig. 6. Reduction of dehydrogenation temperature from LiNH$_2$+LiH system to Li-Mg-N-H system but keeping dehydrogenation amount of 7 mass %.[9]

7. Summary

Because the hydrogen economy is realized in a few ten years of later, funding for the hydrogen related research from the government or public organization is indispensable. Hydrogen is favorable in a weight basis but unfavorable in a volume basis for on board application. Hydrogen storage materials can store hydrogen higher volume density than liquefied hydrogen and pressurized hydrogen of a realistic pressure. As shown above, Japan has a wide range of research and development on hydrogen storage materials. It covers from fundamentals of material as absorber and structural one for container to codes and standards.

Acknowledgments

The author would like to thank Dr. Y. Nakamura and other colleagues of the same research group at AIST for providing information.

References

1. D. Mori, N. Kobayashi, T. Shinozawa, T. Matsunaga, H. Kubo, K. Toh, M. Tsuzuki, *J. Japan Inst. Metals*, **69**, 308 (2005).
2. D. Mori, K. Hirose, N. Haraikawa, N. Kobayashi, T. Shinozawa, T. Matsunaga, H. Kubo, K. Toh, M. Tsuzuki, *2006 MRS Spring Meetings*, EE1.2 (2006).
3. http://unit.aist.go.jp/hydrogenius/html/en-index.html.
4. http://unit.aist.go.jp/energy/hydro-star.
5. Y. Zhang, Y. Tsushio, H. Enoki, E. Akiba, *J. Alloys Comp.*, **393**, 147 (2005).
6. P. Chen, Z. Xiong, J. Luo, J. Lin, K. L. Tan, *NATURE*, **420**, 302 (2002).
7. W. Luo, *J. Alloys Comp.*, **381**, 284 (2004).
8. Z. T. Xiong, G.. T. Wu, J. J. Hu, P. Chen, *Adv. Mater.* **16**, 1522 (2004).
9. Y. Nakamori, G. Kitahara, K. Miwa, S. Towata, S. Orimo., *Appl. Phys. A.*, **80**, 1 (2005).
10. H. Leng, T. Ichikawa, S. Hino, N. Hanada, S. Ishobe, H. Fujii, *J. Phys. Chem. B.*, **108**, 8763 (2004).
11. M. Aoki, T. Noritake, Y. Nakamori, S. Towata, S. Orimo, *J. Alloys Comp.*, **446-447**, 328 (2007).

CHAPTER 11

HYDROGEN STORAGE RESEARCH AND DEVELOPMENT IN KOREA

Jong Won Kim, Sang Sup Han[*] and Kwang Bok Yi[*]

Hydrogen Energy R&D Center,[*] Chemical Process Research CenterKorea Institute of Energy Research 71-2 Jang-dong, Yuseong-gu, Daejeon, 305-343 The Republic of Korea E-mail: jwkim@kier.re.kr, sshan@kier.re.kr[*], cosy32@kier.re.kr[*]

In order to implement fuel cell vehicle (FCV) to the commercial market, an affordable and stable hydrogen storage technology must be preceded. However, storing sufficient hydrogen in various on and off-board sites, while satisfying all the specific requirements without compromising space, is extremely difficult. Korean government launched Hydrogen Energy R&D Center (HERC) as a member of the 21st Century Frontier R&D program, which is led by the Ministry of Science and Technology (MOST) in October 2003. The HERC has conducted research on the key technologies for the production, storage, and use of hydrogen energy for expediting realization of hydrogen economy based on renewable energy sources. The current status of hydrogen storage technology in Korea is reviewed and specific achievements in each research program associated with hydrogen storage are presented. Specifically, the performance of hydrogen gas cylinders is briefly discussed. Five different metal hydride groups (simple metal hydride, Ti-Cr-Mo and Ti-Cr-V systems, Mg-based materials, complex metal hydrides, and carbonaceous nano materials) are also discussed in terms of preparation method and performance. In addition, modeling of organometallic nano-material and measurement technique for hydrogen storage/release is covered in this report.

1. Introduction

As efforts to reduce severe fossil fuel dependence derived by growing transportation, the researches on the efficient and clean fuel alternative has been recently intensified and the demand for the alternatives is expected to become more pronounced due to the recent soaring oil price and the expected shortage of fossil fuels in the near future. Among the alternatives suggested for the past decades, hydrogen has been spotlighted due to its high specific energy density, pollution-free characteristics, and various production resources such as renewables (biomass, hydro, solar, geothermal) as well as fossil fuels.

A typical example of hydrogen utilization system is fuel cell which has a wide range of applications: stationary electricity generator, automobiles, and electronic devices. Particularly, the competition for the commercialization of a fuel cell vehicle has become quite intense due to the recent announcement of automotive industries for the potential appearance of commercially available fuel cell vehicles by 2010. In order to achieve the effective hydrogen utilization, economical hydrogen storage technology should be developed as well as hydrogen production. The hydrogen storage technology is essential not only for hydrogen utilization systems but also for matching the supply and demand sides of hydrogen.

HERC was organized for fundamental technical development of clean and non-pollutant energy for the hydrogen economy in the Republic of Korea. Successful establishment of hydrogen energy system technology would induce the hydrogen market of 300 million US dollars worth in the Republic of Korea. R&D of HERC on the hydrogen as new energy carrier was initiated on October 2003 and will last till March 2013. The first stage of R&D is mainly for basic research on hydrogen production, storage, and utilization, and was carried out from 2003 to 2005. During the second stage from 2006 to 2008, securing core technologies developed in the first stage and demonstrations of viable technologies is carried out. In the third stage, finally, the optimization and establishment of technologies will be attempted for further approach to hydrogen economy. Table 1 shows the detailed programs of HERC are well described according to the fiscal year.

Hydrogen storage is one of the key obstacles to the successful commercialization and market acceptance of hydrogen powered vehicle. Besides the efficiency of power system, it is an extremely challenging technology to store sufficient hydrogen on the vehicle without compromising consumer requirement such as safety, space, driving range, and cost.

There are three main hydrogen storage methods which include compression, liquefaction and hydrogen storage materials.

Table 1. Research & development scheme in HERC.

Stage	1st stage	2nd stage	3rd stage
Goal	Basic R&D	Demonstration	Establishment
Year	2003 2004 2005	2006 2007 2008	2009-2012
Hydrogen Production	Steam reforming of natural gas	Demonstration of 20 Nm^3/h scaled hydrogen gas station	-
	Thermochemical water splitting	Securing core technology	Establishment of hydrogen production technology using natural energy
	Fundamental water electrolysis technology	Securing core technology	
	Fundamental biological technology	Systemization of proven technology	1. Demonstration of 5 Nm^3/h scaled system
	Photocatalytic water splitting	Securing of material synthesis	2. Optimization of hydrogen system operation
Hydrogen Storage	Compressed gas tech & building infrastructure	-	Securing technology of practical application & commercial use
	Complex metal hydrides	Securing core technology	
	Chemical hydrides	Hydrogen storage for transportation & mobile power	(2 ~ 6 wt%, Max. 45 kg/m^3)
	Nano materials	Securing core technology	
Hydrogen Utilization & Policy	Hydrogen-fueled power system	Demonstration of high performance power system	Establishment of high efficient hydrogen-fueled power system
	High sensitive & high leak detecting sensor	Securing technology of sensor production & safety	Technology of the sensor & safety control
	Analysis of hydrogen safety	-	-

Among the technologies currently under development, the hydrogen storage as a highly pressurized gas is the most prominent candidate for the hydrogen powered vehicle. The advanced automobile industries have already demonstrated the highly pressurized hydrogen system on fuel cell vehicles for past several years. The hydrogen storage materials have some advantages such as high volumetric storage capacity, little energy loss, longer storage time and highest safety. Various carbonaceous and non-carbonaceous hydrogen storage materials have been worldwide studied over the past few decades. However, they have low specific storage capacity and poor cyclic properties.

Korea has recently made a great deal of investment on hydrogen storage research through Hydrogen Energy R&D Center launched in 2003, and some of the researches recently started to bear the fruits.

2. HERC's Targets for Hydrogen Storage in the 2nd Stage

HERC hosted the open competition for research quest for hydrogen storage to the scientific community in Korea. HERC is now conducting the 2nd stage research program including the researches on metal hydrides, chemical hydrides, new materials for the storage of hydrogen and characteristics of hydrogen adsorption/desorption.

The effective hydrogen storage capacity target of simple metal hydride in the stage is 3 wt% (150 kg-hydrogen/m^3) with more than life span of 2000 cycles and hydrogen absorption of more than 98% within two minutes. In case of alkali-metal complex hydrides, the effective hydrogen storage capacity target is more than 5 wt%. A process which can produce about 100 grams per batch will be developed and basic thermochemical properties of these hydrides will also be investigated. Using simple and complex metal hydrides, small scale hydrogen storage system is constructed and tested by the operation with fuel cell at Hyundai Motors Company. These results would be used to set up the development direction of hydrogen storage system for FCV.

Mg-based hydrogen storage alloy with the effective hydrogen storage capacity of 5.0 wt% and the sorption/desorption rates of 0.5 wt% at 300°C is developed. Moreover these alloys will be applied to the small scale hydrogen storage system.

An efficient hydrogen storage and generation system (a modular hydrogen generation system for 1 kW with the energy density of 50 kg/m^3) using sodium borohydride is developing. The thermochemical $NaBH_4$ regeneration process is optimized. The hydrogen storage system using hydrazine hydrate having hydrogen capacity of 4 wt% is also studied.

To develop carbonaceous hydrogen storage materials, various surface treatment techniques is used and nano-sized metal particles is introduced on carbon surfaces in order to gain optimal hydrogen storage capacities. The fabrication of carbon nanotubes which can store more than 4 wt% in the condition of 60 ~ 80°C and 50 bar is also tried.

The porous nanostructured materials having the hydrogen sorption capacity of 4.0 wt% at room temperature and 100 bar is being developed. The establishments of mechanism for hydrogen sorption on these kinds of materials are also studied.

HERC also try to establish the innovative methods for design of hydrogen storage materials using molecular modeling and to develop new hydrogen storage materials with high capacity, using the organometallic nanoporous materials and inorganic nanotubes.The DOE's assessment for on-board hydrogen storage options is cited in Table 2.

Table 2. DOE's assessment of challenges for on-board hydrogen storage options.[1]

Key 2010 Targets	Tanks	CH	MH	S
Volume (1.5 kWh/L)	H	M	M	M/H
Weight (2.0 kWh/kg)	M	M	M/H	M
Cost ($4/kWh)	M/H	M/H*	M/H	M/H
Refueling Time (3 min., for 5 kg)	L	L	M/H	M
Discharge Kinetics (0.02 g/s/kW)	L	M	M	M
Durability (1000 cycles)	L	M	M	M

Note: Tanks (to 10,000 psi), CH (Chemical hydrides), Metal Hydride (MH), S (Carbon/Sorbents), H; High (significant challenge), M/H; Medium/High, M; Medium, L; Low (minimum challenge).

3. Current R&D in Korea

3.1. *High pressure hydrogen storage system*

The ultimate goal of research on high pressure hydrogen system is the development of affordable storage system for fuel cell vehicle.

Through this project, development of high pressurized hydrogen system components and its integration technology have been. Some prototype regulators were manufactured and their performance was characterized by leakage, pressure resistance and durability tests.

The system integration technology of high pressurized storage system components such as high pressure tank, regulator, and valves was achieved through fabrication of a large number of fuel cell vehicles. A hydrogen storage system test bench which is able to characterize the performance of high pressure tank and regulator was designed and fabricated. System control algorithm was established by the hydrogen filling test. Some domestic standards of high pressure hydrogen tanks and regulators were suggested based on test results.

The results of this research can be applied to the localization of high pressure hydrogen storage system components as well as the establishment of system design and quality management standard. When the authentication is needed for production process of high pressure hydrogen storage system and its components, these results can be also used as a guideline.

3.1.1. *Type 4 high pressure hydrogen gas cylinder for fuel cell vehicle and charging station*

HERC have recently made great progress in efficient use of hydrogen gas through the development of all composite cylinder (Type 4), which requires ultra-light liner with good gas interrupting characteristic as well as optimized manufacturing process for economical efficiency. Through this project, HERC has developed liner that consists of Polyethylene-Clay (PE-Clay) nano composite with good gas interrupting characteristic and metallic end nozzle with special form for high pressure hydrogen gas cylinder (Fig. 1).

Fig. 1. Photos of Type 3 and Type 4 hydrogen storage vessel for fuel cell car.

Finite Element Method (FEM) program was used to determine the layer-built angle, winding pattern, and optimize them to secure design ability for high pressure hydrogen gas cylinder.

The design qualification of developed high pressure hydrogen gas cylinder was verified through various tests such as permeation test, hydrostatic test, burst test, ambient temperature cycling test, extreme temperature cycling test, environment test, drop test, accelerated stress rupture test, etc.

Using the PE-Clay nano composite liner developed, gas interrupting characteristic increased 30% compared to those the predecessors and the liner thickness also could be reduced. It increased the water capacity and decreased weight so the cylinder can satisfy the development target that hydrogen storage weight percent of the system should be maximum 4.5 wt% and volumetric density of the system should be minimum 20 kg/m^3.

In order to solve the leak problem of existing composite cylinders, HERC have developed, so called, 'leak free plastic liner', which have metallic liner with special shape and triple interrupting lines. The leak free liner has proved its high quality through the extremely harsh condition test performed by internationally authorized inspection body.

The ultra-light composite cylinder was developed and it gained an internationally authorized approval, NGV2-2000, which is the qualification measure for hydrogen gas vehicle fuel cylinders. In order to realize the prototype production by 2007, the troubles that might be caused in actual use will be identified and brought to a settlement through the various field tests.

3.1.2. *Type 3 high pressure hydrogen gas cylinder* [2-3]

Like the projects described above, the objective of this project is to develop safe, efficient, and commercially viable on-board fuel storage system for the hydrogen fuel cell vehicles or natural gas vehicles that use highly compressed gaseous fuels such as hydrogen and natural gas. In this project, the key technologies for high pressure composite vessels have been established and verified, and test facilities for validation have been constructed. Based on the foundations, 207 bar composite pressure vessel for automobile has been developed successfully and deployed for commercial use. There was presented the entire procedure of development and certification of a Type 3 composite cylinder of 207 bar service pressure and 70 l water capacity, which includes design/analysis, processing of filament winding, and the validation through various tests. The design methods of liner configuration and winding patterns were showed. Three dimensional, nonlinear finite element analysis techniques were used to predict burst pressure and failure mode. Design and analysis techniques were verified through burst and cycling tests. In the development of the 350 bar composite cylinders for hydrogen fuel cell vehicle, the key technologies on design, analysis and processing of high-pressure composite cylinders were developed, and the methods were verified through manufacturing and burst testing of prototype cylinders. Using developed technologies, a composite cylinder of 350 bar for FCV was developed and selected qualification testing was conducted to validate the cylinders according to the international standards.

The smart health monitoring system for composite high pressure hydrogen storage vessels was developed and verified. Analytic approaches of thermal stress of the vessel, gauge length selection and proper locations of the sensors were carried out. Fiber Bragg Grating (FBG) sensor system using wavelength swept fiber laser (WSFL) has been developed and verified for a long term on-line monitoring. For the environment test of the vessel, strain measurement using FBG sensors was verified at the extreme temperatures. FBG sensors showed their capability in the various temperatures. For the reliability of sensor embedding, reinforcement tubes for ingress/egress parts of FBG sensor lines were designed and demonstrated. As a feasibility test of FBG sensors in Type 3 composite tanks, the small sized tank with embedded FBG sensors were made and tested. Then, the 350 bar hydrogen storage vessel with 22 embedded FBG sensors was fabricated and tested. Some of the sensors failed during the cure and more sensors failed during the auto-frottage pressure test up to 760 bar and 0.8% strain. The remaining FBG sensors showed good agreement with attached strain gauges and measured internal strains in the cyclic pressure test. For the damage evaluation using embedded FBG sensors, the high frequency multipoint demodulation system using tunable Fabry-Perrot filters was developed. With this system, high frequency vibration signals by external impacts were measured by the embedded FBG sensor in the composite pressure vessel.

In 2006, the control of this project was transferred to the MOCIE and R&D on the topic is being continuously conducted for the next goal.

3.2. *Hydrogen storage system using metal hydride*

HERC also attempted to develop high performance hydrogen storage materials using metal hydrides, which are one of the most promising candidates as hydrogen storage materials due to their high storage capacity, reversibility and secured safety. Since the requirements for hydrogen storage technology differ from other technologies depending on its application fields, the optimum storage media should be chosen.

Metal hydrides and complex metal hydrides were developed for both room and high temperature usages. High performance hydrogen storage materials were developed through the intensive research programs considering various aspects such as chemical composition, microstructure, preparation process, hydrogen behavior, thermodynamics and kinetics.

Several materials have been suggested as hydrogen storage media including metal hydrides, complex metal hydrides and carbonaceous materials. Generally, the metal hydrides have better reversibility in hydrogen absorption and desorption, safety and volumetric storage capacity over other materials.

3.2.1. *Metal hydride*

Metal hydrides have great potentials for reversible on-board hydrogen storage and release at low temperatures and pressures. The optimum "operating P-T window" for proton exchange membrane (PEM) fuel cell vehicular applications is in the range of 1 ~ 10 bar and 25 ~ 120°C. This is based on using the waste heat from the fuel cell to "release" the hydrogen from the media. In near-term, waste heat less than 80°C is available, and moreover, if high temperature membranes are successfully developed, there will be another potential for waste heat at higher temperatures. A simple metal hydride such as $LaNi_5H_6$, that incorporates hydrogen into its crystal structure, can be functioning in this range, but its gravimetric capacity is too low (~ 1.3 wt%) and its cost is expected to be beyond the consideration for vehicular applications.

Researches on the complex metal hydrides are aiming the development of advanced metal hydride materials including light-weight complex hydrides, destabilized binary hydrides, intermetallic hydrides and other on-board reversible hydrides. Engineering studies are also underway to understand the system level issues and to facilitate the design of the optimized packaging and interface systems for on-board transportation applications.

Metal hydrides possess solid form, thus, occupy relatively small space. This is one of the important factors for qualification as a hydrogen storage and release material for the fuel cells. Metal hydrides can also be used for various other applications: Storage tanks for the hydrogen produced by atomic energy or surplus electricity at night, heat pumps, hydrogen purifiers, actuators, static compressors and so on.

Metal hydrides are capable of reversible hydrogen absorption and desorption with varying temperature and pressure. Using this property, hydrogen gas at room temperature and atmospheric pressure may be shrunken to one thousandth of its original volume, which is more effective than liquefying hydrogen in volume-wise. Metal hydrides are relatively safe since they can be used at temperatures and pressures not much higher than the ambient temperature and pressure.

Unfortunately, however, there are still several obstacles to be cleared to put metal hydrides on practical use; their gravimetric storage density is not much higher than that of the pressurized hydrogen gas, and the cyclic life span associated with the reversibility of the reactions with hydrogen is not yet long enough. Therefore, the demand for the development of a high performance hydrogen storage material using metal hydrides is still rapidly growing.

Hydrogen storage material must possess an excellent storage density in terms of both weight and volume. In addition, it must react fast with hydrogen and have a long cycle life span at a reasonably low temperature. In this work of 1st stage from 2003 to 2006, we developed the hydrogen storage material possessing the following properties:

- Effective gravimetric hydrogen storage density > 1.5 wt%,
- Cyclic life span > 1000 cycles,
- Reaction temperature < 100°C,
- Dissociation pressure between 0.05 and 0.5 MPa,
- Reaction rate > 0.05[H/M]/min.

3.2.1.1. Ti-Cr-Mo system

Among the thirteen alloy compositions tested in this study, several alloys exhibited total hydrogen storage capacities higher than 3.5 wt% at 303 K, but their effective hydrogen storage capacities were less than 1 wt%. In this system, unlike the Ti-Cr-V system mentioned below, it was difficult to find a relation between the storage capacity and the lattice parameter. When the radius of hydrogen site was between 0.425~0.450 Å, these alloys were found to show the maximum effective hydrogen storage capacity under the equilibrium hydrogen pressure lower than 5 MPa.

Analyzing the P-C isotherms, the maximum effective hydrogen storage capacity was expected when the titanium content was between 35 and 55 wt%. Although some extra works still need to be done, putting together the above results and the formulas obtained in the present study may help to determine the composition range which will give the maximum effective hydrogen storage capacity.

3.2.1.2. Ti-Cr-V system[4-5]

The alloys whose compositions were represented by the formula, $Ti_{(0.22+X)}Cr_{(0.28+1.5X)}V_{(0.5-2.5X)}$ ($0 \leq X \leq 0.12$), had the total hydrogen storage capacity higher than 3 wt% and the effective hydrogen storage capacity higher than 1.4 wt%. Particularly, among all the tested alloys, the $Ti_{0.32}Cr_{0.43}V_{0.25}$ alloy exhibited the best effective hydrogen storage capacity of 1.65 wt%.

Furthermore, the reversible bcc $\leftrightarrow$ fcc structural transition was observed with hydrogenation and dehydrogenation, which indicated the possibility of pressure cycling. The rate of hydrogen absorption was higher than 0.05[H/M]/min, which is the target value in the present study. Approximately 95% of the hydrogen was absorbed within the first 10 minutes. Energy Dispersive Spectrometer (EDS) analysis revealed micro-segregation. This indicated that microstructure homogenization by heat treatment might be necessary for the achievement of fast hydrogenation rate (Fig. 2).

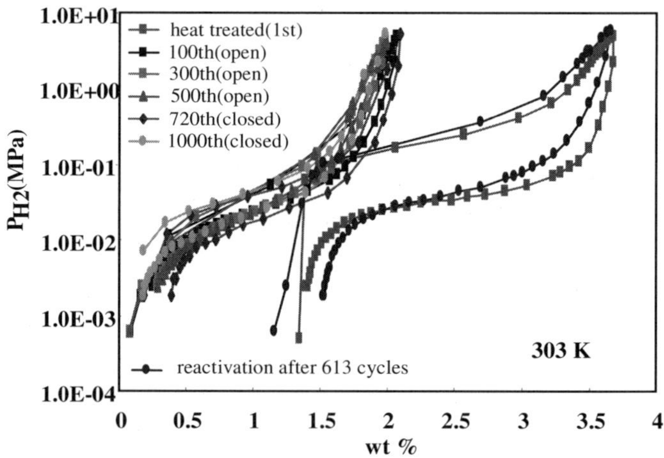

Fig. 2. P-C isotherms of $Ti_{0.32}Cr_{0.43}V_{0.25}$ alloy with cycling and its recovery at 303 K.

The changes in the physical properties and the hydrogen storage properties of the $Ti_{0.32}Cr_{0.43}V_{0.25}$ alloy following the heat treatment were measured and analyzed. The results showed that the total and the effective hydrogen storage capacity increased to 3.7 wt% and 2.3 wt%, respectively. The flatness of the plateau region was also greatly improved.

To investigate the pressure cycling features, pressure swing cyclic tests were carried out. The alloy showed the effective hydrogen storage capacity higher than 2.0 wt% even after 1000 cycles. The hydrogen absorption rate was much higher than 0.05[H/M]/min, and more than 98% of the hydrogen was absorbed within the first 2 minutes.

The heat of hydride formation was determined to be approximately -36 kJ/mol-H_2. The storage capacity was found to be recovered nearly to the original level by reactivating the specimen during the cyclic test. As the number of cycles increased the alloy wore down to fine particles and showed the tendency to turn into an amorphous state.

The material developed in the present study absorbs and desorbs hydrogen at temperatures and pressures not much higher than the ambient ones. Also, other storage properties such as storage density (particularly, volumetric storage density), cycle life, reaction rate with hydrogen were excellent. The gravimetric storage density, however, was not yet enough for fuel cell vehicle application. In the 2nd stage of this work, the goal is to develop hydrogen storage material with effective hydrogen storage capacity of 3 wt%, 150 kg-H_2/m^3 and cycle life span of 2000 cycles.

The successful improvement of the gravimetric storage density to the level of 3 ~ 4 wt% may enable the use of the alloy for on-board applications.

3.2.2. *Mg-based materials*[6-8]

Mg-based materials, especially Mg_2Ni, have the highest specific storage capacity among metal hydrides, which makes them strong candidates for hydrogen storage media.

Magnesium, however, has strong ionic bonding with hydrogen. Due to this, MgH_2 is very stable, and higher temperature and longer period are needed to absorb and desorb the hydrogen. Nevertheless, the hydrogen storage properties of Mg-based materials can be improved by microstructure control.

In this study, HERC developed the preparation method for Mg-based hydrogen storage materials possessing 4.5 wt% of effective hydrogen storage capacity and 4.4 wt% in 20 minutes of hydrogen sorption rate at 320°C through microstructures modification by varying alloy composition and processing routes (Fig. 3).

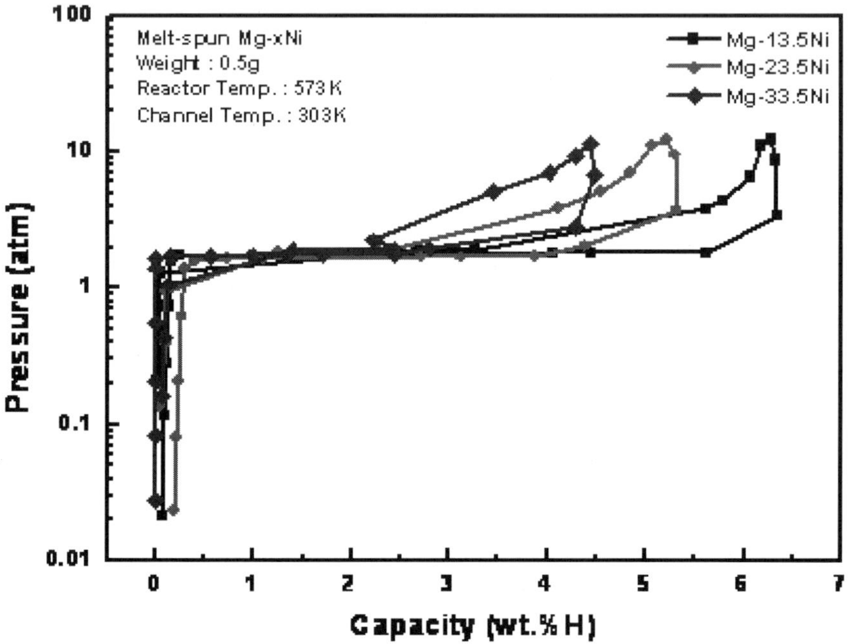

Fig. 3. Hydrogen absorption and desorption properties of Mg-xNi alloy powder.

The microstructure of the alloy was mainly affected by processing routes. In this study, the Mg-based materials were produced by gravity casting, and rapid solidification followed by reactive mechanical grinding. The relation between the microstructure and processing variables was analyzed systematically through Optical Microscopy (OM), Scanning Electron Microscopy (SEM), Transmission Electron Microscopy (TEM) and X-ray Diffraction (XRD) observations. Alloying elements addition to Mg also allowed the control of microstructure. The hydrogen storage properties of Mg-based materials can be also improved by addition of catalyst. In the previous studies, the oxides of transition metals are most effective for improving of hydrogen storage properties. We also tried to develop the preparation method for nanostructured metal oxide by spray conversion and calcination. And the processing technology of composite powder consisted of Mg alloy and metal oxide was developed with success.

Mg-23.5wt%Ni eutectic alloy showed the best hydrogen storage properties among the gravity cast Mg-xNi binary alloys (Fig. 3). But in the case of RSP alloys, hypoeutectic alloy with nano-crystalline structure had superior hydrogen storage properties to amorphous eutectic and hypereutectic alloys. The hydrogen storage properties of amorphous alloys were improved by crystallization with heat treatment because the nanostructured crystalline phases were distributed into the matrix. The eutectic alloy produced by RSP and crystallization with heat treatment had the best hydrogen storage properties among Mg-xNi binary alloys. La and Cu were added into Mg-23.5 wt% eutectic alloy in order to improve the hydrogen storage properties of Mg-Ni binary alloy. Mg-23.5 wt%Ni-aX ternary alloy had faster hydrogen sorption rate, but the effective hydrogen storage capacities were decreased.

The nano-structured metal oxides produced by spray pyrolysis had the superior catalytic effect to commercial metal oxides. The basic processing condition for Mg-oxide composite powders with reactive mechanical grinding under hydrogen atmosphere was established and the Mg-5Ni-10Fe$_2$O$_3$ composite powder had the best hydrogen storage properties.

A hydrogen storage system, heat pump, thermal storage system and high purity hydrogen refining system are possible application fields of Mg-based hydrogen storage materials. Especially if the hydrogen production by nuclear power will be commercialized, Mg-based materials can be applied to huge hydrogen storage system because the high temperature exhausted heat can be utilized economically for hydrogen absorption and desorption of Mg-based materials. Mg-based alloys also have the largest hydrogen storage capacity, high thermal storage capacity, little hysteresis and low price. Therefore, they may fit the role of potential thermal storage media for heat exhaust in industry or solar heat.

3.2.3. *Complex metal hydrides*[9-10]

Complex metal hydrides such as alanate (AlH_4) materials have the potential for higher gravimetric hydrogen capacities in the operational window than simple metal hydrides. Alanates can store and release hydrogen reversibly when catalyzed with titanium dopants, according to the following 2-step reaction for sodium alanate:

$$NaAlH_4 = 1/3\ Na_3AlH_6 + 2/3Al + H_2 \tag{1}$$

$$Na_3AlH_6 = 3\ NaH + Al + 3/2H_2 \tag{2}$$

At atmospheric pressure, the first reaction becomes thermodynamically favorable at temperatures above 33°C and can release 3.7 wt% of hydrogen, while the second reaction takes place above 110°C and can release 1.8 wt% of hydrogen. The amount of hydrogen that a material can release is the key parameter used to determine system (net) gravimetric and volumetric capacities.

Issues with complex metal hydrides include low hydrogen capacity, slow absorption and desorption kinetics, and cost. The maximum material gravimetric capacity of 5.5 wt% hydrogen for sodium alanate is still less than the 2010 DOE system target of 6 wt%. So far, 4 wt% reversible hydrogen content has been experimentally demonstrated with alanate materials. Also, hydrogen release kinetics is too slow for vehicular applications. Furthermore, the packing density of these powders is low (roughly around 50%) and the system-level volumetric capacity is another challenge. Although sodium alanates may not meet the 2010 target, it is envisioned that their continued study will lead to fundamental understanding that can be applied to the design and development of improved types of complex metal hydrides.

The purpose of the study in the 1st stage from 2003 to 2006 was to develop technologies for synthesis of complex metal hydrides for hydrogen storage that can be applied to fuel-cell vehicles, which were characterized by hydrogen storage capacity larger than 4.0 wt%, hydrogen sorption/desorption within 1 hour and over 10 cycle life.

Thermodynamic properties of alkali-metal alanates with over 4.0 wt%

hydrogen have been predicted by computational thermodynamics. Alanates that are predicted to attain thermodynamic equilibrium at moderate temperature and pressure condition were synthesized by mechanochemial methods. In order to enhance the hydrogen sorption/ desorption rate of the alanates, candidate materials for catalysts were selected and dispersed into alkali-metal alanates with the amount of 5 ~ 10 wt% by high-energy ball milling. Thermal decomposition of catalyzed alanates was characterized by Differential Scanning Calorimetry (DSC) and Thermogravimetry/Mass Spectrometry (TG/MS) and was compared to the alanates without catalysts. Nanopowders of materials exhibiting a good catalytic effect were synthesized by mechanochemical methods to maximize their catalytic effect.

Sorption/desorption of the alanates catalyzed with nanopowders is characterized by a pressure-composition isotherm (PCI) measurement apparatus and repeated in several cycles to confirm if the performance is maintained in a long life (Fig. 4).

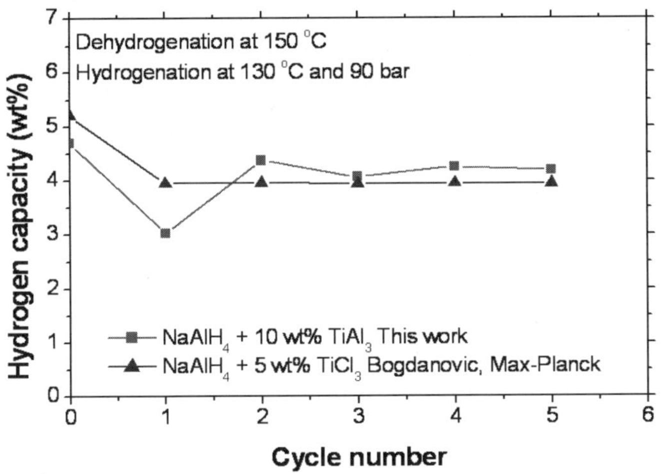

Fig. 4. Cyclic performance with nano-catalyst.

Organic adduct-free Li_3AlH_6 having theoretical hydrogen storage capacity of 5.6 wt% was successfully synthesized by a mechanochemical reaction between LiH and $LiAlH_4$. The synthesized Li_3AlH_6 released about 4.8 wt% hydrogen at 150°C. Although the addition of TiCl$_3$ catalyst enhanced the desorption rate, the hydrogen storage capacity was reduced. It was found that the reaction between TiCl$_3$ and Li$_3$AlH$_6$ forms TiAl$_3$. TiAl$_3$ nanopowder with about 100 nm particle size was mechanochemically synthesized and it presented a good catalytic effect in Li$_3$AlH$_6$ and NaAlH$_4$, inducing less loss in hydrogen storage capacity. Also, NaAlH$_4$ catalyzed with TiAl$_3$ presented reversible hydrogen storage with about 4 wt% hydrogen at 150°C. Thermodynamic database of the Na-Li-Al-H system was built in the framework of the calculation of phase diagrams (CALPHAD) activity.

One of the major issues with complex metal hydride materials, due to the reaction enthalpies involved, is thermal management during refueling. Depending on the amount of hydrogen stored and refueling times required, megawatts to half a gigawatt must be handled during recharging on-board vehicular systems with metal hydrides. Reversibility of these and new materials also needs to be demonstrated for over a thousand cycles.

On the 2nd stage from 2006 to 2008, a feasibility study for application to fuel-cell vehicles will be done in collaboration with an industrial company. Eventually, this technology will also be applicable for hydrogen internal combustion engines and energy storage systems for renewable energy.

3.3. *Carbonaceous nano materials*[11-15]

Carbonaceous materials are chemically stable and have high thermal/electric conductivity, and good mechanical strength and elasticity. In 21st century, carbon materials have been focused as one of the promising materials for hydrogen storage. The surfaces of carbon have high energy density, thus it is relatively easy to modify for introducing hydrogen-friendly functional groups. And the porous structures of carbon also can be advantageous in hydrogen storage.

In this work, studies on new nano-materials, such as nano-structured and nanoporous carbons and nanofibers, were conducted. Also, the hybridization of materials with heterogeneous properties for the hydrogen storage was investigated.

Through heat treatments, ultra-fine fibers were prepared by melt or electro spinning of catalysts, dispersed precursors and carbon materials with targeted storage capacity. Preparation and surface reformation of super active carbon were carried out using the electro spinning system for pitch based nanofibers, activation, chemical or physical surface treatment of graphite nanofibers (GNFs), and alkali treatment of GNFs and CNTs.

1) Electrospun polymer (using polyacrylonitrile and poly(vinylidene fluoride)) nanofibers were carbonized with or without iron(III) acetylacetonate(IAA). The IAA induced catalytic graphitization within the temperature range 1000-1800°C. Carbonization without catalyst produced carbon nanofibers (CNFs) and GNFs were produced in the presence of the catalysts. Their structural properties and morphology were investigated. GNFs with a high surface area of 298 ~ 473 m^2/g showed the typical curve for micro- and meso-porous carbon in nitrogen adsorption/desorption isotherms. CNFs has varied surface area. The hydrogen storage capacity of these CNFs and GNFs was evaluated by the gravimetric method using a magnetic suspension balance (MSB) at room temperature and 100 bar. The hydrogen storage capacity of CNFs was 0.16 ~ 0.5 wt% and GNFs was 0.14 ~ 1.0 wt%. The effective pore size for hydrogen storage compared to the kinetic diameter of the hydrogen molecule is discussed

2) The carbonaceous adsorbent has moisture and thermal stability and adsorbs hydrogen by physical adsorption. Physical adsorption hydrogen storage system does not need extra adsorption/desorption system and the storage material does not degrade by repetition of adsorption/desorption. In this study in order to prepare hydrogen storage material, pitch and metal catalysts had hybridized and spun to superfine pitch fiber by electrospinning method. The pore controlled activated carbon fibers (ACFs) were prepared from pitch fibers by heat treatment. The pore properties of ACFs were measured by adsorption analyzer and changes of surface morphology were also observed by SEM and TEM. The hydrogen adsorption amount was measured with high pressure thermogravimetric analyzer. The pore controlled various ACFs for hydrogen storage were prepared and characterized by further surface characterization with acid and electric decomposition. The hydrogen

storage capacity of ACFs were ranged 0.4 ~ 1.1 wt% at 10MPa and maximum 2.5 at 40MPa.

3) A study is presented on the preparation of high porous super activated carbon. Bituminous coal and lignocellulosic materials such as coconut-shell are chemically activated by KOH, resulting in activated carbons with different porosities. All of prepared activated carbons were microporous and showed the same adsorption properties. The pore structure of the prepared activated carbons possessed the maximum specific surface area of 3,200 m^2/g and total volume of 2.2 cc/g in total pore volume. The hydrogen adsorption capacity was measured by isothermal gravimetric analysis. Hydrogen adsorption on activated carbons was investigated up to 100 bars at 298K. The complete reversibility and fast kinetics of hydrogen adsorption show that most of adsorbed quantity is due to physical adsorption. A linear relation between hydrogen adsorption capacity and pressure is obtained for all samples regardless of their porosities. Hydrogen adsorption capacities are linear function of porosities such as specific surface area, micropore surface area, total pore volume, and micropore volume. The maximum hydrogen adsorption capacity is 1.0 wt% at 100 bars, 298K. Due to the fast adsorption and desorption kinetics and almost complete reversibility, physical adsorption by activated carbon is a promising methods for hydrogen storage at moderate temperature, however, the highest measured hydrogen adsorption capacity is less than 1 wt% at 100 bars, 298K, in spite of their highly developed porosities reaching up to 2,800 m^2/g in specific surface area. New materials with ultra-high porosity, even higher than existing activated carbons, should be investigated to achieve sufficient hydrogen storage capacities at room temperature.

4) In this study, various carbon materials such as CNT (SWCNT, MWCNT), Activated Carbon (AC), ACFs, electro-spun CNFs have been modified by Ni doping and fluorination for developing hydrogen storage capacity. These modified carbon materials were characterized by BET surface area, Scanning Electron Microscopy/Energy Dispersive Spectrometer (SEM/EDS), TEM, XRD, and Raman spectroscopy. Hydrogen storage capacity of SWNTs fluorinated at 423K was remarkably increased 2.6 times than that of pristine SWNTs. For SWNTs

fluorinated at 573K, the amount of hydrogen adsorbed wasn't increased compared with SWNTs fluorinated at 423K. After fluorination treatment, although the micropore volume of ACF was decreased, their amounts of hydrogen storage were found to be increase. We also found that the micropore volume on ACFs was nearly unchanged with Ni doping and hydrogen storage capacities increased considerably due to the catalytic activation of nickel. Though fluorination of carbon materials makes higher hydrogen affinity, the effect of catalytic activation of nickel was more prominent, and thus led to more storage of hydrogen.

In the 2nd stage of this work from 2006 to 2008, various surface treatment techniques will be used on several porous carbon materials. And nano size metal particles, such as Ti, Cu, CR, Ag and Pt will be introduced on carbon surfaces in order to gain optimal hydrogen storage capacity (Fig. 5).

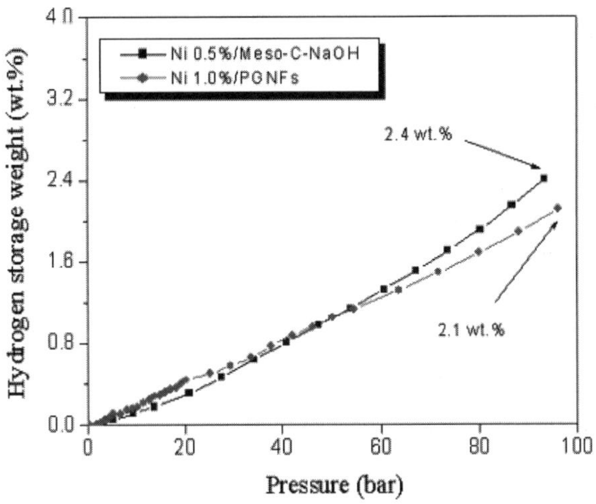

Fig. 5. Hydrogen storage behaviors in nanosized metal-loaded carbon materials.

3.4. *Non-carbonaceous nano naterials*[16-19]

In this research, efficient and mass hydrogen storing technologies were developed by increasing hydrogen bonding. The characteristics of the selected materials were examined in terms of their efficient hydrogen storage capabilities. The method of synthetically creating non-carbonic nanomaterials was discovered. Moreover, the infra-technology that provides stabilizing and economic hydrogen storage is developed.

Metal nanostructured materials, such as metal nanoparticles, nanowires and nanoarrays have been extensively investigated because of their interesting electronic and optical properties and potential application in nanodevices.

A template technique is a promising method to synthesize nanosized materials with ordered structures. Especially using uniform and straight nanochannels of an anodic aluminum oxide film as a template, many researchers have fabricated a variety of one-dimensional nanomaterials, e.g., nanotubes, nanowires or nanorods.

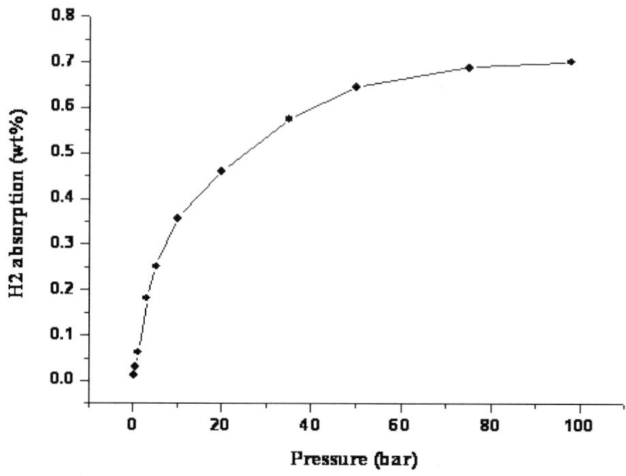

Fig. 6. Hydrogen storage capacity in $Li_4Ti_5O_{12}$ nanotube at 298 K.

The fluorinated Li dispersed nickel oxide nanotubes showed 0.06 wt% at 298K and 1.65 wt% at 77K of the hydrogen adsorption capacities at 47 atm, respectively. And Li dispersed silica nanotubes showed 0.02 wt% at 298K and 2.16 wt% at 77K of hydrogen adsorption capacities at 47 atm, respectively.

Especially, $Cu_2(OH_3)Cl$ nanoplatelets sphere represented 2.35 wt% of hydrogen uptake at 293K under 47 atm.

3.5. *Hydrogen storage using chemical hydrides*[20-22]

Technologies that utilize hydrogen as a clean energy source are beginning to assume a significant position in overcoming energy and environmental problems. As part of the hydrogen storage technologies outcomes of this project contribute to creating an almost pollution-free energy (Fig. 7).

Fig. 7. Photo of fuel cell scooter operated by hydrogen storage system using NaBH$_4$.

Liquid hydrogen storage is expensive and hazardous. Compressed gas storage systems are bulky. These systems consume a large amount of energy. On the other hand, chemical hydrides, such as NaBH$_4$, are safe and practical. Therefore, the hydrogen storage using chemical hydrides is the most promising technology as a source of hydrogen for fuel cells. In this project, we investigate into highly efficient hydrogen storage and hydrogen-generating systems that use alkali chemical hydrides, NaBH$_4$ in particular.

Metal borohydrides such as NaBH$_4$, LiBH$_4$, KBH$_4$ have received considerable attentions as high efficiency hydrogen storage materials. Among them, sodium borohydride(NaBH$_4$) provides a safe and practical mean of hydrogen storage and production at an economical cost compared with other borohydrides. NaBH$_4$ provides the high hydrogen storage density of 10.8 wt%, and it can produce double amount of its stored hydrogen by a hydrolysis reaction as follows:

$$NaBH_4 + 2H_2O \rightarrow NaBO_2 + 4H_2 \qquad\qquad (3)$$

with $\Delta H = 217$ kJ.

In addition, it stores hydrogen safely in an alkaline solution at room temperature. For generation of hydrogen from alkaline $NaBH_4$ solution, suitable catalysts such as Ru and Pt are needed to promote the hydrolysis reaction of $NaBH_4$. Since the cost of such noble metal catalysts are very expensive, it is necessary to develop an economical alternative catalyst with an excellent efficiency of hydrogen generation. Also it is necessary to find a recycling process from by-product, $NaBO_2$, to $NaBH_4$ in order to reduce the cost of $NaBH_4$.

In this work, we investigated the effects of Co and Ni alloy catalysts on the hydrogen generation properties from hydrolysis of alkaline $NaBH_4$ solution and developed high performance Co and Ni alloy catalysts. We examined the cycle properties of the catalysts to elucidate their degradation mechanism in alkaline $NaBH_4$ solution and improve their cycle ability. Finally, $NaBH_4$ recycling processes by ball-milling and thermochemical methods were investigated to develop economical hydrogen storage and production system using $NaBH_4$.

In order to achieve the development of hydrogen storage and production technologies using $NaBH_4$ which is our final goal, it is necessary to develop an economical alternative catalyst with an excellent efficiency of hydrogen generation and to find a recycling process from $NaBO_2$ to $NaBH_4$.

3.5.1. *Optimization of NaBH4 liquid fuel system*

The hydrolysis of $NaBH_4$ depended on the pH of an alkaline solution and the concentration of $NaBH_4$. With increasing pH, hydrogen generation rate decreased due to the increase of the activation energy for hydrolysis of $NaBH_4$. With increasing $NaBH_4$ concentration up to 15 wt%, hydrogen generation rate also increased. However, with increasing $NaBH_4$ concentration more than 15 wt%, hydrogen generation rate decreased due to the blocking of catalytic sites by precipitation of by-product, $NaBO_2$, on catalyst.

3.5.2. *Development of Co and Ni alloy catalysts by a paste method*

Effects of Co and Ni alloy catalysts, prepared by a paste method, on the hydrogen generation properties from hydrolysis of alkaline $NaBH_4$ solution were examined to select the optimum catalyst material. Co catalyst showed the fastest hydrogen generation rate of 211 ml/min.g-catalyst in 10 wt% $NaBH_4$ solution (pH 13) at 25°C, demonstrating that Co has higher intrinsic catalytic activity for the hydrolysis of $NaBH_4$ than Ni alloys.

Hydrogen generation rate of Filamentary Ni catalyst, prepared by a paste method, gradually decreased with cycle number. The degradation of the catalyst resulted from an agglomeration of the catalyst particles and a formation of oxide film, which consists of $Na_2B_4O_7 \cdot 10H_2O$, B_2O_3 and KB_xO_y, on catalyst.

3.5.3. *Development of high performance Co-P catalysts by an electroplating method*

The microstructures of electroplated Co and Co-P catalysts were polycrystalline HCP structure and amorphous structure respectively. Hydrogen generation rates of the Co and Co-P catalysts were determined to be 52 and 954 ml/min.g-catalyst respectively in 1 wt% NaOH + 10 wt% $NaBH_4$ solution at 30°C. It clearly showed that the Co-P catalyst exhibited much higher catalytic activity than did the Co catalyst for hydrolysis of $NaBH_4$.

The catalytic activity of the electroplated Co-P catalyst for hydrolysis of $NaBH_4$ was found to be a function of both cathodic current density and plating time, that is, parameters determining concentration of P in Co-P catalyst. Especially, Co-13 wt% P catalyst electroplated on Cu in the Co-P bath at a cathodic current density of 0.01 A/cm^2 for 1080 s showed the best hydrogen generation rate of 954 ml/min.g-catalyst in 1 wt% NaOH + 10 wt% $NaBH_4$ solution at 30°C.

The sudden change of hydrogen generation performance for electroplated Co-P catalyst was not observed as the cycle number increased. After 50 cycles, Co-P catalyst remained 97% versus initial performance.

3.5.4. *Development of NaBH4 recycling process*

$NaBH_4$ conversion efficiency for ball-milling process was very low, less than 5%. However, for thermochemical process, $NaBH_4$ conversion efficiency increased up to 35% in hydrogen pressure of 6 MPa for 18 h at 550°C.

$NaBH_4$ conversion efficiency for thermochemical process largely depends on the hydrogen pressure. For the case of the hydrogen pressure of 3 MPa, conversion efficiency was 16%, but with increasing the hydrogen pressure increased to 4 MPa, conversion efficiency increased up to 32% very rapidly. Further increase of hydrogen pressure to 6 MPa only slightly increased the conversion efficiency to 35%.

In this study, the Co-P catalyst, which exhibits about 20 times higher catalytic activity for hydrolysis of $NaBH_4$ than did the Co catalyst, was developed successfully by an electroplating method. Also the fabrication process of the electroplated Co-P catalyst is very simple and cost effective. Therefore, it was expected that high performance Co-P catalyst can alternate the noble metal catalysts such as Ru and Pt. HERC will investigate the effects of the composition of a Co-P bath and additives such as glycine on microstructures of Co-P catalysts and their hydrogen generation properties from hydrolysis of alkaline $NaBH_4$ solution in order to increase the hydrogen generation rate from current 954 ml·min^{-1}g-catalyst^{-1} to 1600 ml·min^{-1}g-catalyst^{-1}, which is the hydrogen generation rate of Ru catalyst.

In order to develop recycling process of $NaBH_4$ with high efficiency, HERC will focus on the recycling process by the thermochemical method and examine the effects of reducing agent not only Ca alloys but also carbon materials such as active carbon and acetylene black on conversion efficiency.

During this study aiming at development of high-performance catalyst for hydrogen generation form $NaBH_4$ solution, main effort was devoted to develop Co-B catalysts. The followings summarize conclusions and research directions.

1) Co-B catalyst powders were synthesized by a chemical reduction method. The catalyst was mixture of Co-B and its oxides, showing hydrogen generation rates of 1,000~2,000 ml·min^{-1}g-catalyst^{-1} depending on the experimental conditions (Fig. 8).

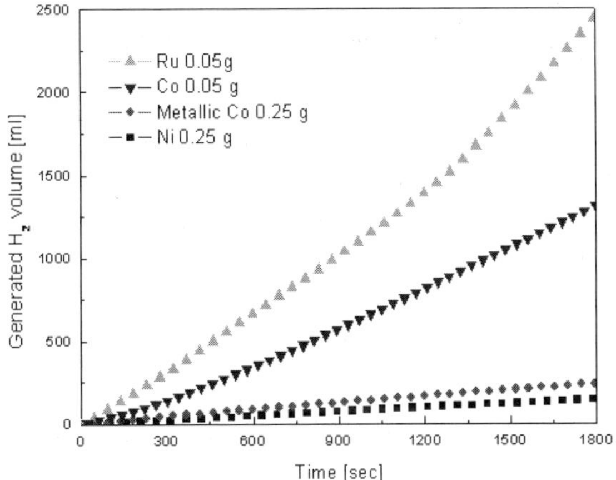

Fig. 8. Hydrogen generation rate of Co-B catalysts.

2) Effects of experimental conditions such as temperature, amount of catalysts, and concentration of reactants on hydrogen generation rates were investigated. With increasing reaction temperature, amount of catalysts, and concentration of NaOH, hydrogen generation rates increased, while the rates decreased with increasing $NaBH_4$ concentration.

3) Preliminary tests using zeolites, alumina and carbon molecular sieves as catalyst supports showed hydrogen generation rates of 50~500 $ml \cdot min^{-1}$g-catalyst^{-1}.

4) Finally, the technology has been developed to coat Co-B catalysts in the pores of Ni foam by a chemical reduction method. Effects of catalyst preparation conditions were investigated. It was revealed that amorphous Co-B catalyst coated on Ni foam showed higher generation rates of hydrogen compared to crystalline one. The optimum temperature of calcining the structured catalysts was in the range of 200 to 250°C with hydrogen generation rates up to 7,000 $ml \cdot min^{-1}$g-catalyst^{-1}.

5) The amorphous Co-B catalysts showed higher generation rates of hydrogen, but they are more prone to be lost during the reaction. Further study should be carried out to improve the stability of the catalysts on the support.

3.6. *Modeling and developing organometallic nanoporous materials*[23-25]

The goal of this research is to develop new hydrogen storage materials with improved capacities using the molecular modeling. Organometallic nanoporous materials were used at the developmental stage.

This research was divided into 3 stages. First, an expert system having the prediction accuracy of 80% for the hydrogen capacity was established using the virtual screening based on the computer aided molecular design for the MOF. Then, the nanoporous metal organic frameworks (MOFs) having the hydrogen capacity of 5.0~6.0 wt% are developed. Lastly, the development of storage materials capable of hydrogen adsorption of 4.5~6.5 wt% is enabled by the molecular modeling and the synthesis of organic zeolite.

The MOF are known to be the most promising and effective nano-structured hydrogen storage for transportation because of their high porosity and surface area, as well. The major advantage of nano-structured materials for hydrogen storage is that specific architectures can be designed into these materials to enhance the weight percentage of stored hydrogen and to control the kinetics involved in the adsorption/release of hydrogen. One of the weak aspects of these materials is the possible complication involved in multi-step synthetic process to diversify the structure.

In our photochemical study using laser photons in UV range, we are trying to overcome the complexity of the multi-step chemical reaction processes and to provide more opportunity to select better structure from the variety of MOF structures. Furthermore, we are hopping to find a way to selectively photo-excite the various reactive site of the hollow MOF structure to enhance the effective storing capability of hydrogen gases.

As a preliminary result, we were able to restore the 3-D structure from the intermediate materials in 2-D structure obtained from the conventional synthetic procedure of MOFs. Although this result was obtained unexpectedly during the photochemical approach for changing the properties of MOFs, it seemed to open the possibility of selecting the 3-D structure from the conventionally provided 2-D structure by introducing photon energies in UV range. The detailed analysis to explain the exact process for initiating the linkage between 2-D structures is under way but the structural analysis using XRD clearly showed the completely new material having cross linkages to form 3D configuration from 2-D structures.

MOFs is microporous crystalline material that has potential applications to hydrogen storage because they can be prepared in principle by design and, in addition, have a large structural diversity due to the many possibilities of available molecular building blocks. Research works on the synthesis and hydrogen sorption behavior of MOFs have been carried out for last 2.5 research years by us. Twenty new MOFs were prepared, and their crystal structures and physical properties have been elucidated. The main research focus was to find out the correlation factors between the structural factors of MOFs and their hydrogen sorption amounts based on the hydrogen sorption experiments at 77K and less than 1 bar. The noticeable results were large variation in the sorption amounts around apparent Langmuir surface area of 2000 m^2/g and pore volume of 0.6 cm^3/g among the known MOFs. This implies that MOFs having these porosity conditions may have an increased hydrogen storing amounts when their physical and chemical environments of pores are to be adjusted. For example, IRMOF-3 that has an electron-donating functional group, $-NH_2$ in the organic linker showed larger hydrogen storage capability than IRMOF-1 that has a un-substituted organic linker. However, considering that the minimum operating conditions of hydrogen storage materials is around room temperature and less than 200 bar, the hydrogen sorption data at low temperature may not be so informative for the developing applicable materials in everyday use because the observed and calculated hydrogen adsorption enthalpies of MOFs are within ~8kJ/mol which should be increased significantly for the everyday use. In this text it is required to obtain plentiful amounts of hydrogen sorption data at room temperature, and to find and explain the adsorption mechanism by aid of various analytical methods including molecular modeling studies. In addition developments of MOFs containing potent adsorption sites should be explored using the investigation results.

The purpose of this study is to improve the capability of hydrogen adsorption capacity of zeolite derivatives dramatically which applicable for hydrogen storage materials. Organic zeolites provide an excellent structural platform for storage of gas molecules since they contain the regularly repeating size of cavities and channels which form hydrogen adsorption site via bonding interaction with metal ions and organic molecules. The organic zeolites, provided by combination of organic molecules, may elevate the capability of hydrogen adsorption by increasing its available void volume. Zeolites can be purchased with chief price and modified easily by simple chemical reaction giving lots of varieties in their structures.

Organic molecules were adsorbed on A, X and Y-type zeolites, which have been substituted with Li^+, Mn^{2+}, and Ca^{2+} cations by ion exchange and adsorption methods. The estimation of their capability for hydrogen storage was carried out. Also, hydrogen adsorption mechanisms at the active sites have been studied using molecular simulations techniques to identify the organic molecules, which increase the hydrogen torage capability. Molecular dynamics, Grand Canonical Monte-Carlo simulation have been carried out. The measurement of hydrogen storage capacity revealed that the adsorption capacity is 0.75 wt% at room temperature and 20 bar, which is ten times higher than that of the ordinary zeolites and may over 1.2 wt% at room temperature and 100 bar.

In order to develop new type of hydrogen storage material efficiently and systematically, we established the expert system of virtual screening based on computer aided molecular modeling with the accuracy of 80%. This expert system will cover from the downstream to upstream, such as compound database, prediction of adsorption capacity of hydrogen, investigation of charging/discharging kinetics and durability of material during the cycle life. On the other hand, we also investigated the adsorption mechanism of hydrogen. Based on this mechanism study, we could find the crucial factor and functional group to improve adsorption capability. Based on virtual screening technology and mechanism study, the research and development of hydrogen storage material with high capacity will be guided.

Virtual screening technology based on molecular modeling was proven as an efficient tool for the development of new material. It predicts and analyzes the property of material before experiment and guides the direction for research and development. Using this technology, we developed organometallic nanoporous material for hydrogen storage which has high capacity of hydrogen adsorption, excellent property in charging/discharging kinetics and high durability during the cycle life. In order to improve the hydrogen adsorption, the polarization of electrons was increased by the introduction of electron donating and withdrawing group in storage molecules (Fig. 9).

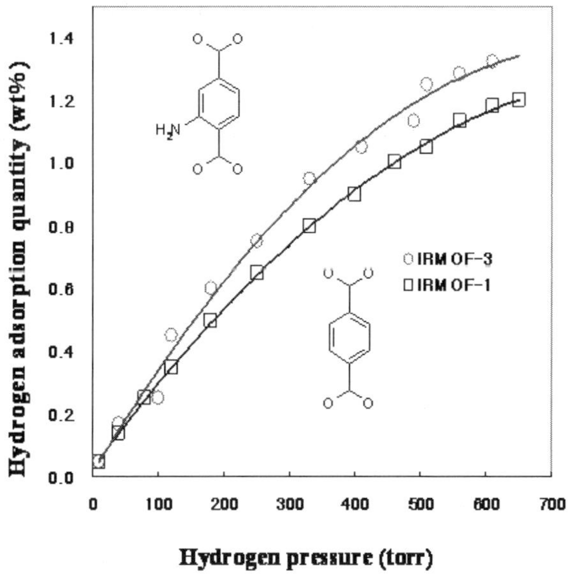

Fig. 9. Hydrogen adsorption isotherms for IRMOF-1 and IRMOF-3 at 78K.

1) Construction of commercially available chemical compound database for the design of organometallic nonoporous material.

2) Quantum mechanical calculation for the investigation of adsorption mechanism of hydrogen.

3) Grand canonical Monte Carlo simulation for the prediction of adsorption capacity of hydrogen.

4) Molecular dynamics simulation for the prediction of hydrogen diffusivity and durability of host material.

5) Quantitative Structure-Uptake Relationship study to estimate the correlation between structure and hydrogen uptakes.

3.7. *Other materials*[26]

Furthermore, a team of scientists, headed by Professor Lee Huen in Korea Advanced Institute of Science and Technology, discovered an affordable way to store hydrogen in ice in April 2005.

Purified water does not have space to embed hydrogen, but Professor Lee found that water combined with organics creates a nano-space to stably store hydrogen at about 0 degrees Celsius, when water turns to ice.

Professor Lee expected his team's new technology of condensing hydrogen inside ice will become the industry standard for hydrogen storage in the future.

3.8. *Measurement techniques for hydrogen storage/release materials*

This study focuses on the development of more accurate measurement technologies for solid-state hydrogen storage/discharge materials. Evaluation techniques for hydrogen storage/release materials belong to the basic technology for securing hydrogen storage technology using solid hydrogen storage materials. Through the development of evaluation techniques for hydrogen storage/release characteristics, the accurate performance evaluation of hydrogen storage material can be acquired and the research period for developing realizable hydrogen storage materials can be shortened. The evaluation technique can be applied for economic feasibility test and insuring confidence for hydrogen storage materials as well. There already exist various technologies and apparatuses for evaluating hydrogen storage materials. None of technologies nor methodologies yet are matured enough to attract general confidence.

With concerns for volumetric hydrogen storage/release characteristics evaluation apparatus, two evaluation units were developed. One (KIER I) is for covering from extremely low to high temperature. The other one (KIER II) is operational only from room temperature to mid-high temperature. KIER II is equipped with compressed gas container, connecting tubes, and sample holder, which consist of same material so that same temperature can be applied for enhanced precision.

Analysis error caused by impurities was minimized by adding sample pre-treatment unit, in which samples are heat treated in an inert gas atmosphere or vacuum status. Characteristics investigation of the evaluation apparatus and its parts has confirmed that the error, caused by contraction of sealing materials for valves, diminished and the evaluation apparatus was controlled precisely within leaking rate of less than 10^{-8} sec^{-1} and temperature standard deviation of less than ±0.0115K.

With concerns for gravimetric hydrogen storage/release characteristics evaluation apparatus, establishment of the gravimetric performance evaluation system was completed. The system consists of raw hydrogen gas pre-treatment unit (continuous 2-beds adsorption purification process unit), hydrogen dosing unit, measurement unit (Magnetic Suspension Balance), data acquisition & control unit, and others.

The system was designed to enable performance evaluation for hydrogen storage materials in the temperature range of -40K ~ 300K and pressure range of 10^{-5} Torr ~ 120 atm. The intrinsic stability of the gravimetric evaluation apparatus is very important factor. However, it was found that buoyancy force correction is a critical factor as well when it comes to hydrogen storage amount calculation from measured data. Therefore, standard procedure for buoyancy force correction was systematically established. Systematizing the gravimetric evaluation apparatus delivered reproducibility within error range of 6.67×10^{-3} wt%, and reliability of the apparatus was acquired through performance evaluation of standard hydrogen storage materials.

Standardizations of measurement technologies of hydrogen storage capacities are applied to various materials to establish an infrastructure in this field.

4. Conclusions and Outlook

In October 2003, The Republic of Korea initiated the systematic research program on hydrogen storage, production, and utilization with launching Hydrogen Energy R&D Center (HERC). The HERC is systematically covering wide range of research programs on hydrogen storage including high pressure hydrogen storage cylinder, various metal hydrides, carbonaceous/non carbonaceous nano materials, and many others. In addition, hydrogen absorption/desorption modeling is covered as well. Detailed research plans for establishment of hydrogen storage system by

2012 is already set up and on its half way now. There are three stages in the master plan of HERC, which are basic R&D, demonstration, and establishment. The main strategy of HERC is upgrading to the next stage by merging or integrating technologies developed in prior stage based on competitively selected portfolio. In this report, the current status of hydrogen storage technologies in the Republic of Korea was reviewed in detail. This report will be used as a milestone for hydrogen storage researches in the Republic of Korea and encouraging international research network formation with other countries.

References

1. S. Satyapal, J. Petrovic, C. Read, G. Thomas and G. Ordaz, *Catalysis Today*, **120**, 246 (2007).
2. D. H. Kang, S. O. Park, C. S. Hong and C. G. Kim, *NDT & E International*, **38**, 712 (2005).
3. D. H. Kang, C. U. Kim and C. G. Kim, *NDT & E International*, **39**, 109 (2005).
4. S. W. Cho, G. Shim, G. S. Choi, C. N. Park, J. H. Yoo and J. Choi, *J. of Alloys and Compounds*, **430**, 136 (2007).
5. S. W. Cho, C. N. Park, J. H. Yoo, J. Choi, J. S. Park, C. Y. Suh and G. Shim, *J. of Alloys and Compounds*, **403**, 262 (2005).
6. K. S. Jung, D. H. Kim, E. Y. Lee and K. S. Lee, *Catalysis Today*, **120**, 270 (2007).
7. C. D. Yim, B. S. You, Y. S. Na and J. S. Bae, *Catalysis Today*, **120**, 276 (2007).
8. M. Y. Song, S. N. Kwon, S. H. Hong, C. G. Park, S. H. Baek, H. S. Kim, D. R. Mumm and J. S. Bae, *Catalysis Today*, **120**, 281 (2007).
9. J. H. Shim, G. J. Lee, B. J. Lee, Y. J. Oh and Y. W. Cho, *Catalysis Today*, **120**, 292 (2007).
10. B. M. Lee, J. W. Jang, J. H. Shim, Y. W. Cho and B. J. Lee, *J. of Alloys and Compounds*, **424,** 370 (2007).
11. H. Jin, Y. S. Lee and I. Hong, *Catalysis Today*, **120**, 399 (2007).
12. J. H. Cho and C. R. Park, *Catalysis Today*, **120**, 407 (2007).
13. S. E. Hong, D. K. Kim, S. M. Jo, D. Y. Kim, B. D. Chin and D. W. Lee, *Catalysis Today*, **120**, 413 (2007).
14. Y. S. Lee, Y. H. Kim, J. S. Hong, J. K. Suh and G. J. Cho, *Catalysis Today*, **120**, 420 (2007).
15. R. Zacharia, K. Y. Kim, S. W. Hwang and K. S. Nahm, *Catalysis Today*, **120**, 426 (2007).
16. S. C. Lee, S. H. Park, S. M. Lee, J. B. Lee and H. J. Kim, *Catalysis Today*, **120**, 358 (2007).
17. J. B. Lee, S. C. Lee, S. M. Lee and H. J. Kim, *Catalysis Today*, **120**, 363 (2007).

18. M. K. Song and K. T. No, *Catalysis Today*, **120**, 374 (2007).

19. S. H. Jhi, *Catalysis Today*, 120, 383 (2007).

20. S. U. Jeong, R. K. Kim, E. A. Cho, H. J. Kim, S. W. Nam, I. H. Oh, S. A. Hong and S. H. Kim, *Journal of Power Sources*, **144**, 129 (2005).

21. K. W. Cho and H. S. Kwon, *Catalysis Today*, **120**, 298 (2007).

22. J. Lee, K. Y. Kong, C. R. Jung, E. Cho, S. P. Yoon, J. Han, T. G. Lee and S. W. Nam, *Catalysis Today*, **120**, 305 (2007).

23. D. Kim, J. Kim, D. H. Jung, T. B. Lee, S. B. Choi, J. H. Yoon, J. Kim, K. Choi and S. H. Choi, *Catalysis Today*, **120**, 317 (2007).

24. J. H. Yoon, S. B. Choi, Y. J. Oh, M. J. Seo, Y. H. Jhon, T. B. Lee, D. Kim, S. H. Choi and J. Kim, *Catalysis Today*, **120**, 324 (2007).

25. T. B. Lee, D. Kim, D. H. Jung, S. B. Choi, J. H. Yoon, J. Kim, K. Choi and S. H. Choi, *Catalysis Today*, **120**, 330 (2007).

26. D.Y. Kim, Y. Park and H. Lee, *Catalysis Today*, **120**, 257 (2007).

CHAPTER 12

DISCOVERING AND DESIGNING BULK METALLIC GLASSES

Srinivasa Ranganathan[a], Tripti Biswas[a] and Anandh Subramaniam[b]

[a]*Department of Materials Engineering*
Indian Institute of Science, Bangalore, 560012, India
[b]*Department of Materials and Metallurgical Engineering*
Indian Institute of Technology, Kanpur, 208016 85, India
E-mail: rangu@materials.iisc.ernet.in

The discovery of metallic glasses and bulk metallic glasses (BMG) were important landmarks in materials research in recent times. Starting with an overview of the techniques for the synthesis of BMG, this paper will critically review the literature on glass forming criteria, glass forming compositions and structural and thermodynamic models describing glasses. With an explosion in the number of compositions forming BMG, their classification becomes important, which will help understand the relationship amongst the various BMG. Following the standard classification scheme based on compositions, Pettifor's approach using Mendeleev number and the importance of bond orbitals in the classification of BMG is highlighted.

1. Introduction

The first metallic glass alloy (Au- 25 at.% Si) was produced by Pol Duwez,[1] while developing rapid quenching techniques for chilling metallic liquids at very high rates of 10^6 K/s. Due to rapid solidification the thickness of the sample was limited to micron dimensions. Chen and Turnbull[2] showed that the samples demonstrated glass transition and thus were true glasses.

The challenge of producing thick metallic glasses was met, when Kui *et al.*[3] produced cm thick bulk metallic glass (BMG) in $Pd_{40}Ni_{40}P_{20}$ alloy. Even though their paper carried the word bulk, it did not cause the excitement that came with the production of metallic glasses in bulk

413

form that came in 1988 with the discovery of Mg-Cu-Y BMG by Inoue *et al.*[4] Intense research has led to the synthesis of many new BMG in different metallic systems. These include La, Zr, Pd, Ca and Mg. Special mention must be made the contribution of Peker and Johnson,[5] who added 22.5 at.% Be in Zr-based BMG and produced a very large sample.

BMG have many unique properties, which can not be obtained in crystalline alloys. The ease of formation of a variety of shapes in the supercooled liquid region makes them more amenable in engineering practice. Most of the glass-forming compositions have been experimentally found. As finding new bulk metallic glass compositions through experiments requires a lot of resources, theoretical approaches are being developed to this end.

Many criteria for BMG formation, established by various scientists will be reviewed. All these BMG-forming criteria are mainly based on: (i) atomic size factor, (ii) electronegativity, (iii) multiple components in the alloy and (iv) orbital effects. The contribution of all these criteria in discovering Mg and Al based bulk metallic glasses will be discussed in detail as illustrative case studies. The relation to eutectic composition given by Turnbull[6] Yavari[7] and pseudo-quaternary double glass formation by Ranganathan[8] will be highlighted.

The progress in increasing thickness with the passage of time has been captured in vivid fashion in a diagram (Fig. 1) due to Loeffler.[9] La-, Mg-, Zr-, Pd- and Ca- based amorphous alloy systems form in bulk with a large thickness up to 100 mm.

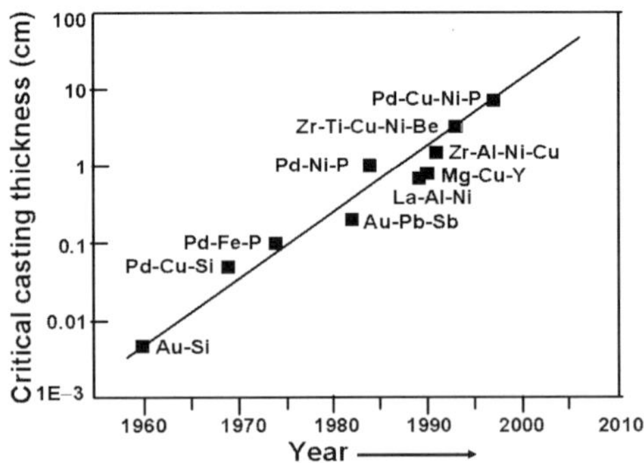

Fig. 1. Critical casting thickness for glass formation as a function of the year the corresponding alloy has been discovered.[9]

A number of books[10,11] reviews[12,16] and several conference proceedings[17-20] in this promising field are already published. The field continues to be exciting with new results reported at frequent intervals. The list of representative BMG is given in Table 1.

Table 1. List of bulk metallic glasses.

Class	Alloy System	Composition	T_{rg}	t_{max} (mm)	R_C (K/s)	Reference
I (ETM/Ln-LTM/BM-Al/Ga)	Zr-LTM-Al Ln-LTM-Al	$Zr_{60}Ni_{20}Al_{20}$	0.64	30	1	Inoue et al.[25]
		$La_{55}Ni_{20}Al_{25}$	0.68	7	10^2	Inoue et al.[21]
		$La_{55}Ni_5Cu_{10}Co_5Al_{25}$	0.70	9	10^2	Inoue et al.[25]
II (ETM/Ln-LTM/BM-Metalloid)	Fe-ETM-LTM-Metalloid	$Fe_{61}Zr_{10}Mo_5W_2Co_7B_{15}$	0.63	6	10^3	Inoue et al.[26]
III (Al/Ga-LTM/BM-Metalloid)	Fe-(Al,Ga)-Metalloid	$Fe_{72}Al_{15}P_{11}C_5B_4Ge_2$		2	10	Inoue et al[27]
IV (IIA-ETM/Ln-LTM/BM)	Mg-Ln-LTM	$Mg_{65}Cu_{25}Y_{10}$	0.59	4		Inoue et al.[28]
		$Mg_{65}Cu_{15}Ag_{10}Y_{10}$		6		Kang et al.[29]
		$Mg_{54}Cu_{26.5}Ag_{8.5}Gd_{11}$		25		Ma et al.[30]
	Zr-IIA-ETM-LTM	$Zr_{41.2}Be_{22.5}Ti_{13.8}Cu_{12.5}Ni_{10}$		25	1-10	Peker and Johnson[5]
V (LTM/BM-Metalloid)	Pd-LTM-Metalloid	$Pd_{40}Ni_{40}P_{20}$		5	1.28	Drehman et al.[16]
		$Pd_{40}Ni_{40}P_{20}$ (fluxed)		100		
		$Pd_{40}Cu_{30}Ni_{10}P_{20}$ (fluxed)	0.71	72	0.1	Kui et al.[3] Inoue et al.[26]
VI (ETM/Ln-LTM/BM)	Ti-LTM-BM	$Ti_{50}Ni_{15}Cu_{32}Sn_3$	0.53	1		Kim et al.[32]
		$Ti_{40}Zr_{25}Ni_8Cu_9Be_{18}$	0.63	5		
VII (IIA-LTM/BM)	Ca-IIA-LTM	$Ca_{57}Mg_{19}Cu_{24}$	0.60	4		Amiya and Inoue[33]

2. Synthesis

Liquid metal can be transformed into glass by avoiding the nucleation of a crystalline phase during cooling, i.e. the cooling curve should not intersect the nose of the TTT curve (Fig. 2).

Liquid metal can be transformed into glass by avoiding the nucleation of a crystalline phase during cooling, i.e. the cooling curve should not intersect the nose of the TTT curve (Fig. 2). Many metallic alloys have been amorphized so far by using rapid solidification techniques. Since the discovery of metallic glass formation by ultra-rapid melt quenching,[1] it was thought that metallic glasses can be processed only as very thin ribbons or fine powders, due to the required high cooling rate. This had severely limited the technological applications of metallic glasses.

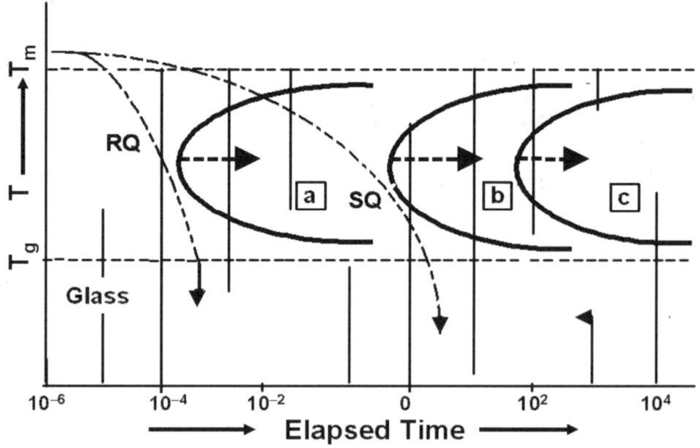

Fig. 2. CCT diagram for glass formation. Critical cooling rate continues to decrease for (a) metallic glasses, (b) bulk metallic glasses and (c) fluxed bulk metallic glasses.[22]

BMGs can be prepared by two different paths: by solidification or by solid state processing. Several techniques have been used for the solidification of the melt that leads to the formation of BMG. The different techniques are: water quenching[21] arc melting, copper mold casting[22] and mechanical alloying.[23] Water quenching is the oldest technique to be used for preparing BMG. In this technique, the alloys were melted by induction melting inside a water cooled quartz tube under a protective atmosphere. The cooling rates depend on the diameter of the tube but in general it varies from 10 to 100 K/s. $La_{55}Al_{25}Ni_{20}$ BMG rods up to 1.6 mm diameter were produced using this technique.[21] Peker and Johnson[5] produced $Zr_{41.2}Ti_{13.8}Cu_{12.5}Ni_{10}Be_{22.5}$ BMG rods up to 14 mm diameter using induction melting on a water cooled silver boat under Ti-gettered argon atmosphere. $Pd_{40}Ni_{10}Cu_{30}P_{20}$ BMG rods with 40 mm diameter were produced using this technique.[24]

It is possible to synthesize glasses using arc melting of the alloys on a copper hearth. Zr-based quaternary and quinary Zr-Al-LTM (Co, Ni, Cu) alloys were synthesized using this technique.[31] Copper surface influences easy heterogeneous nucleation sites and the Zr_2Co, Zr_2Ni, Zr_2Cu phases form at the surface contacted with the copper hearth. Single glassy phase was achieved by re-melting followed by water quenching of the arc melted ingot.

Copper mold casting is one of the important and useful techniques for the production of BMG. This technique has many variants. Even though the final solidification takes place inside a copper mould, the methods differ in the way the melt is fed to it. The various methods to feed the melt to copper mould are: electromagnetic levitation, suction casting and high pressure die casting. The first method consists of electromagnetic levitation of the melt followed by its being dropped into a water cooled copper mould. $Zr_{57}Ti_5Al_{10}Cu_{20}Ni_8$ BMG rods with 10 mm diameter were prepared using this method.[34] Several advantages of this technique include a homogeneous melt and minimum contamination.

Suction casting is another important method to synthesize BMG. In this method, a moving piston generates a pressure difference between the castings and melting chambers, which causes the molten alloy to be drawn into the copper mould. $Zr_{55}Cu_{30}Ni_5Al_{10}$ BMG rods with 30 mm dimension were produced using this technique.[35] This technique offers a high casting rate. Cavities and holes free BMG can be produced using it. This is due to the sucking force generated during the process.

High pressure die casting is another useful method. Inoue *et al.*[36] have produced $Mg_{65}Cu_{25}Y_{10}$ BMG up to 7 mm diameter using this technique and could produce Zr-based BMG using this method. This method is particularly attractive due to its high solidification rate.

While BMG forming processes discussed above belong to solidification route where the liquid is cooled to a glassy solid by suppressing crystallisation, mechanical alloying (MA) belongs to solid state route. In mechanical alloying technique, alloying is achieved through the diffusion of the constituent elements. It can give rise to amorphous powders. The amorphous powders can be consolidated into a bulk form at temperatures above Tg and below Tx, i.e. in the supercooled liquid region.[32] Glass-formation of different alloy systems using this procedure has been reported by Murty and Ranganathan.[37] $Zr_{54.5}Ti_{7.5}Cu_{20}Ni_8Al_{10}$ and $Zr_{59.52}Nb_4Cu_{19.2}Ni_{7.68}Al_{9.6}$ alloys are amorphised by this technique.[38]

3. Structure

Since the glass transition and crystallisation behaviour are connected with the glass structure, investigations of the atomic distribution in the amorphous state is of high importance to the understanding and control of the phase transformations. There are several model-based approaches to understanding the structure of metallic glasses. Bernal's[39] dense random packing (DRP) model describes the atomic arrangement in monatomic liquids. Bernal proposed five different types of polyhedral holes to describe the topology of DRP. They are the tetrahedron, the octahedron, the trigonal prism, the archimedean antiprism, and the tetragonal dodecahedron. The five different polyhedral holes are shown in Fig. 3.

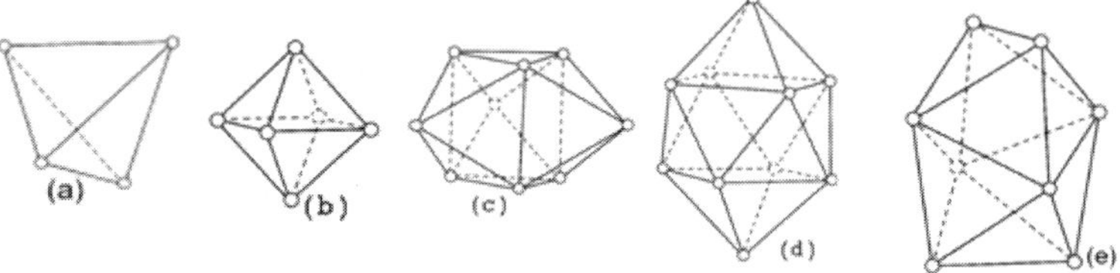

Fig. 3. Idealised holes found by Bernal to describe the topology of DRP: (a) tetrahedron; (b) octahedron; (c) trigonal prism capped with three half-octahedral; (d) Archimedean antiprism capped with two half-octahedral; (e) tetragonal dodecahedron.[39]

Bernal's model applies to monatomic glasses. Therefore, it was desired to construct a model to understand the structure of the binary glasses. Polk[40] proposed a model based approach to understand the structure of metal-metalloid glasses. In his random packing model, the atomic arrangement of metal atoms are similar to Bernal's[39] hard sphere packing and the metalloid atoms occupy the larger holes inherent in such structures. Polk's model is restricted to binary glasses.

A major advance in this field is due to Miracle.[41] His model fulfills the deficiency of understanding of the atomic structure of multicomponent glassy alloys. In the model a glass forming system can contain a maximum of three different solutes. He has considered the fcc and hcp as the cluster packing unit because of their high packing density in comparison with bcc. In this model α is the primary cluster-forming solute species, β is the secondary solute that occupies cluster octahedral interstices and γ is the tertiary solute that occupies cluster tetrahedral interstices (Fig. 4).

Fig. 4. Illustrations of portions of a single cluster unit cell for the dense cluster packing (DRP) model. (a) A 2D representation of a DRP structure in a (100) plane of clusters illustrating the features of interpenetrating clusters and efficient atomic packing around each solute. (b) A portion of a cluster unit cell of a <12-10-9> model system representing a Zr-(Al,Ti)-(Cu,Ni)-Be alloy. The three solutes with progressively smaller atomic size are α, β and γ.[41]

Miracle *et al.*[42] have shown that the cluster configurations do not necessarily form regular polyhedra. Their model shows twenty different efficiently packed clusters (Fig. 5). The above-mentioned models are applicable for short range ordering (SRO) of glassy structure.

The structure of glasses can also contain medium range order (MRO). Sheng *et al.*[43] proposed a model based approach to describe both short range order and medium range order structure of glasses. They have analyseda range of model binary systems which have different chemistry and atomic size ratio. They have described the different types of short range and medium range order.

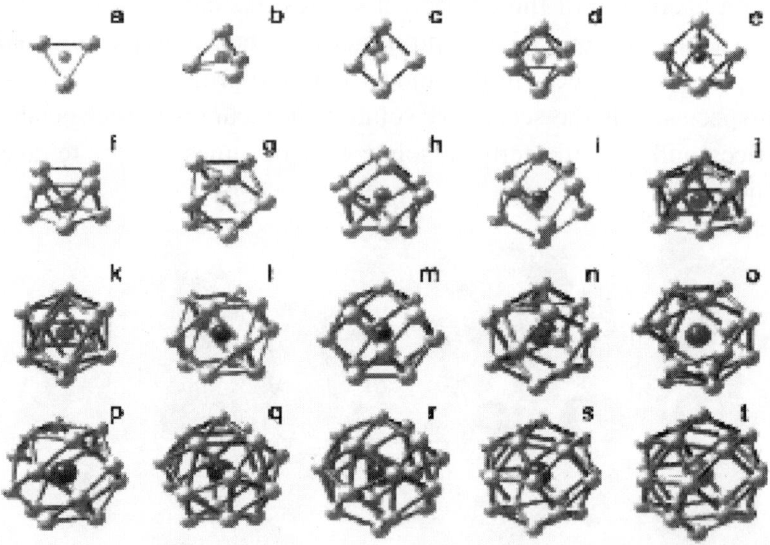

Fig. 5. Efficiently packed clusters with Foppl notations of (a) 3; (b) 1, 3; (c) 1, 3, 1; (d) 1, 4, 1; (e) 1, 3, 3; (f) 4, 4; (g) 3, 3, 3; (h) 2, 4, 2, 2; (i) 1, 4, 4, 1; (j) 1, 5, 5; (k) 1, 5, 5, 1; (l) 4, 4, 4; (m) 1, 4, 4, 4; (n) 1, 4, 2, 2, 4, 1; (o) 3, 3, 3, 3, 3; (p) 4, 4, 4, 4; (q) 1, 4, 2, 2, 4, 2, 2; (r) 1, 4, 4, 4, 4, 1; (s) 1, 3, 3, 6, 5, 1; and (t) 1, 3, 3, 6, 3, 3, 1.[42]

4. Eutectics and Metallic Glasses

It was Cohen and Turnbull[44] who first reported that glass formation occurs near the eutectic compositions. They reported that Duwez's Au-Si glass with 25 at.% Si and its nearby eutectic compositions (31 at.% Si) favour the glass formation.

Dubey and Ramchandrarao[45] have carried out a classic study on the tendency of glass formation in binary alloys with respect to eutectic compositions. They have studied the relationship between glass formation and phase diagram features: slopes of liquidus and solidus lines, extent of solid solubility, existence of intermetallic phases in the light of the driving force for nucleation in a binary melt. In case of binary pure metal eutectics, the glass forming compositions are symmetric with respect to the eutectic compositions. In other cases, the binary eutectics consist of solid solution and an intermetallic phase leading to an asymmetry in the glass forming composition range (GFCR) with respect to the eutectic compositions. In case of asymmetric GFCR, the glass formation prefers towards the side where intermetallic phase is the primary phase.

Tan *et al.*[46] pointed out that the exact eutectic composition may not always be the best for BMG formation. The best glass forming composition can be achieved at the off-eutectic (<1 at.%) composition.[46,47] The maximum thickness of the glass can be achieved only at the off-eutectic composition and in a very narrow composition range. Figure 6 shows the schematic diagram of eutectic coupled zone and its relation to the glass forming ability.[46]

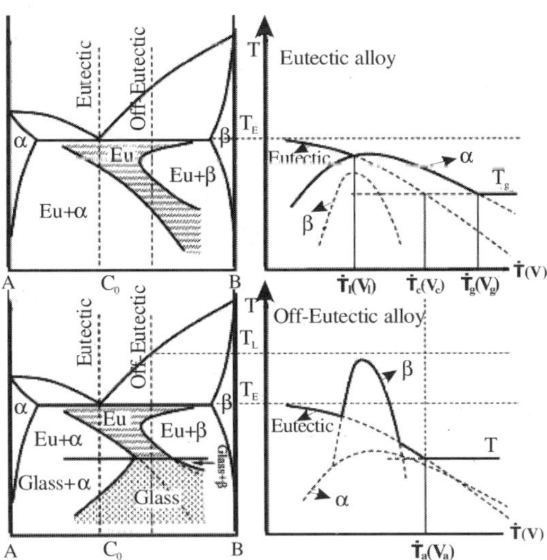

Fig. 6. Schematic diagram showing skewed eutectic coupled zone and its relation to the glass forming ability: (a) a eutectic system with a skewed coupled zone; (b) glass forming and composite forming regions related to the skewed coupled zone; (c) growth temperature of the constituents as a function of cooling rate (growth rate) for the eutectic alloy and (d) growth temperature of the constituents as a function of cooling rate (growth rate) for an off-eutectic alloy.[46]

In some ways, the search for the best BMG composition turns into the search for the eutectic composition. Stockdale[48] has suggested that the compositions of eutectic points correspond with simple whole number ratios of the two kinds of atoms: solvent (A>50 at.%) and solute

(B<50 at.%). At that time, there was insufficient data to confirm this prediction. An insight was gained, when Hume-Rothery and Anderson[49] plotted the frequency of occurrence of binary eutectic compositions versus alloy compositions from Hansen's[50] text book on 'The constitution of binary alloys'. They showed that certain regions of compositions confirm the Stockdale hypothesis. These compositions correspond to A_8B_1, A_5B_1, A_3B_1, A_2B_1 and A_3B_2 (Fig. 7), where A and B are the solvent and solute atoms, respectively.

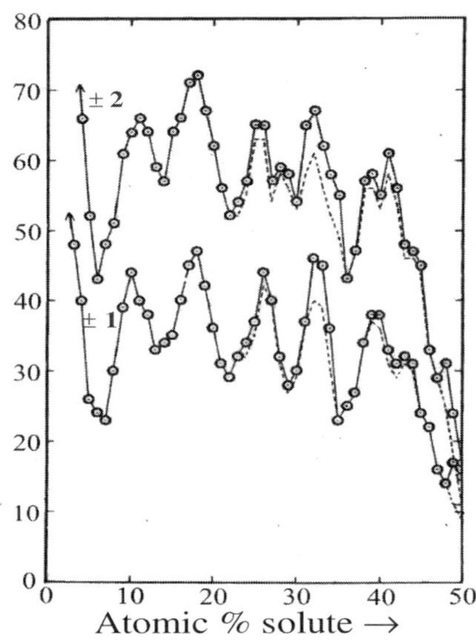

Fig. 7. The frequency of occurrence of binary eutectic compositions versus alloy compositions.[49]

Gilman[51] has plotted eutectic composition distribution curves for the special class of eutectics consisting of transition metals and metalloids: B, C and P to verify Stockdale's suggestion. He has chosen transition metal-metalloid systems because they form many eutectics and these eutectics have a clear tendency to form glasses. He found a clear maximum around eutectic compositions that corresponded to metal/ metalloid atomic ratios 6/1 and 5/1.

Yavari[7] correlated the eutectic compositions with 'Miracle Glasses'. He re-examined the eutectic composition distribution curves plotted by

Hume-Rothery and Anderson.[49] He considered Miracle's fcc cluster packing to correspond to glass composition: $A_6B_1C_1D_2$ (mentioned in Sec. 3) and used three schemes to reach the binary glass forming compositions. The first scheme includes the replacement of C and D atoms with additional B atoms that leads to the composition A_3B_2. In the second scheme, the D site has been filled with additional B atoms and the C site has been left empty, which leads to the composition A_2B_1. The third scheme includes the filling of C site with additional B atoms and leaving D sites empty, leading to the composition A_3B_1. He showed that these compositions correspond to the maxima of Hume-Rothery and Anderson's eutectic composition distribution curves (Fig. 7). He picked up the plot for binary glassy alloys versus their radius ratios RA/RB, and showed that[52] the compositions of maxima for whole numbers correspond to efficient packing of A around B atoms. Therefore, the distribution of experimental eutectic points provided an added support to the Miracle's model for metallic glasses with eutectic and near eutectic compositions.

Apart from binary eutectics, the ternary and quaternary eutectic compositions also lead to the formation of BMG with larger dimensions. The ternary eutectic $Pd_{40}Ni_{40}P_{20}$ alloy gives rise to BMG with 1 cm in diameter.[53] The substitution of Ni with 30 at.% Cu in $Pd_{40}Ni_{40}P_{20}$ alloy leads to the quaternary eutectic $Pd_{40}Ni_{10}Cu_{30}P_{20}$ composition that gives rise to BMG with 40 mm in dimension.[24]

5. Glass Forming Criteria

Several parameters have been used to measure the glass-forming ability (GFA) of metallic alloys. The reduced glass transition temperature (T_{rg}) defined by Turnbull[6] as the ratio between the glass transition (T_g) and liquidus temperatures (T_l) has been successfully used to evaluate the GFA of various metallic glasses. The interval between glass transition temperature (T_g) and crystallisation onset temperature (T_x) (i.e., $\Delta T_x = T_x - T_g$), referred to as the supercooled liquid region, is one of the important parameters for determining the GFA of glass-forming alloys.[12] T_g, T_x and T_l values can be obtained from DSC plots (Fig. 8). Lu and Liu[54] have proposed another parameter to determine the GFA, known as

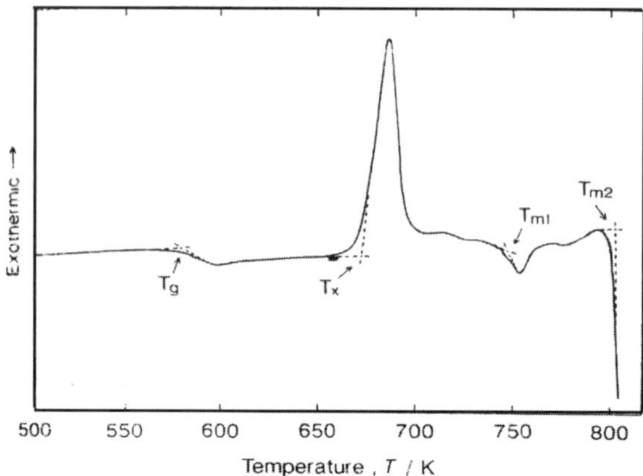

Fig. 8. DSC curve of the cylindrical $Pd_{40}Ni_{10}Cu_{30}P_{20}$ glassy alloy with a diameter of 40 mm. The glass transition temperature (T_g), crystallisation onset temperature (T_x) and the liquid onset temperature (T_{m1} and T_{m2}) can be obtained from this plot.[58]

the γ parameter ($\gamma = T_x/(T_g + T_l)$). This is obtained by the simple additive assumption of a devitrification tendency of a glass and suppression of crystallization during solidification, to predict the GFA for various glass forming systems. Chen *et al.*[55] defined a new parameter $\delta = T_x/(T_l - T_g)$ to predict the GFA of various glass forming systems. This parameter δ has been derived from the classical nucleation and growth theory. These criteria are based on experimental observations.

Ramchandrarao[56] and Egami and Waseda[57] have shown that the atomic size difference between constituent elements should be greater than 10% to form a glassy phase. The other two inverse Hume-Rothery criteria are: a) if electronegativity difference is large between the solute and the host, then it is more likely to form a compound, its solubility in the host will be limited and b) a valency difference between elements is more likely to form intermetallics, favouring glass formation. Therefore, a violation of Hume Rothery's rules for solid solution formation is in favour of glass formation.

Inoue's[12] well-known three empirical criteria of bulk glass formation are: (1) multicomponent alloy systems consisting of more than three elements, (2) significant difference in atomic size ratios between constituent elements (>12%), and (3) high negative heat of mixing among the constituent elements. These criteria were deduced from a number of experimental observations and theoretical calculations. Inoue's second empirical rule can be correlated with Hume-Rothery's first rule for solid solution formation. The inverse of Hume-Rothery's first rule results in Inoue's second criterion for bulk metallic glass formation. An 'inverse Hume-Rothery' criterion states that the size of atoms present must differ by at least 12-15%.[12]

Senkov and Miracle[59] have studied the effect of atomic size distribution on glass forming ability of various amorphous and BMG alloys. They have studied the atomic size distribution for both ordinary amorphous alloys as well as bulk amorphous alloys. The composition of glass forming alloys is used to get information about the most important topological parameters: the relative atomic sizes of the atoms present and the relative numbers of atoms. Using this information, one can plot atomic size distributions (composition versus atomic radius) (Figs. 9, 10) for a number of alloys with different GFA, and can distinguish between marginal glasses and bulk metallic glasses.

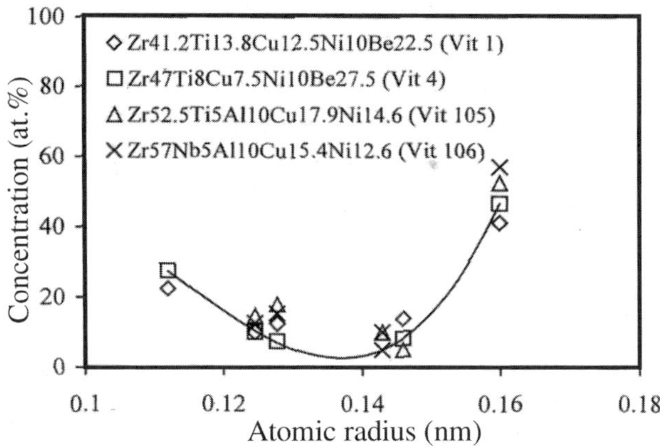

Fig. 9. Atomic size distribution of several Zr-based bulk metallic glasses.[59]

The atomic size distribution plots for marginal glass forming alloys show a concave downward curve with a single peak; a concave upward shape with a single-peak is the signature of bulk glass forming alloys. The general trend shows that the marginal glass forming systems contain at least one element with smaller atomic radius than that of the base

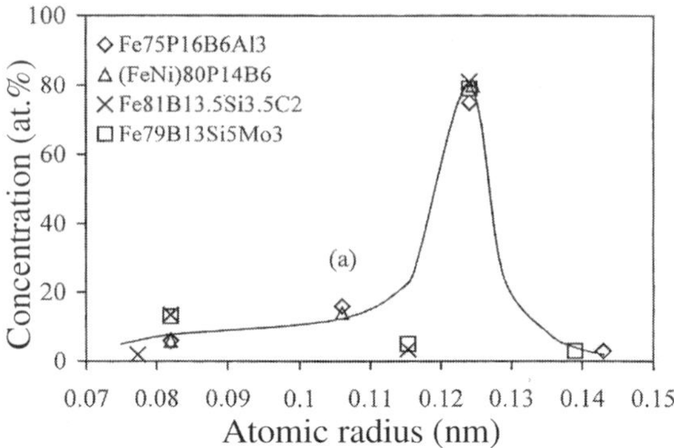

Fig. 10. Atomic size distribution of several Fe-based marginal metallic glasses.[59]

element, and at least one element with larger atomic radius. The concentration of the alloying element decreases with an increase in the differences in atomic size between alloying element and base element. In the case of a bulk glass forming system, the atomic radius of the base element is greater than that of the other alloying elements present in the system.

Mondal and Murty[60] have proposed another GFA parameter called α (Tx/Tl) which can be used for glass forming alloys those do not show any glass transition. They have shown that this new parameter has very good correlation with critical cooling rate for glass formation.

6. Multicomponent Alloys

Alloying is a powerful technique to improve the properties of engineering materials. Alloys containing a number of elements i.e. multicomponent alloys open up numerous possibilities.[61] It is possible to obtain a variety of structure types: bulk metallic glasses, multifunctional, superplastic, superelastic alloys (gum alloys), and nanostructured high entropy alloys[8] using multicomponent alloys.

In a study of intermetallics, Mackay[62] has found an interesting relationship between the number of intermetallics and the number of component elements present in the system. He plotted the frequency of occurrence of intermetallics as a function of number of component. An updated version is shown in Fig. 11.

The frequency of occurrence of intermetallics increases from two to three component system. The frequency decreases with a further increase of the number of components.

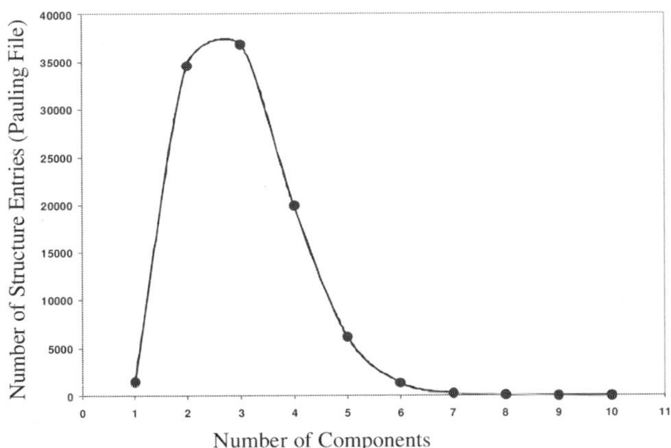

Fig. 11. The frequency of occurrence of structures as a function of number of components.

Inoue's first criterion for BMG-formation: multicomponent can be correlated with Greer's 'confusion principle'.[63] This principle states that, if a system contains many elements with different atomic sizes, then, during rapid cooling from the melt to a solid state, a random structure results.

Multicomponent alloys can be broadly categorised into two types with respect to elemental composition present in the alloy. They are: multicomponent alloy based on a single element whose concentration exceeds 50 at.% i.e. the conventional alloys. The second category of multicomponent alloys are those in which elements are present in substantial proportions. In the first type, the selection of base element depends on the primary property requirement, and other elements are added to improve the secondary properties. The second type of multicomponent alloys can be called as high entropy alloys (HEAs) due to their higher mixing entropies compared to conventional alloys.[61] High entropy alloys consist of five or more major elements, in which the concentration of each element does not exceed 50 at.%. HEAs could be designed based on equal-mole and non-equal mole ratios.

Several effects of multi-element on microstructure include high entropy effect, lattice distortion effect, sluggish diffusion effect and cocktail effect. It was expected that the addition of many elements into an alloy system will result in complicated microstructure. It has been found experimentally that the high mixing entropy of a multicomponent system enhances simple multiphase solid solution or even a single phase instead of the formation of complicated intermetallics. The high mixing entropy of the system prevents the phase separation into terminal solid solution phases or complicated intermetallics.[64] Many elements in a system could result an extended crystal structure of bcc or fcc due to lattice distortion effect.[65] Lattice distortion takes place because all the atoms are solute atoms with different atomic size. Multi-element systems containing elements with different atomic size could give rise to BMG due to combined effect of lattice distortion and slow diffusion rate.[65]

Eskandarany and Inoue[66] were able to prepare multicomponent $Ti_{60}Al_{15}Cu_{10}W_{10}Ni_5$ glassy alloy powder which exhibits the glass transition temperature at 733 K using low energy ball milling technique. The average particle size of 0.38 mm has been achieved using this technique. The binary to hexanary high entropy alloys in Al-Fe-Ti-Cr-Zn-Cu with excellent homogeneity were synthesized by Varalakshmi *et al.*[67] using mechanical alloying technique. The crystallite size of 10 nm was achieved using this technique.

7. Classification of Bulk Metallic Glasses

Glass forming systems can be broadly classified into two categories: metal-metal and metal-metalloid. They can also be classified as ferrous and nonferrous BMG. These classifications are based on the constituent elements present in the system. These are convenient classifications but do not give any insight into their structure and stoichiometry.

Though it was not explicitly recognised at that time, Inoue[12] classified ternary BMG into five classes with each glass drawing its components from five groups of elements. This was later refined by Takeuchi and Inoue[68] into seven classes with a slight change in the grouping of elements (Fig. 12).

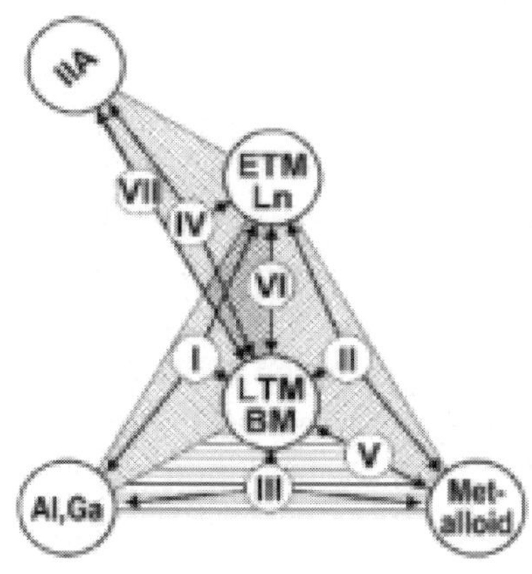

I	Zr-Al-Ni,Ln-Al-Ni
	Zr-Al-Cu,Ln-Al-Cu
	Zr-Al-Ni-Cu,Ln-Al-Ni-Cu
	Zr-Ti-Al-Ni-Cu
	Zr-Ga-Ni,Ln-Ga-Ni,Ln-Ga-Cu
II	Fe-Zr-B,Fe-Hf-B
	Fe-Zr-Hf-B
	Fe-Co-Ln-B
	Co-Zr-Nb-B
	Co-Fe-Ta-B
III	Fe-(Al,Ga)-Metalloid
IV	Mg-Ln-Ni,Mg-Ln-Cu
	Zr-Ti-Be-Ni-Cu
	Ti-Cu-Ni-Sn-Be-Zr
V	Pd-Ni-P
	Pd-Cu-Ni-P
	Pt-Ni-P
VI	Cu-Zr-Ti
	Ni-Nb-Ta,Ni-Nb-Sn
	Ti-Zr-Cu-Ni
	Ti-Ni-Cu-Sn
	Ti-Cu-Ni-Mo-Fe
VII	Ca-Mg-Cu
	Ca-Mg-Zn

IIA: Alkaline Metal
ETM: Early transition metal (IIIB-VIIB)
Ln: Lanthanide metal
LTM: late transition metal (VIII-IIB)
BM: IIIA-IVA metal (In, Sn, Tl, Pb)

Fig. 12. Seven groups of BMG consisting of groups of elements: ETM/Ln, LTM/BM, Al, Ga, IIA and metalloid (modified from previous classification by Inoue[12]).[68]

The elements are recognized as early transition metal (ETM)/ Lanthanides (Ln), late transition metal (LTM), (Al,Ga), (Mg-Be). The first class consists of (ETM) or (Ln), Al, and (LTM). The examples of first class are Zr-Al-Ni and Ln-Al-Ni. The second class consists of LTM, ETM and metalloid (e.g. Fe-Zr-B, Co-Nb-B). The third group contains LTM, (Al,Ga) and metalloid systems (e.g. Fe-(Al,Ga)-B). The fourth group is composed of Mg-Ln-LTM and ETM (Zr,Ti)-Be-LTM systems. The fifth group consists of LTM and metalloid (e.g. Pd-Cu-Ni-P, Pd-Ni-P). It appears convenient to divide class I into IA and IB based on Zr and Ln. A brief discussion on these seven classes is given below.

Three varieties of atoms are required to form seven groups of BMG.[68] Definite stoichiometric composition for each group favours the glass formation. The definite stoichiometry for each group leads to a crystalline approximant for glasses like rational approximant for quasicrystals. Therefore, the multicomponent glasses can be considered as pseudo-ternary.[69]

La is one of the important basic elements to form BMG. Excellent glass forming ability of La-(Al,Ga)-LTM alloys can be achieved in La-rich compositions. Inoue *et al.*[21] reported the first La-based BMG with composition $La_{55}Al_{25}Ni_{20}$. Partial substitution of Ni with other two LTM-Cu and Co, gives rise to 9 mm thick $La_{55}Al_{25}Cu_{10}Ni_5Co_5$ BMG70 with wide ΔTx (of 98 K) and a high Trg of 0.70. Y based Y-Al-LTM (Co) system with composition $Y_{56}Al_{24}Co_{20}$ gives rise to a BMG with 1.5 mm dimension, ΔTx of 45 K and a Trg of 0.61. Substitution of Y with 20 at.% Sc for $Y_{56}Al_{24}Co_{20}$ alloy results 2.5 cm diameter BMG. The addition of 20 at.% Sc to this system results higher Trg of 0.66 compared to that of $Y_{56}Al_{24}Co_{20}$ alloy and ΔTx of 115 K, which is the largest value reported for metallic glasses.[71]

Zr-based BMG are one of the most widely studied systems. Zr, Ti and Hf are in the same group (ETMs). They can be alloyed with LTM such as Ni or Cu.[72] Addition of Al to these systems (ETM-LTM) results better glass forming ability. Addition of Be (very small size atom with 1.11 Å radius) to this system improves the glass-forming ability. $Zr_{41.2}Ti_{13.8}Cu_{12.5}Ni_{10.0}Be_{22.5}$ alloy gives rise to BMG with an exceptional glass forming ability.[5]

Mg is another important basic element for bulk glass formation. Mg-Ln-LTM forms bulk metallic glasses. Mg65Cu25Y10 alloy, a well-known BMG, is particularly important for its high GFA.[28] Substitution of Cu with LTMs (Ag and Pd) leads to further increase in glass forming ability. The simultaneous addition of Ag and Pd element is very effective for the improvement of glass forming ability of $Mg_{65}Cu_{15}Ag_5Pd_5$ BMG.[33] Mg-based glass forming systems are discussed in detail in Sec. 11.

Pd-based Pd-LTM-Metalloid system is another important BMG former. Drehman *et al.*[31] showed that $Pd_{40}Ni_{40}P_{20}$ alloy gives rise to 5 mm size metallic glass. Kui *et al.*[3] produced the first BMG with 1 cm dimension for the same composition, by cooling it's melt in B_2O_3 fluxes. Inoue *et al.*[24] showed that partial substitution of Ni with Cu leading to $Pd_{40}Cu_{30}Ni_{10}P_{20}$ gives rise to 40 mm diameter BMG at a cooling rate of 1.57 K/s. Fluxing led to a still smaller cooling rate.[73] This alloy crystallised with one peak that indicates proximity to eutectic composition. Nishiyama and Inoue[74] have modified the composition of this alloy to get exact quaternary eutectic which can be solidified at the slowest possible rate.

Fe-based BMG can be formed in three varieties of Fe-based multi-component systems. The alloy systems comprise Fe-(Al,Ga)-metalloids,[27] Fe-ETM (Zr,Hf,Nb)-B[26] and Fe-Co-Ln-B.[11] Most of the ferrous glasses were mainly developed for their magnetic properties whereas Fe-Al-metalloid type glasses were produced for non-ferromagnetic applications. The other ferrous glass systems comprise (Co,Ni)-ETM (Zr,Hf,Nb)-B.[26] Fe-based multicomponent Fe-Mn-Mo-Cr-C-B leads to glass dimension with 4 mm diameter. The Trg for these alloys vary between 0.60-0.61. A minor addition of Y or Ln improves the glass forming ability. Ponnambalam *et al.*[75] showed that it is possible to produce 12 mm diameter BMG rods with a small addition of Ln to this multicomponent alloys though the Trg reduces to 0.58. Lu *et al.*[76] showed that addition of Y makes it possible to produce bulk amorphous steel with 12 mm size by drop casting technique. The high glass forming ability of this alloy is attributed to its proximity to a deep eutectic.

Al is an essential element in many glass-forming systems. It is important to note that search for Al based metallic glasses led to the discovery of BMG. Al-based binary Al-Ge,[77] Al-Ni[78] and Al-Pd[79] glasses were synthesized by Indian researchers but was not pursued further. The details of Al-based glass forming systems are described in Sec. 10.

8. Thermodynamic Approach

There are many physical metallurgical factors including Inoue's[80] empirical rules for the glass formation (multicomponent, atomic size mismatch and negative heat of mixing), that have been used in conjunction with proximity of the composition to the eutectic composition, to estimate the GFA of alloys. The principal thermal parameters which have been developed to estimate the GFA are: reduced glass transition temperature ($T_{rg} = T_g/T_l$) and supercooled liquid region ($\Delta T_x = T_x - T_g$). To estimate the value of these parameters, the value of glass transition temperature (T_g), the crystallisation onset temperature (T_x) and the liquid onset temperature (T_l) need to be known. These are measured from experiments and hence T_g and T_x are not predictable quantities. It is important to note that none of the above mentioned factors are consistently applicable to a wide range of glass forming alloys.

The CALPHAD method and Miedema model[81] can be used to estimate the glass forming composition range for any metallic system. Saunders and Miodownik[82] and Bormann[83] have used the CALPHAD approach to evaluate phase diagrams and to determine thermodynamic quantities. It is possible to calculate the glass forming composition range (GFCR) by extrapolating the properties of liquid phase to lower temperature.[82] They have calculated GFCR without taking into account the glass transition. Palumbo *et al.*[84] calculated GFCR using the CALPHAD approach, where they have also considered the glass transition for obtaining a better result. Bormann[83] showed that the thermodynamic functions of the undercooled liquid are strongly influenced by the excess specific heat, which reflects the chemical and topological ordering of the liquid on cooling. This method is mainly applicable for the binary transition metal systems with a large heat of mixing between constituent elements. Shao[85] has proposed a model based approach for the prediction of GFCR considering glass transition temperature as a second order phase transformation.

Midema's model[86] could be used to calculate the GFCR for binary metallic systems. This model based approach has been extensively used by Nagarajan and Ranganathan,[87] Murty,[88] Basu[89] and Takeuchi and Inoue[90] to evaluate the GFCR in various alloy systems. Murty *et al.*[91] have calculated the GFCR for binary Ti-Ni, Ti-Cu and ternary Ti-Cu-Ni systems. Extended Miedema model has been used to calculate the GFCR for ternary alloy systems.[92,88]

It is possible to calculate the mixing enthalpy and mismatch entropy of any binary or multinary alloy systems, to quantify Inoue's empirical criteria of BMG formation.[93] Takeuchi and Inoue93 have proposed a model-based approach (shown in Fig. 13) for the estimation of the GFA by deducing thermodynamical functions from Inoue's well-known empirical criteria of bulk glass formation.

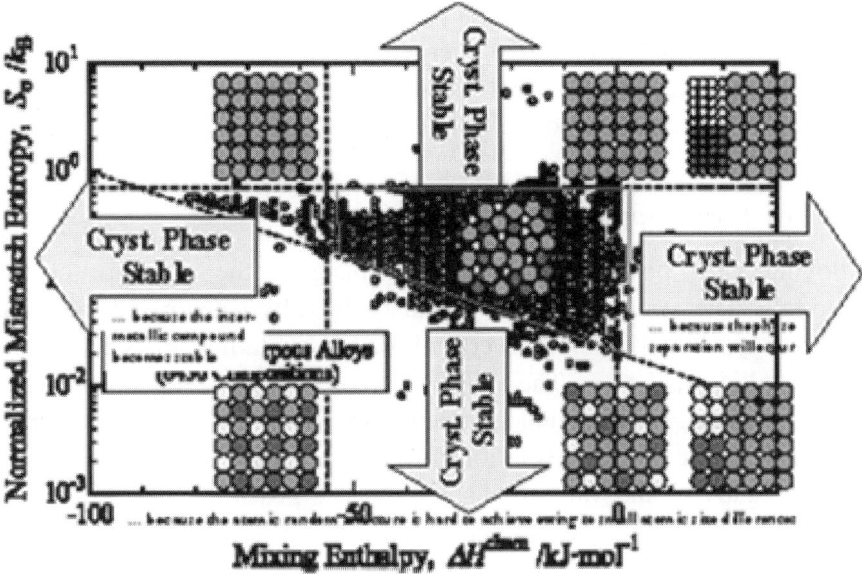

Fig. 13. Mixing enthalpy versus normalized mismatch entropy plot for BMG compositions.[93]

They calculated the mismatch entropy (Sσ), arising from the atomic size difference between constituent elements, which gives a thermodynamic measure of the second empirical criterion for bulk glass formation. The enthalpy of mixing (ΔHmix) has also been calculated by using this model. The enthalpy of mixing is a quantitative measure of the third empirical criterion of bulk glass formation. In this model, the mismatch entropy and mixing enthalpy were defined as functions of composition in a multicomponent system satisfying the first empirical criterion for bulk glass formation. Takeuchi and Inoue[21] have calculated the mismatch entropy and mixing enthalpy for 6450 alloys in 351 ternary amorphous systems. According to their calculations, the values of mixing enthalpy for glass forming compositions lies within −55 and 0 kJ/mol. The normalized mismatch entropy for glass forming alloys can vary between 0.12 to 0.99.

Rao *et al.*[94] calculated the free energy change from solid solution to amorphous transformation for Zr-based metallic glasses using Miedema's model,[95] Miracle[96] and Monsoori's[97] model. They have shown that the reduced glass transition temperature of this system has strong correlation with the free energy change during the process. Bhatt *et al.*[98] have calculated the best glass forming composition for Zr-Cu-Al system using Gallego's model.[92]

9.   Structure Mapping

Alloying is a powerful method for changing the structure of the materials. The structure of alloys can be broadly categorised into four different types: solid solution, crystalline intermetallics, quasicrystalline intermetallics, and glass. The ability to predict the structure of a material is one of the fundamental problems in materials research. Hume-Rothery's rules for solid solution formation include three factors: size, electronegativity and valence. Therefore, it is possible to predict extent of solid solutions by considering these three factors. Pettifor[99] has pioneered the structure of binary crystalline intermetallics. He has added a fourth factor of bond orbitals along with size, electronegativity and valence and created a chemical scale. In this chemical scale, each element in the periodic table is assigned by a unique number called Mendeleev number (Fig. 14).

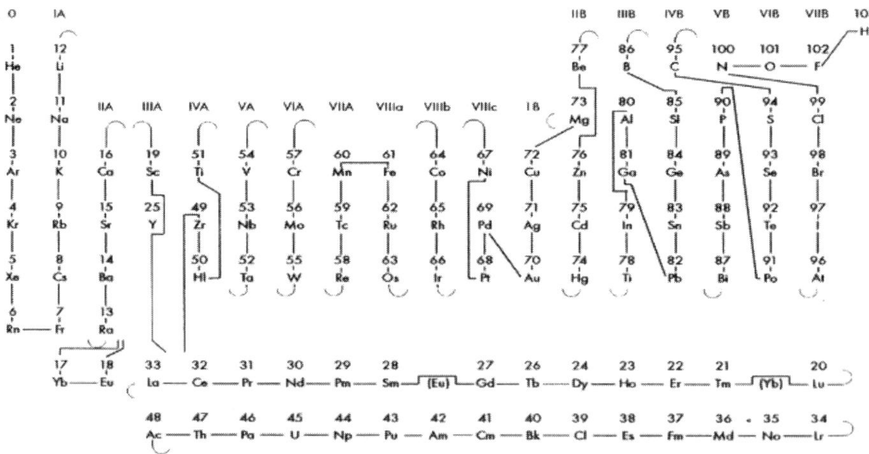

Fig. 14. String running through this modified periodic table puts all elements in sequential order, according to Mendeleev number: note that group IIA elements Be and Mg have been grouped with group IIB, divalent rare-earths have been separated from trivalent, and Y has been slotted between Tb and Dy.[100]

Pettifor[99] first introduced this number based on the phenomenological optimization of the separation of binary AB compounds into different structure types. Therefore, elements which are neighbors in their atomic numbers could have very different Mendeleev numbers and elements with different atomic numbers could be neighbors in their Mendeleev numbers. This single number can be used to differentiate various structures. Figure 15 shows the AB_2 structure map.[100] The boundaries are drawn to separate compounds of different structure type. The bare patches show the region of positive heats of formation, correspond to binary nonformer.

Pettifor[100] has also shown that the structure mapping approach can be used to predict the structure of ternary and quaternary compounds by treating them as pseudo-binaries. He has used the method of calculating average Mendeleev number for this prediction. If A_xB_y is considered as the binary alloy and ternary and quaternary additions C and D preferentially go the A and B sites, respectively, then the alloy can be treated as a pseudo-binary alloy $(A_xC_{1-x})(B_yD_{1-y})$. The pseudo-binary alloy is characterised by the average Mendeleev numbers M*A and M*B are given by M*A = xMA + (1-x)Mc and M*B = yMB + (1-y)MD. This simple scheme shows that the structural domain of pure AB and AB3 binaries are similar to that of the pseudo-binaries.

Nearly twenty years after the advocacy of this structure mapping approach, Villars *et al.*[101,102] used Mendeleev number for the prediction of compound former/nonformer in any binary, ternary and quaternary systems. They reported that the Mendeleev number is a highly effective feature compared to atomic number for this structure prediction method.

Jeevan and Ranganathan[103] used the Pettifor pseudo-binary structure mapping approach for the classification of quasicrystals. They have shown that AB_6 and Al_3B_5 compounds those are related to quasicrystals, are actually the extension of the AB and AB_3 compounds. Ranganathan and Inoue[104] used this approach for the identification of pseudo-binary quasicrystalline intermetallics. They have shown that Pettifor's average Mendeleev number scheme is also applicable for A_2B_3, A_5B_2, AB_2 and AB_6 stoichiometry and related quasicrystals.

Takeuchi and Inoue[68] were the first to use this approach for the classification of BMG. They have adopted Pettifor's pseudo binary

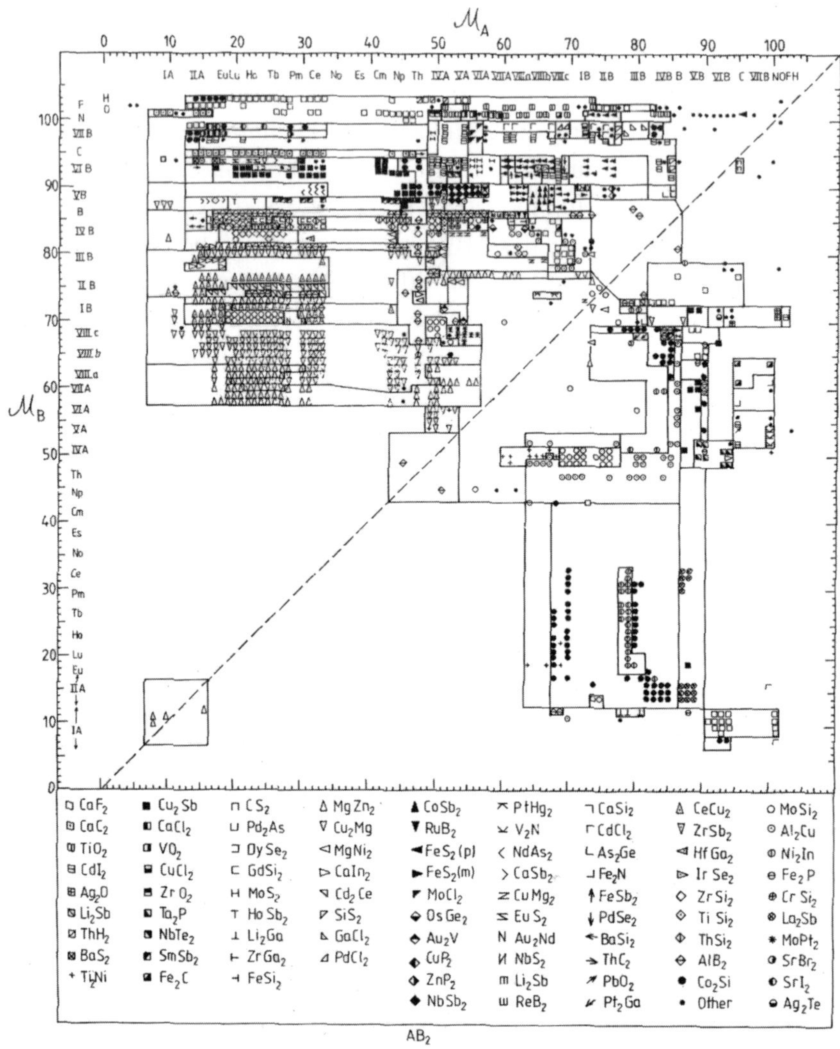

Fig. 15. AB_2 structure maps using Mendeleev number.[100]

structure mapping approach to analyze the characteristic of atomic pairs in ferrous BMG.

They have adopted Pettifor's pseudo binary structure mapping approach to analyze the characteristic of atomic pairs in ferrous BMG. They have described the multicomponent BMG forming systems as the sum of pseudo binary systems. For an example, a ternary $A_aB_bC_c$ (A, B and C are the constituent elements and a, b and c are the composition in at.%) BMG can be described as the sum of three pseudo binaries: Aa/(a+b)Bb/(a+b), Bb/(b+c)Cc/(b+c), and Cc/(c+a)Aa/(c+a). Multicomponent BMG systems also can be described as the sum of many pseudo binaries.

Ranganathan[8] has suggested a new idea for choosing the systems with better GFA. The metallic elements can be categorised in four different groups depending upon their outer electron shell configuration. They are *s*, *p*, *d* and *f* block elements. (*s/p*)-*d*-*f* is the common combination of glass forming systems. Figure 16 shows the plot of atomic radius versus Mendeleev number.[104] The general trend shows that elements far away from each other in this diagram favour glass formation.

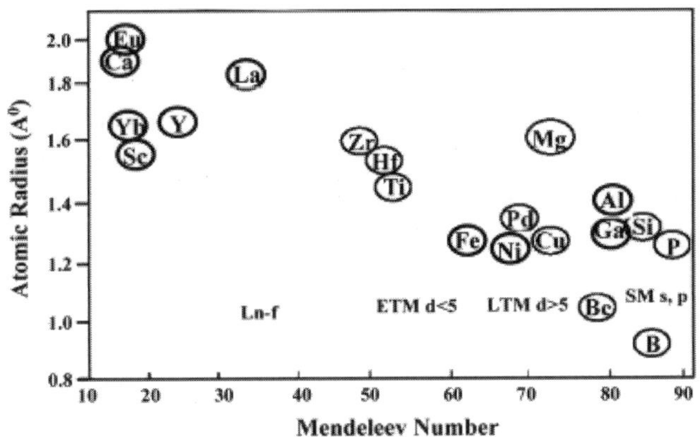

Fig. 16. Relative position of Al, TM and RE with respect to Mendeleev number and atomic size. It is seen that atomic size difference with RE elements is very high and contributes to glass formation.[104]

Takeuchi *et al.*[105] explicitly analysed BMG with a tetrahedron composition diagram. It is easy to find out all possible multicomponent systems and the bonding nature among the constituent elements. Takeuchi and Inoue's[68] seven classes of BMG could be related to four constituent classes: *s*- (*s* block element), *d*E*f*- (E: early transition metal and *f* block element), *d*L*p*- (L: late transition metal and *p* block element) and *p*- (metalloid) depending up on their orbital electrons.

10. Al-Based Glass Forming Systems

The preliminary study of Al based metallic glasses were started by Indian investigators. Al-based binary Al-Ge, Al-Ni and Al-Pd glasses were synthesized by Ramchandrarao et al.,[77] Chattopadhyay et al.[78] and Sastry et al.,[79] respectively, but have not been pursued in vigorous fashion. The development of Al-based metallic glasses has actually been improved due to superlative contribution of Inoue et al.[72] The first single phase amorphous structure has been achieved in Al (50 at.%)-Fe-B and Al-Co-B ternary alloys.[106] Subsequently, glass formation in melt spun Al-Fe-Si, Al-Fe-Ge and Al-Mn-Si alloys have been discovered by Suzuki et al.[107] and Inoue et al.[108] Glass formation in Al-rich alloys consisting of Al-ETM-LTM (Al-Zr-Cu, Al-Zr-Ni and Al-Nb-Ni) has been found by Tsai et al.[109,110] Several other ternary amorphous alloys with 50 at.% Al have been discovered. All above alloys are extremely brittle in nature. An amorphous Al-Ni-Si and Al-Ni-Ge alloys with comparatively good bending ductility was discovered by Inoue et al.[111] These alloys contain above about 80 at.% Al. Sufficient amount of ductility of these alloys can be achieved due to presence of high Al concentration. The maximum thickness of 780 μm has been achieved in $Al_{86}La_5Ni_9$ alloy.[112] Table 2 shows the list of Al-based glasses.

Table 2. Al-based metal-metal glass forming compositions.

Type	System	Composition
Binary	Al-AEM	$Al_{93}Ba_7$, $Al_{93}Ca_7$
	Al-RE	$Al_{92}La_8$, $Al_{92}Ce_8$, $Al_{91}Y_9$, $Al_{90}La_{10}$, $Al_{92}Sm_8$
Ternary	Al-RE-LTM	$Al_{90}Ce_{10-x}Fe_x$, $Al_{90}Ce_7Co_3$, $Al_{86}La_5Ni_9$, $Al_{88}Y_7Fe_5$, $Al_{90}Nd_5Ni_5$, $Al_{88}Y_8Ni_4$, $Al_{88}Y_7Fe_5$, $Al_{90}Nd_5Fe_5$, $Al_{89}La_6Ni_5$, $Al_{87}Ce_3Ni_{10}$, $Al_{85}Y_{10}Ni_5$, $Al_{75}Y_4Ni_{21}$, $Al_{70}Si_{17}Fe_{13}$, $Al_{85}Ce_5Mg_{10}$, $Al_{84}Ni_{12}Zr_4$, $Al_{89}Fe_{10}Zr_1$,
Quaternary	Al-RE-ETM-LTM	$Al_{85}Nd_8Ni_5Co_2$, $Al_{88}Ce_2Ni_9Fe_1$, $Al_{88}Nd_5Ni_6Fe_1$, $Al_{87}Nd_3Ni_7Cu_3$, $Al_{85}Ce_5Ni_8Co_2$, $Al_{83.5}Y_5Ni_{8.5}(Co,Fe)_3$
	Al-ETM-LTM	$Al_{84.5}Y_{8.5}Ni_5Co_2$ $Al_{84}Ni_{12-x}Zr_4Cu_x$, $Al_{84}Ni_{12-x}Zr_4Co_x$
Quinary	Al-RE-ETM-LTM-LTM	$Al_{85}Y_4Nb_4Ni_5Co_2$, $Al_{85}Y_6Ni_5Co_2Zr_2$ $Al_{84}Y_6Ni_4Co_2Sc_4$, $Al_{85}Y_6Ni_5Co_2Fe_2$ $Al_{85}Y_6Ni_4Co_3Fe_2$, $Al_{85}Y_8Ni_3Co_2Fe_2$ $Al_{85}Y_6Ni_3Co_3Fe_3$, $Al_{84}Y_3Ni_8Co_4Zr_1$ AlYNiCoPd, AlYNiCoCu

Two important characteristics of Al-based glasses are that the alloys contain more than 80 at.% of the base element i.e. Al and do not have a deep eutectic, which is an important parameter for glass formation. Instead, formation of Al-based metallic glasses is driven by the three factors: multicomponent combination of constituent elements, a large atomic size differences among the constituent elements (>12 at.%), and a negative heat of mixing among constituent elements. Therefore, Inoue's empirical rules are very much applicable for Al-based glasses. Al-rich alloys, addition of rare earth (RE) and transition metals (TM) leads to the formation of glasses. Typical range for Al-TM-RE glass forming alloys are 80-90 at.% Al, 3-20 at.% RE and 1-15 at.% TM.[19] A limited glass formation is also possible in binary Al-RE and Al-TM alloys.

It is possible to obtain glasses in Al-Y system with 8-12 at.% Y.[19] Al-rich ternary Al-LTM-RE metallic glasses were found to be formed in Al-(Fe, Co, Ni and Cu)-(Y, La) alloys.[113] It is interesting to note that like all other glasses, Al based glasses also formed at slightly off-eutectic composition.[19] Complete substitution of Ln with alkali metals lower the glass forming ability of Al-LTM-Ln-alloys.[114] This indicates the importance of lanthanides in glass formation of Al-rich alloys.

11. Mg-Based Glass Forming Systems

The experimental evidences show the formation of a variety of interesting metastable quasicrystalline, nanocrystalline, nano-quasicrystalline and glassy phases of Mg-based alloys. Many of these phases can made to co-exist by suitable processing leading to an interesting combination of properties. The first bulk metallic glass was found in the Mg-Cu-Y system with composition $Mg_{65}Cu_{25}Y_{10}$.[115] The thickness of this BMG was 4 mm. Since this discovery a great deal of research work on Mg-based glasses took place throughout the years in various parts of the world. Mg-Cu-Y alloy have a significant supercooled liquid regime, over a wide composition range. Inoue *et al.*[116] have shown that $Mg_{65}Cu_{25}Y_{10}$ alloy has the maximum supercooled liquid region of 69 K among all the Mg-based metallic glasses. Mg-Ln-TM alloys possess high glass forming ability due to their wide supercooled liquid region compared to other Mg-based amorphous alloys. Busch *et al.*[117]

studied the thermodynamics of this system. They have reported strong liquid like nature for this BMG, which correlates the excellent GFA found in this composition. Mg-Cu-Y glasses with 1-7 mm in dimension were synthesized by Inoue *et al.*[36] These glasses possess tensile strength of 600 MPa. Mg-based glassy alloys with high tensile strengths in a number of alloy systems including Mg-Ln, Mg-LTM (Ni, Cu) binary and Mg-Ln-LTM (LTM: Ni, Cu, Zn), Mg-Y-Ln, Mg-Y-Al, Mg-Ca-Al, Mg-Zn-Al and Mg-Ga-Ca ternary have been found.[118] Mg content in all these alloys are more than 50 at.%. Table 3 shows the list of Mg-based glasses.

Mg based glasses having high strength and good bending ductility have been formed in a wide composition range of Mg-Ca-Ni, Mg-Ce-Ni, Mg-Y-Ni, Mg-Y-Cu.[118,119] A wide glass forming composition range was also found in Mg-Y-Mm (Mm: mischmetal) system. The Mg content in Mg-Y-Mm glassy alloys is up to 91%, which is one of the highest Mg content among the Mg-based glassy alloys resulting in this low density. Moreover, Mg-rich Mg-Y-Mm (5 at.% Y and 4 at.% Mm) glassy alloys possess a good bending ductility (i.e. can be bent through 180° without fracture) and quite high tensile strength of 550 MPa.

Table 3. Mg based metal-metal glass forming compositions.

Type	System	Composition
Binary	Mg-TM	$Mg_{75}Zn_{25}$, $Mg_{75}Ni_{25}$, $Mg_{75}Cu_{25}$
	Mg-AEM	Mg-Ca
	Mg-RE	Mg-Y
Ternary	Mg-TM-RE	$Mg_{65}Cu_{25}Y_{10}$, $Mg_{65}Cu_{25}Gd_{10}$, $Mg_{65}Cu_{25}Nd_{10}$, $Mg_{80}Ni_{15}Y_5$, $Mg_{84}Ni_{15}Sr$, $Mg_{80}Cu_{10}Y_{10}$, $Mg_{98}Cu_1Y_1$
	Mg-AEM-(M, TM)	$Mg_{70}Ca_{10}Al_{20}$, $Mg_{87.5}Ca_5Ni_{7.5}$, $Mg_{90}Ca_{2.5}Ni_{7.5}$
	Mg-RE-RE	$Mg_{91}Y_5Mm_4$
Quaternary	Mg-TM-RE-RE	$Mg_{65}Cu_{25}Gd_5Y_5$, $Mg_{65}Cu_{25}Nd_5Y_5$, $Mg_{65}Cu_{25}Y_{10-x}Gd_x$ $Mg_{54}Cu_{26.5}Ag_{8.5}Gd_{11}$, $Mg_{56}Cu_{26.5}Ag_{6.5}Gd_{11}$, $Mg_{54}Cu_{28}Ag_7Gd_{11}$, $Mg_{65}Cu_{20}Zn_5Y_{10}$, $Mg_{75}Cu_5Ni_{10}Gd_{10}$, $Mg_{70}Cu_{15}Ni_5Gd_{10}$
Quinary	Mg-TM-TM-TM-RE	$Mg_{65}Cu_{15}Ag_5Pd_5Y_{10}$

Multicomponent interaction in alloys is an important way to develop glasses with substantial amount of GFA. Partial replacement of Cu with

5 at.% Zn in $Mg_{65}Cu_{25}Y_{10}$ alloy leads to the formation of BMG (6 mm) with a significantly stronger GFA than $Mg_{65}Cu_{25}Y1_0$ glass.[120] The addition of Zn in the $Mg_{65}Cu_{25}Y_{10}$ metallic glass results in multiple stage crystallization and yields a reduction of the supercooled region, but Trg is slightly increased compared to $Mg_{65}Cu_{25}Y_{10}$ glass.

A series of Mg-Cu-RE alloys was studied by Xi *et al.*[121] They have studied the effect of atomic size and electronegativity of rare earth elements on glass forming ability of $Mg_{65}Cu_{25}Re_{10}$ alloys. The replacement of Ce and La with Y is expected to bring about a further increase in the specific tensile strength defined by the ratio of tensile fracture strength (σf) to density (ρ), though the melting temperature of Y is much higher than that of La and Ce so the production of homogeneously mixed alloys is more difficult for the Y-containing alloys.[115]

Countinuous research work has taken place to improve the thickness of Mg-LTM-RE BMG alloys. While $Mg_{65}Cu_{25}Y_{10}$ alloy led to the formation of BMG with 4 mm in diameter[36] a substitution of Cu with 10 at.% Ag results a BMG with 6 mm in diameter.[122] A spectacular improvement took place due to the discovery of inch diameter Mg-LTM-RE glasses by Ma *et al.*[30] They have produced $Mg_{54}Cu_{26.5}Ag_{8.5}Gd_{11}$ BMG with 25 mm in diameter and $Mg_{54}Cu_{28}Ag_7Y_{11}$ BMG with 16 mm in diameter using a new scheme to locate the best glass forming composition in a multicomponent system (Fig. 17a). They have shown that Mg-Cu-Ag-Gd has better glass forming ability than Mg-Cu-Ag-Y alloys. They have also compared their BMG forming alloys with Inoue's[28] and Kim's[122] BMG (Fig. 17b). They have used a completely new strategy to find out the best glass forming composition in a three-dimensional composition space. Initially, they have chosen Inoue's $Mg_{65}Cu_{25}Y_{10}$ BMG, which is an eutectic composition, and made a careful search with small composition steps towards steeper liquidus slope to find out the best BMG composition. Using this method, they could achieve 9 mm thick BMG at $Mg_{58.5}Cu_{30.5}Y_{11}$. Then they have examined the effect of Ag addition on the GFA. They have substituted Cu with Ag and made a systematic examination of a few consecutive compositional planes to find out the best BMG forming composition. Each plane contains a fixed Ag to Cu ratio.

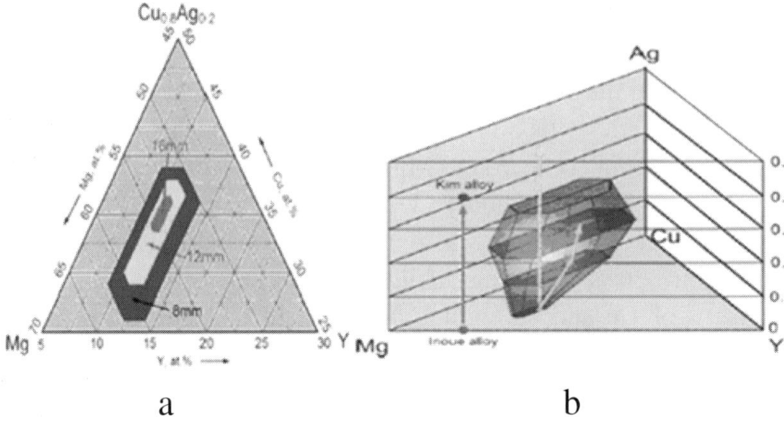

a   b

Fig. 17. (a) Composition ranges for BMG formation (D_c = 8 mm to the maximum) on a representative 2D plane ($x = 0.2$ in Cu$_{1-x}$Ag$_x$); (b) Composition region (the blue gemstone shape body) for the D_c = 8 mm BMG. The Mg$_{65}$Cu$_{25}$Y$_{10}$ (Inoue alloy, D_c = 4 mm) and Mg$_{65}$Cu$_{15}$Ag$_{10}$Y$_{10}$ (Kim's alloy, D_c = 6 mm) are outside of Mg$_{54}$Cu$_{28}$Ag$_7$Y$_{11}$ (Ma alloy, D_c = 16 mm), a straight-up green arrow indicates fixing of the total Ag+Cu content at 30.5 at.%.[30]

12.   Thermal Stability of Glasses

As glasses are metastable, they crystallise above a critical temperature. ΔT_x is known often used as the parameter for development of BMG. Herold and Koster[123] have shown that this transformation can be broadly classified as polymorphous, primary or eutectic crystallisation. In polymorphous crystallisation the product phase has the same composition as the matrix, in primary crystallisation the composition of the precipitate differs from that of the matrix, and in eutectic crystallisation the glassy matrix transforms into two different crystalline phases. These are well documented. Ranganathan and Heimendahil[124] have carried out a classic study on the activation energy calculations during devitrification of commercially available metallic glasses (Metglas: Fe$_{40}$Ni$_{40}$P$_{14}$B$_6$ and Fe$_{32}$Ni$_{36}$Cr$_{14}$P$_{12}$B$_6$). They have correlated the experimental observations with the theoretical results those are in good agreement. They also showed the relationship among the activation energies for nucleation, growth and overall transformation.

The devitrification studies of BMG have brought additional features into focus. These include nanocrystallisation, nanoquasicrystallization and phase separation. These three processes are briefly described below.

Nanocrystallisation takes place during the devitrification of both metallic and bulk metallic glasses. Perepezko and Hebert[125] have reported that the partial crystallization of Al-based metallic glasses leads to an improvement of properties. Basu and Ranganathan[126] have observed the formation of nanocrystallisation during the devitrification of Al-rich Al-LTM (Fe, Ni)-La and Al-ETM (Ti)-LTM (Ni)-La glasses. They have shown that α-Al nanocrystals are precipitated at the initial stage of crystallisation which results a solute rich matrix. They have also reported that the grain growth process is restricted by the precipitated crystallites.

Koester *et al.*[127] observed that Zr-based bulk metallic glasses give rise to quasicrystals on annealing. Since then this subject has invoked considerable interest, as it indicated a strong connection between icosahedral order and the structure of the bulk metallic glasses. In particular, it has been found that melt-spun glassy ternary Zr-TM (TM = Fe, Co, Ni, Cu) -M (M = Pt, Pd, Ag, Au) alloys give rise to quasicrystals on annealing. This led to a modification of the empirical rules of the formation of bulk metallic glasses to state that at least one atomic interaction, namely that between TM and M must lead to slightly negative or even positive enthalpy of mixing. Figure 18 shows such a nanocrystallisation from a Zr-Ti-Ni glass.[128] There is extensive report in the literature for such observations and provide the strongest evidence for the link between metallic glasses and icosahedral quasicrystals.

In glass forming system, phase separation offers an opportunity for designing composites with a hierarchical microstructure at different length scales. Phase separation systems also provide an opportunity to develop materials with unique properties. Park *et al.*[129] have succeeded in synthesizing phase separating (Zr,Ti)-Y-Al-Co and Y-Ti-Al-Co glasses. They have correlated the experimental observation with theoretical results and confirmed the phase separation of both MG and BMG.

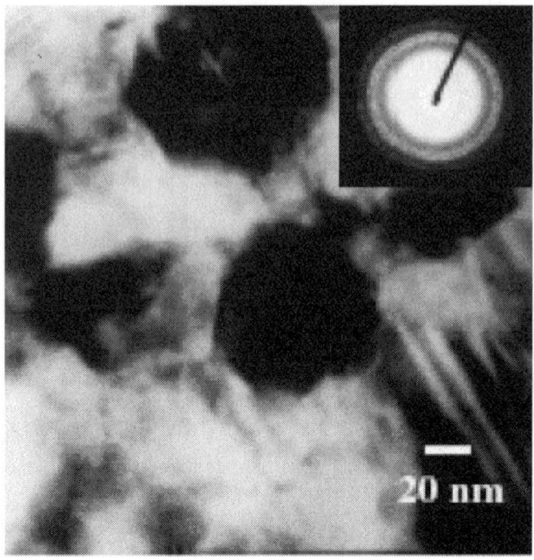

Fig. 18. Nanocrystallization from a Zr-Ti-Ni glass.[128]

13. Summary

Since the pioneering discovery of metallic glasses by Pol Duwez the field of amorphous alloys has made considerable strides. The discovery of alloy compositions forming bulk glasses infused renewed interest in this area, which has led to an explosion in the number of compositions having bulk glass formability, which have been realized by various synthetic routes. Concurrent research has lead to better structural and thermodynamic understanding of metallic glasses, which has further led to insightful classification schemes for BMG. Computational methods are playing an increasing role in the design of BMG, which is expected to lead to a considerable saving in time. Although prospective applications for metallic glasses have seen a leap with the availability of myriad compositions which form BMG; the real applications have been slow to come by. Reviewing the vigorous activity in this area, one can conclude with confidence that amorphous metallic alloys have a profound future.

Acknowledgments

The authors are grateful to Prof. K. Chattopadhyay, Prof. U. Ramamurty, Dr. Eric Lord, Dr. Joysurya Basu (India), Prof. A. Inoue, Prof. A. Takeuchi, Prof. D. Louzguine, Prof. J. Saida (Japan), Prof. A.L. Greer (UK), Prof. Jun Shen (China) and Dr. D. Miracle (USA) for fruitful collaborative research and numerous stimulating discussions. Gratitude is also expressed to the Defence Research and Development Organization and the Board of Research in Nuclear Sciences (Government of India) and the US Air Force office of Scientific Research.

References

1. W. Klement, R.H. Willens, P. Duwez, *Nature*, **187**, 869 (1960).
2. H.S. Chen, and D. Turnbull, *Appl. Phys.Lett.* **10**, 284 (1967).
3. H.W. Kui, A.L. Greer and D. Turnbull, *Appl. Phys. Lett.* **45**, 615 (1984).
4. A. Inoue, K. Ohtera, K. Kita and T. Masumoto, *J. Appl. Phys.* **27**, 615 (1988).
5. A. Peker, and W.L. Johnson, *Appl. Phys. Lett.* **63**, 2342 (1993).
6. D. Turnbull, *Contemp. Phys.* **10**, 473 (1969).
7. A.R. Yavari, *Nature*, **4**, 2 (2005).
8. S. Ranganathan, private communication (2003).
9. J.F. Loffler, Intermetallics, **11**, 529 (2003).
10. A. Inoue, *Bulk Amorphous Alloys: Preparation and Fundamental Characteristics* (Trans Tech Publications, Zurich, 1998).
11. A. Inoue, *Bulk amorphous alloys: Practical characteristics and applications* (Trans Tech Publications, Zurich, 1999).
12. A. Inoue, *Acta Mater.* **48**, 279 (2000).
13. W.L. Johnson, *MRS Bull.* **24**, 42 (1999).
14. W.L. Johnson, *J. Mater.* **54**, 40 (2002).
15. J. Basu and S. Ranganathan, *Sadhana*, **28**, 783 (2003).
16. H. Wang, C. Dong and C.H. Shek, *Mater. Sci. Engg.* **R44**, 45 (2004).
17. W.L. Johnson, A. Inoue and C.T. Liu, Eds., *Bulk metallic glasses* (MRS., Warrendale, 1999).
18. A. Inoue, W. Zhang, T. Zhang and K. Kurosaka, *Acta Mater.* **49**, 2645 (2001).
19. T. Egami, A.L. Greer, A. Inoue and S. Ranganathan, Eds., *Supercooled liquid, the glass transition and bulk metallic glasses* (MRS., Warrendale, 2003).
20. P.K. Liaw and R.A Buchanan, Eds., *Bulk Metallic Glasses*, (TMS, Warrendale 2004).
21. A. Inoue, T. Zhang and T. Masumoto, *Mater. Trans. JIM*, **12**, 965 (1989).
22. A. Inoue, N. Nishiyama, T. Masumoto, *Trans. Mat. Res. Soc. Jpn.* **A16**, 87 (1994).

23.	J. Eckert, *Mater. Sci., Engg.* **A226-228**, 364 (1997).

24.	A. Inoue, N. Nishiyama, T. Matsuda, *Mater. Trans. JIM*, **37**, 181 (1996).

25.	A. Inoue, H. Yamaguchi, T. Zhang and T. Masumoto, *Mater. Trans. JIM*, **31**, 104 (1990).

26.	A. Inoue, T. Zhang, A. Takeuchi, *Appl. Phys. Lett.* **71**, 464 (1997).

27.	A. Inoue and J.S. Gook, *Mater. Trans. JIM*, **36**, 1180 (1995).

28.	A. Inoue, A. Kato, T. Zhang, S.G. Kim and T. Masumoto, *Mater. Trans. JIM*, **32**, 609 (1991).

29.	H.G. Kang, E.S. Park, W.T. Kim, D.H. Kim and H.K. Cho, *Mater. Trans. JIM*, **41**, 846 (2000).

30.	H. Ma, L.L. Shi, J. Xu, Y. Li and E. Ma, *Appl. Phys. Lett.* **87**, 181915 (2005).

31.	A.J. Drehman, A.L. Greer and D. Turnbull, *Appl. Phys. Lett.* **41**, 716 (1982).

32.	Y.C. Kim, W.T. Kim and D.H. Kim, *Mater. Sci. Engg.* **A375-377**, 1337 (2004).

33.	K.Amiya and A. Inoue, *Mater. Trans. JIM*, **41**, 1460 (2002).

34.	L.Q. Xing, P. Ochin, M. Harmelin, F. Faudot, J. Bigot and J.P. Chevalier, *Mater. Sci. Engg.* **A220**, 155 (1996).

35.	A. Inoue and T. Zhang, *Mater. Trans. JIM*, **37**, 185 (1996).

36.	A. Inoue, T. Nakamura, N. Nishiyama and T. Masumoto, *Mater. Trans. JIM*, **33**, 937 (1992).

37.	B.S. Murty and S. Ranganathan, *Int. Mater. Rev.* **43**, 1 (1998).

38.	S. Madge, M.E. Thesis, Indian Institute of Science, Bangalore (1999).

39.	J.D. Bernal, *Nature*, **185**, 68 (1960).

40.	D.E. Polk, *Acta Metall.* **20**, 485 (1972).

41.	D.B. Miracle, *Nature Materials*, **3**, 697 (2004).

42.	D.B. Miracle, E.A. Lord and S. Ranganathan, *Mater. Trans.* 47, 1737 (2006).

43.	H.W. Sheng, W.K. Luo, F.M. Alamgir, J.M. Bai and E. Ma, *Nature*, **439**, 419 (2006).

44.	M.H. Cohen and D. Turnbull, *Nature*, **189**, 131 (1961).

45.	K.S. Dubey, and P. Ramchandrarao, *Intl. J. Rapid Solidification*, **5**, 127 (1990).

46.	H. Tan, Y. Zhang, D. Ma, Y.P. Feng and Y. Li, *Acta Mater.* **51**, 4551 (2003).

47.	D. Wang, Y. Li, B.B. Sun, M.L. Sui, K. Lu and E. Ma, *Appl. Phys. Lett.* **84**, 4029 (2004).

48.	D. Stockdale, *Proc. Roy. Soc.* **A152**, 81 (1935).

49.	W. Hume-Rothery and E. Anderson, *Phil. Mag.* **5**, 383 (1960).

50.	P. Hansen, *The constitution of binary alloys*, (McGraw-Hill, New York, 1958).

51.	J.J. Gilman, *Phil. Mag.* **B 37**, 577 (1978).

52.	D.B. Miracle, W.S. Sanders and O.N. Senkov, *Phil. Mag.* **83**, 2409 (2003).

53.	H.W. Kui, A.L. Greer and D. Turnbull, *Appl. Phys. Lett.* **45**, 615 (1984).

54.	Z.P. Lu and C.T. Liu, *Acta Mater.* **50**, 3501 (2002).

55.	Q. Chen, J. Shen, D. Zhang, H. Fan, J. Sun and D.G. McCartney, *Mater. Sci. Engg.* **A 433**, 155 (2005).

56.	P. Ramachandra Rao, *Z Metallk.* **71**, 172 (1980).

57. T. Egami and Y. Waseda, *J. Non-cryst. Solids*, **64**, 113 (1984).

58. A. Inoue and J.S. Gook, *Mater. Trans. JIM*, **37**, 32 (1996).

59. O.N. Senkov and D.B. Miracle, *MRS Bull.* **36**, 2183 (2001).

60. K. Mondal and B.S. Murty, *J. Non-cryst. Sol.* **351**, 1366 (2005).

61. J. Yeh, private communication (1996).

62. A.L. Mackay, *Cryst. Reports*, **46**, 524 (2001).

63. A.L. Greer, *Nature*, **366**, 303 (1993).

64. B. Cantor, I.T.H. Chang, P. Knight and A.J.B. Vincent, *Mater. Sci. Engg.* **A 375-377**, 213 (2004).

65. J.W. Yeh, S.K. Chen, S.J. Lin, J.Y. Gan, T.S. Chin, T.T. Shun, C.H. Tsau and S.Y. Chang, *Adv. Engg. Mat.* **6**, 299 (2004).

66. M.S. El-Eskandarany and A. Inoue, *Metall. Mater. Trans.* **A37**, 2231 (2006).

67. S. Varalakshmi, M. Kamaraj and B.S. Murty, *J. Alloys Comp.* (in press) (2007).

68. A. Takeuchi and A. Inoue, *J. Optoelectron. Adv. Mater.* **8**, 1679 (2006).

69. T. Biswas and S. Ranganthan, *Ann. Chim. Sci. Mat.* **31**, 649 (2006).

70. A. Inoue, T. Zhang and T. Masumoto, *Sol.* **156-158**, 473 (1993).

71. F.Q. Guo, S.J. Poon and G.J. Shiflet, *Appl. Phys. Lett.* **83**, 2575 (2003).

72. A. Inoue, T. Zhang and T. Masumoto, *Mater. Trans. JIM*, **31**, 177 (1990).

73. N. Nishiyama and A. Inoue, *Mater. Trans., JIM*, **38**, 464 (1997).

74. N. Nishiyama and A. Inoue, *Mater. Trans.* **43**, 1913 (2002).

75. V. Ponnambalam, S.J. Poon, G.J. Shiflet, V.M. Keppens, R. Taylor and G. Petculescu, *Appl. Phys. Lett.* **83**, 131 (2003).

76. Z.P. Lu, C.T. Liu, J.R. Thompson and W.D. Porter, *Phy. Rev. Lett.* **92**, 2455031 (2004).

77. P. Ramchandrarao, S. Suryanarayana and T.R. Anantharaman, *Metall. Trans.* **2**, 617 (1971).

78. K. Chattopadhyay, P. Ramachandra Rao, S. Lele and T.R. Anantharaman, *Proc. 2nd Int. Conf. on Rapidly Quenched Metals* (MIT Press, Cambridge, 1976), p.157.

79. G.V.S. Sastry, C. Suryanarayana, O.N. Srivastava and H.A. Davies, *Trans. Indian Inst. Met.* **31**, 292 (1978).

80. A. Inoue, *Mater. Trans. JIM*, **36**, 866 (1995).

81. A.R. Miedema, F.R. De Boer and R. Boom, *Physica B*, **103**, 67 (1981).

82. N. Saunders, A.P. Miodownik, 1998, *CALPHAD (Calculation of Phase Diagrams): A Comprehensive Guide*, Pergamon Materials Series: Vol. 1, Ed. R. W. Cahn (Elsevier, Oxford, 1998).

83. R. Bormann, *Mater. Sci. Engg.* **A178**, 55 (1994).

84. M. Palumbo, M. Satta, G. Cacciamani and M. Baricco, *Mater. Trans. JIM*, **47**, 2950 (2006).

85. G. Shao, *J. Appl. Phys.* **88**, 4443 (2000).

86. A.R. Miedema, *Philips Tech. Rev.* **36**, 217 (1976).

87. R. Nagarajan and S. Ranganathan, *Mater. Sci. Engg.* **A179-180**, 168 (1994).

88. B.S. Murty, Ph.D. Thesis, Indian Institute of Science, Bangalore (1992).

89. J. Basu, Ph.D. Thesis, Indian Institute of Science, Bangalore (2004).

90. A. Takeuchi and A. Inoue, *Mater. Trans. JIM*, **42**, 1435 (2001).

91. B.S. Murty, M. Rao and S. Ranganathan, *CALPHAD*, **19**, 297 (1995).

92. L.J. Gallego, J.A. Somoza and J.A. Alonso, *J. Phys.: Cond. Matter*, **2**, 6245 (1990).

93. A. Takeuchi and A. Inoue, *Mater. Trans. JIM*, **41**, 1372 (2000).

94. B.S. Rao, J. Bhatt and B.S. Murty, *Mater. Sci. Engg.* **A 449-451**, 211 (2007).

95. A.K. Niessen, F.R. de Boer, R. Boom, P.F. de Chatel, W.C.M. Mattern and A.R. Miedema, *CALPHAD*, **1**, 51 (1983).

96. O.N. Senkov and D.B. Miracle, *J. Non-Cryst. Solids*, **317**, 34 (2003).

97. G.A. Mansoori, N.F. Carnahan, K.E. Starling and T.W. Leland, *J. Chem. Phys.* **54**, 1523 (1971).

98. J. Bhatt, W. Jiang, X. Junhai, W. Quing, C. Dong and B.S. Murty, *Intermetallics*, **15**, 716 (2007).

99. D.G. Pettifor, *Solid. State Commun.* **51**, 31 (1984).

100. D.G. Pettifor, *Physica*, **B 149**, 3 (1988).

101. P. Villars. K. Brandenburg, M. Berndt, S. LeClair, A. Jackson, Y.H. Pao, B. Igelink, M. Oxley, B. Bakshi, P. Chen and S. Iwata, *J. Alloy Compd.* **317-318**, 26 (2001).

102. P. Villars, K. Cenzual, J. Daams, Y. Chen and S. Iwata, *J. Alloy Compd.* **367**, 167 (2004).

103. H.S. Jeevan and S. Ranganathan, *J. Non-cryst. Sol.* **334-335**, 184 (2004).

104. S. Ranganathan and A. Inoue, *Acta Mater.* **54**, 3647 (2006).

105. A. Takeuchi, B.S. Murty, M. Hasegawa, S. Ranganthan and A. Inoue, *Mater. Trans.* **48**, 1304 (2007).

106. A. Inoue, A. Kitamura and T. Masumoto, *J. Mater. Sci.* **16**, 1895 (1981).

107. R.O. Suzuki, Y. Komatsu, K.E. Kobayashi, P.H. Shingu, *J. Mater. Sci.* **18**, 1195 (1983).

108. A. Inoue, Y. Bize, H.M. Kimura, M. Yamamoto, A.P. Tsai, T. Masumoto, *J. Mater. Sci. Lett.* **6**, 811 (1987).

109. A.P. Tsai, A. Inoue and T. Masumoto: *Metall. Trans.* **19 A**, 391 (1988).

110. A.P. Tsai, A. Inoue and T. Masumoto. *Matall. Trans.* **19 A**, 1369 (1988).

111. A. Inoue, H.M. Kimura, T. Masumoto, A.P. Tsai and Y. Bizen, *J. Mater. Sei. Lett.* **6**, 771 (1987).

112. W.S. Sanders, J.S. Warner and D.B. Miracle, *Intermetallics*, **14**, 348 (2006).

113. A. Inoue, K. Ohtera, A.P. Tsai and T. Masumoto, *Jpn. J. Appl. Phys.* **27** L280 (1988).

114. F.Q. Guo, S.J. Poon and G.J. Shiflet, *Scr. Mater.* **43**, 1089 (2000).

115. S.G. Kim, A. Inoue and T. Masumoto, *Mater. Trans. JIM*, **31**, 929 (1990).

116. A. Inoue, T. Nakamura, T. Sugita, T. Zhang and T. Masumoto, *Mater. Trans. JIM*, **34**, 351 (1993).

117. R. Busch, W. Liu and W.L. Johnson, *J. Appl. Phys.* **83**, 4134 (1998).

118. A. Inoue and T. Masumoto, *Mater. Sci. Engg.* **A 173**, 1 (1993).

119. H. Horikiri, A. Kato, A. Inoue and T. Masumoto, *Mater. Sci. Engg.* **A 179/180**, 702 (1994).
120. H. Men, Z.Q. Hu and J. Xu, *Scripta Mater.* **46**, 699 (2002).
121. X.K. Xi, D.Q. Zhao, M.X. Pan and W.H. Wang, *Intermetallics*, **13**, 638 (2005).
122. H.G. Kang, E.S. Park, W.T. Kim, D.H. Kim, H.K. Cho, *Mater. Trans. JIM*, **41**, 846 (2000).
123. U. Herold and U. Koster, *Rapidly Quenched Metals III*, Ed. B. Cantor, (Metals Society, London, 1978), p. 281.
124. S. Ranganathan and M.V. Heimendahl, *J. Mater. Sci.* **16**, 2401 (1981).
125. J.H. Perepezko and R.J. Hebert, *J. Met.* **54**, 34 (2002).
126. J. Basu and S. Ranganathan, *Intermetallics*, **12**, 1045 (2004).
127. U. Koester, J. Meinhardt, S. Ros and H. Liebertz, *Appl. Phys. Lett.* **69**, 179 (1996).
128. J. Basu and S. Ranganathan, *Acta Mater.* **54**, 3637 (2006).
129. B.J. Park, H.J. Chang, D.H. Kim, W.T. Kim, K. Chattopadhyay, T.A. Abinandanan and S. Bhattacharyya, *Phys. Rev. Lett.* **96**, 245503 (2006).